THE BARBARA KRAUS
DICTIONARY OF PROTEIN

Other Books by Barbara Kraus

The Dictionary of Calories and Carbohydrates

The Dictionary of Sodium, Fats, and Cholesterol

The Cookbook of the United Nations

The Cookbook to Serve 2, 6 or 24

THE BARBARA KRAUS DICTIONARY OF PROTEIN

Over 8,000 brand names and basic foods with their protein (and caloric) count.

by Barbara Kraus

HARPER'S MAGAZINE PRESS

Published in Association with Harper & Row

New York

THE BARBARA KRAUS DICTIONARY OF PROTEIN Copyright © 1975, by Barbara Kraus. All rights reserved. Printed in the United States of America. No part of this book may be used or reproduced in any manner whatsoever without written permission except in the case of brief quotations embodied in critical articles and reviews. For information address Harper & Row, Publishers, Inc., 10 East 53rd Street, New York, N.Y. 10022. Published simultaneously in Canada by Fitzhenry & Whiteside Limited, Toronto.

Library of Congress Cataloging in Publication Data

Kraus, Barbara.
  The Barbara Kraus dictionary of protein.
  Bibliography: p.
  1. Proteins—Tables. 2. Food—Composition—Tables.
I. Title. II. Title: Dictionary of protein.
TX553.P7K7   641.1′2′0212   74–27302
ISBN 0–06–125101–1

75 76 77 78 79 10 9 8 7 6 5 4 3 2 1

For
Marge Altman
Rebecca K. Pecot
Shirley Wershba

CONTENTS

PREFACE

WHY THIS BOOK?

As the population of Planet Earth grows inexorably larger, the difficulty of producing sufficient protein for the needs of its inhabitants is of increasing concern. When the protein is gotten from one source in particular, the problem is aggravated. Americans, especially, are used to deriving an extraordinary amount of their protein needs from meat—a cultural habit, that is becoming something of a luxury in our lifetime, since it is essentially wasteful of the planet's protein supply. In this country, cattle are fed much of the inexpensive but valuable plant protein. The amount of meat actually made available to our dinner tables is disproportionately low compared with the amount of protein that the livestock had to consume to produce it.

If we wish to reduce the cost of protein in our diet, we should keep in mind that cheaper cuts of meat are as nutritious as the more expensive steaks and chops. Why buy so much beef, lamb, pork or veal when fish and poultry cost less and are as good for you? Eggs and milk are also excellent sources of protein and are usually cheaper than meat. But for the real bargains, look to the plant products.

Dry beans including lentils, soybeans, soya products, peanuts and peanut butter are all good sources of protein. Whole grain cereals also contribute significant amounts of this nutrient. Nuts, too, are high in protein though, except for peanuts, they are usually expensive.

A small serving of meat a day may satisfy a craving for this food and meat dishes can be extended by combining them with fish or poultry, with cereal products such as macaroni, noodles or rice. The protein in such combinations complement each other and may result in more efficient sources than any single food provides.

It is perfectly feasible to provide a diet adequate in protein content without any animal products. In fact, in experimental work it has been shown that potatoes were satisfactory as the sole source of protein for test subjects. Of course, one should not rely on a single source of protein. A mixed diet of many different foods is still the safest way to good nutrition.

Individuals following a vegetarian diet must make certain that they are getting the proper amount of protein, since vegetable foods generally contain lower total amounts of the nutrient than animal products. For example, "meat-like beef loaf" (p. 334) would provide only 21.8 grams of protein in a 4-ounce serving compared with 32 grams in a 4-ounce serving of lean roast beef (p. 38). Also, because some animal protein is needed to provide vitamin B_{12}, a dietary supplement is recommended. Furthermore, many trace minerals are higher in animal foods than in vegetables. To supply these requirements, some vegetarians include milk and/or eggs in their diets. But in any case, a variety of sources of protein (as well as other nutrients) should be used which would include different kinds of legumes, cereals, nuts and vegetables.

The value of this book is that it dramatizes the wide array of choices one has for acquiring needed protein, and leaves it up to you to make the most economical and pleasing selections. However, the fact is that you don't need as much protein as you may think to keep you healthy. The recommended dietary allowances (RDA) differ for different ages and sexes. Infants require about 1 gram of protein for each pound of body weight. Children require from 23 to 36 grams per day. The adult man needs 56 grams and the adult woman 46 grams. Age and weight affect the requirement to some extent but these figures will serve as a yardstick to remember in your dietary planning.

For the benefit and convenience of those who must watch their weight, the caloric value of the same foods and portions has been listed alongside their protein content.

BARBARA KRAUS

INTRODUCTION

The Greeks had a word for it—*proteios*—meaning primary or holding first place. Protein is present in varying amounts in all living cells, animal and plant; it is indispensable to growth and development. Children need it especially in their rapidly growing years, but it is essential throughout life for replacement of cells and tissues of the body.

Protein is an extremely complex compound and it occurs in different forms in blood, tissue, and cells. The element common to all these forms is nitrogen. In the laboratory, nitrogen is determined by chemical analysis. From these analyses, protein is calculated by appropriate factors. In many foods, nitrogen takes up about 16% of the protein, so a factor of 6.25 ($^{100}/_{16}$) is used to calculate the protein content of food items. However, because of variations in this figure, appropriate adjustments are made where necessary. In most food composition tables, protein is calculated by using specific factors for different classes of foods.

All proteins are not of equal value in human nutrition. Some are better in fostering growth and development than others. Animal products have been recognized as being very efficient sources of the nutrient. This knowledge plus the pleasing taste of meat, milk, eggs, and their products have led in the affluent countries to an excessive consumption of animal products.

Since animals get their protein from plant sources, meat and animal products become a secondary source of the nutrient and, therefore, a more costly source in terms of land, labor, and feed. For this reason, in times of high prices, consumers look to lower-cost sources of supply, i.e., to plant products.

It is perfectly feasible to plan a diet that contains essential nutrients with a limited amount or no animal products. But such a diet needs careful planning, especially for the young growing child, the elderly, and the pregnant or nursing mother. The data in this book will help you to select foods that will provide a pleasing and attractive diet from a wide variety of basic and commercial food items.

Quality of Protein

Amino Acid Theory

But what about the quality of protein from different kinds of foods?

There are two theories as to the relative quality or efficiency of this nutrient. For many years, it has been thought that quality depended on the *amino acid* content of the protein. Amino acids are called the "building blocks" of protein. They combine in different proportions to build the particular protein that the organism requires. Therefore, the amino acid content of the protein in meat is different from that in potatoes or beans.

Altogether, there are eighteen amino acids commonly found in foods. Eight of them are called "essential" to the human diet because the body cannot manufacture them itself in the necessary amounts. The others, though required, can be produced by the body when there is an adequate supply of nitrogen in the food consumed. The amounts of the eight essential amino acids needed are not large and can be supplied by relatively small servings of foods that provide "complete" protein (i.e. those that together contain the eight essential amino acids) with enough other sources to provide the total requirement. These foods are: meat, poultry, fish, eggs, milk, soybeans, and their products. Egg protein has been used as the standard against which to measure the quality of other sources in the diet.

According to the amino acid theory, complete proteins must be consumed at each meal or the protein is not available for its function in cell-building. It will not be carried over for this purpose from one meal to the next.

Total Nitrogen Theory

Another theory of protein efficiency that has only recently been summarized in American scientific publications is based on the total nitrogen content of the diet.

In the studies reported, different plant foods such as potato, soy, wheat, corn, rice, and beans were fed to the test subjects in combination with egg, milk, or meat and all were found to be equal to or better in quality than egg (the standard) alone. The authors concluded that the nonessential proteins were utilized to provide the body with the total nitrogen it required to keep the subjects in nitrogen balance. Bean and corn combinations and potato alone also were found to keep the subjects in balance.

The practical application of this theory is that a diet containing large proportions of plant foods combined with small amounts of animal products, and certain combinations of plant products, are satisfactory to supply the body's nitrogen requirement. It can be concluded that such familiar combinations in American diets as cereal with milk, succotash (corn and beans), cheese sandwiches, and many others can supply the protein needs.

Protein Requirements

Whichever theory prevails in the future, the oft-repeated advice of the nutritionist still holds: eat a variety of foods from many different sources. By so doing the normal individual receives all the nutrients he requires. Only in exceptional circumstances are special supplements recommended.

The tables in this book provide the basis for estimating your protein intake. Values are given in terms of convenient serving size as well as in purchased containers such as packages or cans. Just add the grams of protein in each food you eat during the day to get the total amount in your diet.

It is not necessary to consume more protein than you require because the excess over your daily requirement is converted to energy in terms of calories. If excess weight is a concern, this means an addition to your problem. As has been indicated, protein from animal sources is a costly nutrient to produce. Therefore, to reduce the high cost of foods, selections of cheaper sources of this nutrient can be made from the tables. Even small amounts from many different types of food will easily fulfill your daily allowance.

Arrangement of this Book

Foods are listed alphabetically by brand name or by the name of the food. The singular form is used for entries, that is, blackberry instead of blackberries. Most items are listed individually though a few are grouped, for example, all candies are listed together so that if you are looking for *Mr. Goodbar* bar, you look first under Candy, then under *M* in alphabetical order. But, if you are looking for a breakfast food such as Oatmeal, you will find it under *O* in the main alphabet. Many cross references are included to assist in finding items called by different names.

Under the main headings, it was often not possible nor even desirable to follow an alphabetical arrangement. For basic foods such as apricots, for example, the first entries are for the fresh product weighed with seeds as it is purchased in the store, then the fruit in small portions as they may be eaten or measured. These entries are followed by the processed products, canned (although it may actually be a bottle or jar), dehydrated, dried, and frozen. This basic plan, with adaptations where necessary, was also followed for fruits, vegetables and meats.

In almost all entries where data were available the U.S. Department of Agriculture figures are shown first. The Department values represent averages from several manufacturers and are shown for comparison with the values from individual companies or for use where particular brands are not available.

All brand-name products have been italicized and company names appear in parentheses.

Portions Used

The portion column is a most important one to read and note. Common household measures are used whenever possible. For some items, the amounts given are those commonly purchased in the store, such as 1 pound of meat, or a 21-oz. can of pie filling. These quantities can be divided into the number of servings used in the home and the nutritive values available to each person served can then be readily determined. Of course, any ingredients added in preparing such products must also be taken into account.

The smaller portions given are for foods as served or measured in moderate amounts, such as ½ cup of juice reconstituted, or 4 ounces of meat. Be sure to adjust the amount of the nutrients to the actual portions you use. For example, if you serve 1 cup of juice instead of ½ cup, multiply the amount of the nutrients shown for the smaller amount by 2.

The size of the portions you use is extremely important in controlling the intake of any nutrient. The amount of a nutrient is directly related to the weight of the food served. Also, the weight of a volumetric measure, such as a cup or a pint, may vary considerably depending on many factors. For example, 4 ounces by weight may be very different from ½ cup or 4 fluid ounces. Ounces in the tables are always ounces by weight unless specified as fluid ounces, or fractions of a cup or other volumetric measure. Foods that are fluffy in texture such as flaked coconut and bean sprouts vary greatly in weight per cup depending on how tightly packed they are. Such foods as canned green beans also vary when measured with and without liquid; for instance, canned beans with liquid weigh 4.2 ounces for ½ cup, but drained beans weigh 2.5 ounces for the same ½ cup. Check the weights of your serving portions regularly. Bear in mind that you can reduce or increase the intake of any nutrient by changing the serving size.

It was impossible to convert all the portions to a uniform basis. Some sources were only able to report data in terms of weights with no information on cup or other volumetric measures. We have shown small portions in quantities that might reasonably be expected to be served or measured in the home or institution. Package sizes are useful to show the composition of products as they are purchased and may be divided into the number of serving portions prepared from the entire product, taking into account any added ingredients.

You will find in the portion column the phrases "weighed with bone," or "weighed with skin and seeds" or other inedible parts. These descriptions apply to the products as you purchase them in the markets but the nutritive values as shown are for the amount of edible foods after you discard the bone, skin, seed or other inedible part. The weight given in the "measure or quantity" column is to the nearest gram or fraction of an ounce.

Data on the composition of foods are constantly changing for many reasons. Better sampling and analytical methods, improvements in marketing procedures, and changes in formulas of mixed products, all may alter values for all of the nutrients. Weights of packaged foods are frequently changed. It is essential to read label information and to make intelligent use of food tables.

Other Nutrients

These tables are not intended as a dietary guide. Any drastic change from a normal mixed diet should be undertaken only under the guidance of a qualified physician. Do not forget that other nutrients are extremely important in diet planning—fats, carbohydrates, minerals and vitamins. From a nutritional viewpoint, perhaps the best advice that can be given is to eat a varied diet with all classes of food represented. Meat, fish, chicken, fats and oils, milk, vegetables, fruits, and grain products are all important sources of essential nutrients, and some foods from each of these groups should be included in the diet every day. With the great abundance and variety of foods on the grocer's shelves, there is no reason why the dieter should not enjoy a tasty, nutritious and attractive diet.

Sources of Data

Values in this dictionary are based on publications issued by the U.S. Department of Agriculture and on data submitted by manufacturers and processors. The U.S. Department of Agriculture issued basic tables on food composition for use in the United States. The commercial products from U.S.D.A. publications represent average values obtained on products of more than one company. The figures designated "home recipe" are based on recipes on file with the Department of Agriculture. Data on commercial products listed by brand name in this publication are based on values supplied by manufacturers and processors for their own individual products. Very few supermarket brand names, such as A & P's *Ann Page*, or private labels are included in this book because they are not usually analyzed under these trade names. Every care has been taken to interpret the data and the descriptions supplied by the companies as fully and accurately as possible. Many values have been recalculated to different portions from those submitted in order to bring about greater uniformity among similar items.

Analyses of foods to provide information on nutritive values are extremely expensive to conduct. Many small companies cannot afford to have their products analyzed and thus were unable to provide data for this book or were able to provide only a portion of the data requested. Other companies have simply never gotten around to having the analyses done. New requirements for labeling nutritive values for products may provide information on additional items in the future.

Foods Listed by Groups

Foods in the following classes are reported together rather than as individual items in the main alphabet: Baby Food; Bread; Cake Icing; Cake Icing Mix; Candy; Cheese; Cookie; Cookie Mix; Cracker; Gravy; Salad Dressing; and Sauce.

B. K.

ABBREVIATIONS AND SYMBOLS

(USDA) = United States Department
    of Agriculture
(HEW/FAO) = Health, Education and Welfare/
    Food and Agriculture Organization
1 = prepared as package directs[1]
< = less than
& = and
" = inch
canned = bottled or jars as well as cans
dia. = diameter
fl. = fluid
liq. = liquid
lb. = pound

med. = medium
oz. = ounce
pkg. = package
pt. = pint
qt. = quart
sq. = square
T. = tablespoon
tsp. = teaspoon
Tr. = trace
wt. = weight

Italics or name in parentheses = registered trademark, ®
Blank spaces indicate that no data are available.

EQUIVALENTS

By Weight

1 pound = 16 ounces
1 ounce = 28.35 grams
3.52 ounces = 100 grams

By Volume

1 quart = 4 cups
1 cup = 8 fluid ounces
1 cup = ½ pint
1 cup = 16 tablespoons
2 tablespoons = 1 fluid ounce
1 tablespoon = 3 teaspoons
1 pound butter = 4 sticks or 2 cups

[1]If the package directions call for whole or skim milk, the data given here are for whole milk, unless otherwise stated.

Food and Description	Measure or Quantity	Protein (grams)	Calories

A

ABALONE (USDA):
Raw, meat only	4 oz.	21.2	111
Canned	4 oz.	18.1	91

AC'CENT	¼ tsp. (1 gram)	.1	3

ACEROLA, fresh (USDA):
Fruit	½ lb. (weighed with seeds)	.8	52
Fruit, flesh only	4 oz.	.5	32
Juice	½ cup (4.3 oz.)	.5	28

ALBACORE, raw meat only
(USDA)	4 oz.	28.7	201

ALEWIFE (USDA):
Raw, meat only	4 oz.	22.0	144
Canned, solids & liq.	4 oz.	18.4	160

ALMOND:
In shell:
(USDA)	4 oz. (weighed in shell)	10.8	347
(USDA)	1 cup (2.8 oz.)	7.4	239

Shelled:
Plain:
Whole (USDA)	1 oz.	5.3	170
Whole (USDA)	1 cup (5 oz.)	26.4	849
Whole (USDA)	13–15 almonds (.6 oz.)	3.3	105
Chopped (USDA)	1 cup (4.5 oz.)	23.6	759
(Blue Diamond)	1 cup (5.6 oz.)	39.2	1008

Blanched:
Salted (USDA)	1 cup (5.5 oz.)	29.2	984
Slivered (Blue Diamond)	1 cup (5.6 oz.)	36.8	1008

Chocolate-covered (see
CANDY)
Roasted:
Dry, salted (Flavor House)	1 oz.	5.3	178
Dry, salted (Planters)	1 oz.	5.3	191

(USDA): United States Department of Agriculture

Food and Description	Measure or Quantity	Protein (grams)	Calories
Salted (USDA)	1 oz.	5.3	178
Salted (USDA)	1 cup (5.5 oz.)	29.2	984
ALMOND MEAL, partially defatted (USDA)	1 oz.	11.2	116
ALPHA-BITS, oat cereal (Post)	1 cup (1 oz.)	2.2	113
A.M., fruit juice drink (Mott's)	½ cup	.1	62
AMARANTH, raw (USDA):			
Untrimmed	1 lb. (weighed untrimmed)	10.0	103
Trimmed	4 oz.	4.0	41
AMBROSIA, chilled, bottled (Kraft)	4 oz.	1.0	85
ANALOG (see **TEXTURED VEGETABLE PROTEIN**)			
ANCHOVY, PICKLED, canned, with or without added oil (USDA)	1 oz.	5.4	50
ANGEL FOOD CAKE, home recipe (USDA)	1/12 of 8″ cake (1.4 oz.)	2.8	108
ANGEL FOOD CAKE MIX:			
Dry (USDA)	4 oz.	9.5	437
*Made with water & flavorings (USDA)	1/12 of 10″ cake (1.9 oz.)	3.0	137
*(Betty Crocker):			
1 step	1/16 of cake	2.5	111
2 step	1/16 of cake	2.6	100
Confetti	1/16 of cake	2.6	114
Lemon custard	1/16 of cake	2.6	111
Strawberry	1/16 of cake	2.7	115
*(Duncan Hines)	1/12 of cake (2 oz.)	3.4	131
*(Pillsbury) white or raspberry	1/12 of cake	3.0	140
*(Swans Down)	1/12 of cake (1.8 oz.)	3.2	132

(USDA): United States Department of Agriculture
*Prepared as Package Directs

Food and Description	Measure or Quantity	Protein (grams)	Calories
APPLE, any variety:			
Fresh (USDA):			
Eaten with skin	1 lb. (weighed with skin & core)	.8	242
Eaten with skin	1 med., 2 ½" dia. (about 3 per lb.)	.3	80
Eaten without skin	1 lb. (weighed with skin & core)	.8	211
Eaten without skin	1 med., 2 ½" dia. (about 3 per lb.)	.3	70
Pared, diced	1 cup (3.8 oz.)	.2	59
Pared, quartered	1 cup (4.3 oz.)	.2	66
Canned (see **APPLESAUCE**)			
Dehydrated, sulfured:			
Uncooked (USDA)	1 oz.	.4	100
Cooked, sweetened (USDA)	½ cup (4.2 oz.)	.2	91
Dried, sulfured:			
Uncooked (USDA)	1 cup (3 oz.)	.9	236
Uncooked (Del Monte)	1 cup (3 oz.)	.6	212
Cooked, unsweetened (USDA)	½ cup (4.3 oz.)	.4	94
Cooked, sweetened (USDA)	½ cup (4.9 oz.)	.4	157
Frozen, sweetened, slices, not thawed (USDA)	4 oz.	.2	105
APPLE BROWN BETTY, home recipe (USDA)	1 cup (8.1 oz.)	3.7	347
APPLE BUTTER:			
(USDA)	½ cup (5 oz.)	.6	262
(USDA)	1 T. (.6 oz.)	<.1	33
(Bama)	1 T. (.6 oz.)	.1	27
(Smucker's) Cider	1 T. (.6 oz.)	<.1	37
APPLE CAKE MIX, cinnamon:			
*(Betty Crocker) pudding cake	⅙ of cake	2.1	223
*(Betty Crocker) upside down cake	⅑ of cake	1.9	269
*(Duncan Hines)	1/12 of cake (2.7 oz.)	2.3	202
APPLE CIDER:			
(USDA)	½ cup (4.4 oz.)	.1	58
Cherry or sweet (Mott's)	½ cup	.1	59

(USDA): United States Department of Agriculture
*Prepared as Package Directs

[3]

Food and Description	Measure or Quantity	Protein (grams)	Calories
APPLE DRINK, canned:			
(Del Monte)	½ cup (4.3 oz.)	.1	54
(Hi-C)	½ cup (4.2 oz.)	.5	60
APPLE DUMPLING, frozen			
(Pepperidge Farm)	1 dumpling	2.3	276
APPLE FRITTER, frozen			
(Mrs. Paul's)	8-oz. pkg.	6.5	474
APPLE JACKS, cereal (Kellogg's)	1 cup (1 oz.)	1.3	110
APPLE JELLY, dietetic or low calorie:			
(Diet Delight)	1 T. (.6 oz.)	Tr.	22
(Kraft)	1 oz.	<.1	34
(Smucker's) plain or with fruit	1 T. (.7 oz.)	Tr.	49
(S and W) *Nutradiet*	1 T. (.5 oz.)	<.1	13
(Slenderella)	1 T. (.6 oz.)	Tr.	25
(Tillie Lewis)	1 T. (.8 oz.)	Tr.	11
APPLE JUICE, canned:			
(USDA)	½ cup (4.4 oz.)	.1	58
(Heinz)	5½-fl.-oz. can	.2	88
(Mott's)	½ cup	.1	59
APPLE NECTAR (Mott's)	½ cup	.1	63
APPLE PIE:			
Home recipe, 2 crust (USDA)	⅙ of 9″ pie (5.6 oz.)	3.5	404
(Drake's)	2-oz. pie	1.9	204
(Hostess)	4½-oz. pie	3.4	421
(McDonald's)	1 serving (3 oz.)	2.0	269
(Tastykake)	4-oz. pie	3.4	380
French apple (Hostess)	4½-oz. pie	3.6	447
French apple (Tastykake)	4½-oz. pie	7.1	451
Frozen:			
Baked (USDA)	5-oz. serving	2.7	361
(Banquet)	5-oz. serving	3.5	351
(Morton)	⅙ of 24-oz. pie	2.2	270
(Mrs. Smith's)	⅙ of 8″ pie (4.2 oz.)	1.8	302
Golden deluxe (Mrs. Smith's)	⅛ of 10″ pie (5.6 oz.)	2.3	533
Dutch apple (Mrs. Smith's)	⅙ of 8″ pie (4.2 oz.)	2.3	309

(USDA): United States Department of Agriculture

Food and Description	Measure or Quantity	Protein (grams)	Calories
Dutch apple (Mrs. Smith's)	⅛ of 10″ pie (5.6 oz.)	3.0	412
Natural juice (Mrs. Smith's)	⅙ of 8″ pie (4.2 oz.)	2.0	340
Tart (Mrs. Smith's)	⅙ of 8″ pie (4.2 oz.)	2.0	340
Tart (Pepperidge Farm)	3-oz. tart	2.3	276
APPLE PIE FILLING, canned:			
Sweetened:			
(Comstock)	⅙ of 8″ pie	.1	115
(Lucky Leaf)	8 oz.	.2	248
(Wilderness)	21-oz. can	1.4	709
French (Wilderness)	21-oz. can	2.8	703
Unsweetened (Lucky Leaf)	8 oz.	.5	98
APPLESAUCE, canned:			
Sweetened:			
(USDA)	½ cup (4.5 oz.)	.3	116
(Del Monte)	½ cup (4.6 oz.)	.2	119
(Hunt's)	5-oz. can	.3	94
(Mott's)	½ cup (4.5 oz.)	.2	105
(Stokely-Van Camp)	½ cup (4.2 oz.)	.2	109
Unsweetened, dietetic or low calorie:			
(USDA)	½ cup (4.3 oz.)	.2	50
(Blue Boy)	4 oz.	.2	43
(Diet Delight)	½ cup (4.4 oz.)	.1	58
(Mott's)	½ cup (4.4 oz.)	.2	56
(S and W) *Nutradiet,* low calorie	4 oz.	.2	56
(S and W) *Nutradiet,* unsweetened	4 oz.	.3	57
***APPLESAUCE CAKE MIX:**			
(Duncan Hines)	⅑ of cake (2.7 oz.)	2.3	200
Spice (Pillsbury)	1/12 of cake	3.0	200
APPLE TURNOVER:			
(Pepperidge Farm)	1 turnover (3.3 oz.)	2.9	315
(Pillsbury)	1 turnover	1.0	150
APRICOT:			
Fresh (USDA):			
Whole	1 lb. (weighed with pits)	4.3	217

(USDA): United States Department of Agriculture
*Prepared as Package Directs

Food and Description	Measure or Quantity	Protein (grams)	Calories
Whole	3 apricots (about 12 per lb.)	1.1	55
Halves	1 cup (5.5 oz.)	1.6	80
Canned, regular pack, solids & liq.:			
Juice pack (USDA)	4 oz.	1.1	61
Light syrup (USDA)	4 oz.	.8	75
Heavy syrup:			
Halves & syrup (USDA)	½ cup (4.4 oz.)	.8	108
Halves & syrup (USDA)	4 med. halves with 2 T. syrup (4.3 oz.)	.7	105
(Del Monte)	½ cup (4.4 oz.)	.9	104
(Hunt's)	½ cup (4.5 oz.)	.8	103
(Stokely-Van Camp)	½ cup (4.2 oz.)	.7	103
Extra heavy syrup (USDA)	4 oz.	.7	115
Canned, unsweetened or low calorie, solids & liq.:			
Water pack, halves & liq. (USDA)	½ cup (4.3 oz.)	.9	46
(Diet Delight)	½ cup (4.4 oz.)	.7	60
(S and W) *Nutradiet*	2 whole (3.5 oz.)	.6	32
(Tillie Lewis)	½ of 8-oz. can	.5	54
Dehydrated, sulfured (USDA):			
Uncooked	4 oz.	6.4	376
Cooked, sugar added, solids & liq.	4 oz.	1.5	135
Dried, sulfured:			
Uncooked:			
(USDA)	1 lb.	22.7	1179
(USDA)	14 large halves (½ cup or 2.8 oz.)	4.0	211
(USDA)	10 small halves (¼ cup or 1.3 oz.)	1.9	99
(USDA)	1 cup (4.5 oz.)	6.4	333
(Del Monte)	1 cup (4.5 oz.)	4.7	326
Cooked:			
Sweetened	½ cup with liq. (12-13 halves, 5.7 oz.)	2.3	198
Unsweetened	½ cup with liq. (4.3 oz.)	2.0	104
Frozen, sweetened, not thawed (USDA)	4 oz.	.8	111

(USDA): United States Department of Agriculture

Food and Description	Measure or Quantity	Protein (grams)	Calories
APRICOT-APPLE JUICE & PRUNE, (Sunsweet)	½ cup	.3	63
APRICOT, CANDIED (USDA)	1 oz.	.2	96
APRICOT NECTAR, canned:			
Sweetened:			
(USDA)	½ cup (4.2 oz.)	.4	68
(Del Monte)	½ cup (4.3 oz.)	.4	68
(Heinz)	5½-fl.-oz. can	.7	90
(Sunsweet)	½ cup	.4	75
Low calorie (S and W) *Nutradiet*	4 oz. (by wt.)	.5	35
APRICOT-ORANGE PIE (Tastykake)	4-oz. pie	4.3	377
APRICOT PIE FILLING:			
(Comstock)	1 cup (10¾ oz.)	1.4	314
(Lucky Leaf)	8 oz.	.8	316
(Wilderness)	21-oz. can	2.6	756
APRICOT & PINEAPPLE NECTAR (S and W) *Nutradiet,* unsweetened	4 oz. (by wt.)	.5	35
APRICOT & PINEAPPLE PRESERVE:			
Sweetened (Bama)	1 T. (.7 oz.)	.1	54
Sweetened (Smucker's)	1 T.	.3	50
Dietetic or low calorie:			
(Diet Delight)	1 T. (.6 oz.)	.1	21
(S and W) *Nutradiet*	1 T. (.5 oz.)	<.1	10
(Tillie Lewis)	1 T. (.8 oz.)	.1	12
APRICOT PRESERVE or JAM:			
Sweetened			
(Bama)	1 T. (.7 oz.)	.1	51
(Smucker's)	1 T. (.7 oz.)	.1	49
Low calorie (Slenderella)	1 T. (.7 oz.)	.1	28

(USDA): United States Department of Agriculture

Food and Description	Measure or Quantity	Protein (grams)	Calories
ARTICHOKE, Globe or French (see also **JERUSALEM ARTICHOKE**):			
Raw, whole (USDA)	1 lb. (weighed untrimmed)	5.3	85
Boiled, drained (USDA)	1 large artichoke (15. oz.)	11.9	187
Canned, marinated, drained (Cara Mia)	1 heart	.2	13
Frozen, hearts (Birds Eye)	5-6 hearts (3 oz.)	1.2	22
ASPARAGUS:			
Raw, whole spears (USDA)	1 lb. (weighed untrimmed)	6.4	66
Boiled, whole spears (USDA)	4 spears (½″ dia. at base, 2.1 oz.)	1.3	12
Boiled, 1½″-2″ pieces, drained (USDA)	1 cup (5.1 oz.)	3.2	29
Canned, regular pack:			
Green:			
Spears & liq. (USDA)	1 cup (8.6 oz.)	4.6	44
Spears only (USDA)	1 cup (7.6 oz.)	5.2	45
Spears only (USDA)	6 med. spears (3.4 oz.)	2.3	20
Liq. only (USDA)	2 T. (1.1 oz.)	.2	3
Spears & liq. (Green Giant)	⅓ of 15-oz. can	2.4	24
Spears & liq., (*Le Sueur*)	¼ of 1-lb. 3-oz. can	2.3	23
Solids & liq. (Stokely-Van Camp)	½ cup (3.9 oz.)	2.1	20
Cut spears & liq. (Green Giant)	½ of 10½-oz. can	2.5	25
Drained solids (Del Monte)	1 cup (7.8 oz.)	3.2	34
White:			
Spears & liq. (USDA)	1 cup (8.4 oz.)	3.8	43
Spears only (USDA)	1 cup (7.6 oz.)	4.5	47
Spears only (USDA)	6 med. spears (3.4 oz.)	2.0	21
Liq. only (USDA)	2 T. (1.1 oz.)	.2	3
Spears & liq. (Del Monte)	1 cup (8 oz.)	3.0	34
Spears only (Del Monte)	1 cup (7.6 oz.)	3.7	39

(USDA): United States Department of Agriculture

Food and Description	Measure or Quantity	Protein (grams)	Calories
Canned, dietetic pack:			
Green:			
Spears & liq. (USDA)	4 oz.	2.3	18
Drained solids (USDA)	4 oz.	2.9	23
Liq. only (USDA)	4 oz.	.9	10
Spears & liq. (Blue Boy)	4 oz.	2.3	20
Solids & liq. (Diet Delight)	4 oz.	1.5	19
Solids & liq. (Tillie Lewis)	½ cup (4.2 oz.)	2.4	22
(S and W) *Nutradiet,*			
unseasoned	5 spears (3.5 oz.)	1.9	16
White:			
Spears & liq. (USDA)	4 oz.	1.6	18
Drained solids (USDA)	4 oz.	2.2	22
Liq. only (USDA)	4 oz.	.7	9
Frozen:			
Cuts & tips, not thawed (USDA)	4 oz.	3.7	26
Cuts & tips, boiled, drained (USDA)	½ cup (3.2 oz.)	2.9	20
Cuts (Birds Eye)	½ cup (3.3 oz.)	3.0	20
Cut spears in butter sauce (Green Giant)	⅓ of 9-oz. pkg.	1.9	45
Spears, not thawed (USDA)	4 oz.	3.7	27
Spears, boiled, drained (USDA)	4 oz.	3.6	26
Spears (Birds Eye)	⅓ of 10-oz. pkg.	3.0	22
Spears with Hollandaise sauce (Birds Eye)	⅓ of 10-oz. pkg.	3.2	97
ASPARAGUS SOUP, Cream of, canned:			
Condensed (USDA)	8 oz. (by wt.)	4.5	123
*Prepared with equal volume water (USDA)	1 cup (8.5 oz.)	2.4	65
*Prepared with equal volume milk (USDA)	1 cup (8.5 oz.)	6.7	144
*(Campbell)	1 cup	2.2	80
AVOCADO, peeled, pitted (USDA):			
All commercial varieties:			
Whole	1 lb. (weighed with seed & skin)	7.1	568

Food and Description	Measure or Quantity	Protein (grams)	Calories
Diced	½ cup (2.6 oz.)	1.6	124
Mashed	½ cup (4.1 oz.)	2.4	194
California varieties, mainly Fuerte:			
Whole	½ avocado (3⅛″ dia.)	2.4	185
½″ cubes	½ cup (2.7 oz.)	1.7	130
Florida varieties:			
Whole	½ avocado (3⅝″ dia.)	2.0	195
½″ cubes	½ cup (2.7 oz.)	1.0	97
*AWAKE (Birds Eye)	½ cup (4.4 oz.)	Tr.	55
AYDS, vanilla or chocolate	1 piece (7 grams)	.2	26

B

BABY FOOD:

Apple:			
& apricot, junior (Beech-Nut)	7¾ oz.	.4	212
& apricot, strained (Beech-Nut)	4¾ oz.	.3	123
& cranberry, junior (Heinz)	7¾ oz.	.3	191
& cranberry, strained (Heinz)	4¾ oz.	.1	124
& honey, junior (Heinz)	7½ oz.	1.4	155
& honey with tapioca, strained (Heinz)	4½ oz.	.8	89
& pear, junior (Heinz)	7¾ oz.	.5	180
& pear, strained (Heinz)	4½ oz.	.3	108
Dutch, dessert (Gerber):			
Junior	7⁸⁄₁₀ oz.	.3	207
Strained	4⁷⁄₁₀ oz.	.1	126
Apple-apricot juice, strained (Heinz)	4½ fl. oz.	.2	92
Apple Betty (Beech-Nut):			
Junior	7¾ oz.	.7	234
Strained	4¾ oz.	.5	147
Apple-cherry juice:			
Strained (Beech-Nut)	4⅕ fl. oz. (4.4 oz.)	.1	74
Strained (Gerber)	4⅕ fl. oz.	.3	59
Strained (Heinz)	4½ fl. oz.	.2	86
Apple-grape juice:			
Strained (Beech-Nut)	4⅕ fl. oz. (4.4 oz.)	.1	81
Strained (Heinz)	4½ fl. oz.	.2	89

*Prepared as Package Directs

Food and Description	Measure or Quantity	Protein (grams)	Calories
Apple juice:			
Strained (Beech-Nut)	4⅕ fl. oz. (4.4 oz.)	0.	57
Strained (Gerber)	4⅕ fl. oz. (4.6 oz.)	.2	65
Strained (Heinz)	4½ fl. oz.	.1	88
Apple pie (Heinz):			
Junior	7¾ oz.	.7	219
Strained	4¾ oz.	.4	133
Apple-pineapple juice, strained (Heinz)	4½ fl. oz.	.2	92
Apple-prune & honey (Heinz):			
Junior	7½ oz.	1.3	181
With tapioca, junior	1 jar	1.3	181
With tapioca, strained	4½ oz.	.9	107
Apple-prune juice, strained (Heinz)	4½ fl. oz.	.3	89
Applesauce:			
Junior (Beech-Nut)	7¾ oz.	.2	197
Junior (Gerber)	7⁸⁄₁₀ oz.	.4	185
Junior (Heinz)	7¾ oz.	.4	182
Strained (Beech-Nut)	4¾ oz.	.1	121
Strained (Gerber)	4⁷⁄₁₀ oz.	.2	113
Strained (Heinz)	4½ oz.	.2	98
& apricots, junior (Gerber)	7⁸⁄₁₀ oz.	.7	197
& apricots, junior (Heinz)	7¾ oz.	.8	174
& apricots, strained (Gerber)	4⁷⁄₁₀ oz.	.4	120
& apricots, strained (Heinz)	4¾ oz.	.3	98
& cherries, junior (Beech-Nut)	7¾ oz.	.4	206
& cherries, strained (Beech-Nut)	4¾ oz.	.3	127
& pineapple, junior (Gerber)	7⁸⁄₁₀ oz.	.5	171
& pineapple, strained (Gerber)	4⁷⁄₁₀ oz.	.3	109
& raspberries, junior (Beech-Nut)	7¾ oz.	.4	243
& raspberries, strained (Beech-Nut)	4¾ oz.	.1	143
Apricot with tapioca:			
Junior (Beech-Nut)	7¾ oz.	.7	188
Junior (Gerber)	7⁸⁄₁₀ oz.	.8	181
Junior (Heinz)	7¾ oz.	.6	224
Strained (Beech-Nut)	4¾ oz.	.4	103
Strained (Gerber)	4⁷⁄₁₀ oz.	.5	109

Food and Description	Measure or Quantity	Protein (grams)	Calories
Strained (Heinz)	4¾ oz.	1.2	141
Banana:			
Strained (Heinz)	4½ oz.	.5	105
Dessert, junior (Beech-Nut)	7¾ oz.	.7	204
Pie, junior (Heinz)	7¾ oz.	1.3	207
Pie, strained (Heinz)	4¾ oz.	.8	116
& pineapple, junior (Heinz)	7¾ oz.	.6	167
& pineapple, strained (Heinz)	4¾ oz.	.2	102
& pineapple with tapioca:			
Junior (Beech-Nut)	7¾ oz.	.7	204
Junior (Gerber)	7⁸⁄₁₀ oz.	.6	183
Strained (Beech-Nut)	4¾ oz.	.7	129
Strained (Gerber)	4⁷⁄₁₀ oz.	.4	114
Pudding, junior (Gerber)	7⁸⁄₁₀ oz.	2.1	215
With tapioca:			
Strained (Beech-Nut)	4¾ oz.	.3	119
Strained (Gerber)	4⁷⁄₁₀ oz.	.4	118
Bean, green:			
Junior (Beech-Nut)	7¼ oz.	2.7	62
Strained (Beech-Nut)	4½ oz.	1.5	40
Strained (Gerber)	4½ oz.	1.7	42
Strained (Heinz)	4½ oz.	1.8	37
Creamed with bacon, junior (Gerber)	7½ oz.	3.9	145
In butter sauce (Beech-Nut):			
Junior	7¼ oz.	2.7	94
Strained	4½ oz.	1.7	58
With potatoes & ham, casserole, toddler (Gerber)	6⅕ oz.	4.2	142
Beef:			
Junior (Beech-Nut)	3½ oz.	12.9	87
Junior (Gerber)	3½ oz.	14.9	95
Strained (Beech-Nut)	3½ oz.	12.9	101
Strained (Gerber)	3½ oz.	13.4	90
Beef & beef broth (Heinz):			
Junior	3½ oz.	14.5	99
Strained	3½ oz.	14.0	92
Beef & beef heart, strained (Gerber)	3½ oz.	12.3	85
Beef dinner:			
Junior (Beech-Nut)	4½ oz.	6.7	120
Strained (Beech-Nut)	4½ oz.	6.8	134
& noodles, junior (Beech-Nut)	7½ oz.	4.2	138

Food and Description	Measure or Quantity	Protein (grams)	Calories
& noodles, junior (Gerber)	7½ oz.	5.8	109
& noodles, strained (Beech-Nut)	4½ oz.	2.3	79
& noodles, strained (Gerber)	4½ oz.	3.6	63
& noodles, strained (Heinz)	4½ oz.	2.6	59
With vegetables:			
Junior (Gerber)	4½ oz.	8.4	106
Strained (Gerber)	4½ oz.	7.8	106
Strained (Heinz)	4¾ oz.	8.0	109
With vegetables & cereal, junior (Heinz)	4¾ oz.	8.8	110
Beef lasagna, toddler (Gerber)	6⅕ oz.	7.6	137
Beef liver, strained (Gerber)	3½ oz.	14.0	92
Beef liver soup, strained (Heinz)	4½ oz.	4.6	57
Beef stew, toddler (Gerber)	6⅕ oz.	10.2	122
Beet:			
Strained (Gerber)	4½ oz.	1.7	52
Strained (Heinz)	4½ oz.	1.5	60
Blueberry buckle (Gerber):			
Junior	7⁸⁄₁₀ oz.	.5	187
Strained	4½ oz.	.2	106
Butterscotch pudding (Gerber):			
Junior	7½ oz.	3.6	198
Strained	4½ oz.	2.1	130
Caramel pudding (Beech-Nut):			
Junior	7¾ oz.	2.8	215
Strained	4¾ oz.	1.7	131
Carrot:			
Junior (Beech-Nut)	7½ oz.	2.1	81
Junior (Gerber)	7½ oz.	1.5	69
Junior (Heinz)	7¾ oz.	1.5	86
Strained (Beech-Nut)	4½ oz.	1.3	47
Strained (Gerber)	4½ oz.	.9	41
Strained (Heinz)	4½ oz.	.9	52
& pea, junior (Gerber)	7½ oz.	4.9	90
In butter sauce (Beech-Nut):			
Junior	7½ oz.	2.1	119
Strained	4½ oz.	1.3	72
Cereal, dry:			
Barley (Gerber)	3 T. (7 grams)	.8	27
Barley, instant (Heinz)	1 oz.	3.1	101
High protein (Gerber)	3 T. (7 grams)	2.5	27

Food and Description	Measure or Quantity	Protein (grams)	Calories
High protein, instant (Heinz)	1 oz.	10.1	99
Hi-protein (Beech-Nut)	1 oz.	9.9	104
Mixed (Beech-Nut)	1 oz.	4.1	106
Mixed (Gerber)	3 T. (7 grams)	.8	27
Mixed (Heinz)	1 oz.	3.7	100
Mixed, honey (Beech-Nut)	1 oz.	3.4	106
Mixed, with banana (Gerber)	3 T. (7 grams)	.8	28
Oatmeal (Beech-Nut)	1 oz.	3.8	109
Oatmeal (Gerber)	3 T. (7 grams)	1.1	28
Oatmeal, instant (Heinz)	1 oz.	4.0	112
Oatmeal, honey (Beech-Nut)	1 oz.	3.1	109
Oatmeal, with banana (Gerber)	3 T. (7 grams)	.9	28
Rice (Beech-Nut)	1 oz.	1.6	107
Rice (Gerber)	3 T. (7 grams)	.5	26
Rice, instant (Heinz)	1 oz.	2.4	101
Rice, honey (Beech-Nut)	1 oz.	1.4	106
Rice, with strawberry (Gerber)	3 T. (7 grams)	.5	28
With applesauce & banana:			
Junior (Gerber)	7⁸⁄₁₀ oz.	3.0	180
Strained (Gerber)	4⁷⁄₁₀ oz.	1.6	113
Strained (Heinz)	4¾ oz.	2.1	108
With egg yolks & bacon:			
Junior (Beech-Nut)	7½ oz.	5.3	201
Junior (Gerber)	7½ oz.	4.7	159
Junior (Heinz)	7½ oz.	5.7	164
Strained (Beech-Nut)	4½ oz.	3.5	120
Strained (Gerber)	4½ oz.	2.8	92
Strained (Heinz)	4½ oz.	2.9	112
With fruit, strained (Beech-Nut)	4¾ oz.	1.6	113
High protein with apple & banana, strained (Heinz)	4¾ oz.	5.8	132
Oatmeal, with applesauce & banana, junior (Gerber)	7⁸⁄₁₀ oz.	3.5	169
Oatmeal, with applesauce & banana, strained (Gerber)	4⁷⁄₁₀ oz.	1.9	101
Oatmeal, with fruit, strained (Beech-Nut)	4¾ oz.	1.9	99
Rice, with applesauce & banana, strained (Gerber)	4⁷⁄₁₀ oz.	.5	94

Food and Description	Measure or Quantity	Protein (grams)	Calories
Cheese:			
Cottage, creamed with pineapple:			
Junior (Beech-Nut)	7¾ oz.	5.7	186
Strained (Gerber)	4⁷⁄₁₀ oz.	8.0	179
Cottage, creamed with pineapple juice, strained (Beech-Nut)	4¾ oz.	3.6	200
Cottage, dessert, with pineapple (Gerber):			
Junior	7⁸⁄₁₀ oz.	6.8	199
Strained	4½ oz.	4.0	116
Cottage, with banana (Heinz):			
Junior	7¾ oz.	4.8	169
Strained	4½ oz.	2.4	97
Cherry vanilla pudding (Gerber):			
Junior	7⁸⁄₁₀ oz.	.8	191
Strained	4⁷⁄₁₀ oz.	.4	117
Chicken:			
Junior (Beech-Nut)	3½ oz.	13.4	97
Junior (Gerber)	3½ oz.	14.4	133
Strained (Beech-Nut)	3½ oz.	12.1	97
Strained (Gerber)	3½ oz.	13.5	131
Chicken & chicken broth (Heinz):			
Junior	3½ oz.	13.0	103
Strained	3½ oz.	13.7	114
Chicken dinner:			
Junior (Beech-Nut)	4½ oz.	6.8	105
Strained (Beech-Nut)	4½ oz.	6.9	110
Noodle:			
Junior (Beech-Nut)	7½ oz.	3.4	95
Junior (Gerber)	7½ oz.	4.3	95
Junior (Heinz)	7½ oz.	4.2	123
Strained (Beech-Nut)	4½ oz.	2.0	58
Strained (Gerber)	4½ oz.	2.9	62
Strained (Heinz)	4½ oz.	2.9	70
With vegetables, junior (Beech-Nut)	7½ oz.	3.8	98
With vegetables, junior (Gerber)	4½ oz.	8.1	114

Food and Description	Measure or Quantity	Protein (grams)	Calories
With vegetables, junior (Heinz)	4¾ oz.	9.5	128
With vegetables, strained (Beech-Nut)	4½ oz.	2.4	61
With vegetables, strained (Gerber)	4½ oz.	8.0	111
With vegetables, strained (Heinz)	4¾ oz.	8.9	124
Chicken soup:			
Junior (Heinz)	7½ oz.	3.5	113
Strained (Heinz)	4½ oz.	2.4	65
Cream of, junior (Gerber)	7½ oz.	5.2	118
Cream of, strained (Gerber)	4½ oz.	3.2	74
Chicken stew, toddler (Gerber)	6 oz.	8.2	130
Chicken sticks:			
Junior (Beech-Nut)	2½ oz.	9.9	146
Junior (Gerber)	2½ oz.	11.1	134
Junior (Heinz)	1 jar	6.9	79
Cookie, animal-shaped (Gerber)	1 cookie (6 grams)	.8	29
Cookie, assorted (Beech-Nut)	½ oz.	1.6	61
Corn, creamed:			
Junior (Gerber)	7½ oz.	3.2	135
Junior (Heinz)	7½ oz.	1.9	153
Strained (Beech-Nut)	4½ oz.	2.0	123
Strained (Gerber)	4½ oz.	1.9	83
Strained (Heinz)	4½ oz.	1.1	92
Custard:			
Junior (Beech-Nut)	7¾ oz.	3.3	210
Junior (Heinz)	7¾ oz.	3.8	213
Strained (Beech-Nut)	4½ oz.	1.9	125
Strained (Heinz)	4½ oz.	2.2	122
Chocolate, junior (Gerber)	7⁸⁄₁₀ oz.	4.5	212
Chocolate, strained (Beech-Nut)	4½ oz.	1.9	138
Chocolate, strained (Gerber)	4½ oz.	2.4	126
Vanilla, junior (Gerber)	7½ oz.	4.0	207
Vanilla, strained (Gerber)	4½ oz.	2.2	116
Egg yolk:			
Strained (Beech-Nut)	3⅓ oz.	9.0	181
Strained (Gerber)	3³⁄₁₀ oz.	9.2	187
Strained (Heinz)	3¼ oz.	9.6	189
& bacon, strained (Beech-Nut)	3⅓ oz.	10.1	174

Food and Description	Measure or Quantity	Protein (grams)	Calories
& ham, strained (Gerber)	3³⁄₁₀ oz.	9.3	182
Fruit (Heinz):			
Mixed, & honey, junior	7½ oz.	1.4	215
Mixed, & honey, with			
tapioca, strained	4½ oz.	.9	105
Fruit dessert:			
Junior (Heinz)	7¾ oz.	.6	207
Strained (Heinz)	4½ oz.	.3	124
Tropical, junior (Beech-Nut)	7¾ oz.	.2	208
With tapioca:			
Junior (Beech-Nut)	7¾ oz.	.4	221
Junior (Gerber)	7⁸⁄₁₀ oz.	.6	204
Strained (Beech-Nut)	4¾ oz.	.3	139
Strained (Gerber)	4⁷⁄₁₀ oz.	.4	121
Fruit juice:			
Mixed, strained (Beech-Nut)	4⅕ fl. oz. (4.4 oz.)	.2	79
Mixed, strained (Gerber)	4⅕ fl. oz. (4.6 oz.)	.3	77
Ham:			
Junior (Gerber)	3½ oz.	14.8	116
Strained (Beech-Nut)	3½ oz.	11.6	115
Strained (Gerber)	3½ oz.	13.7	113
Ham dinner:			
Junior (Beech-Nut)	4½ oz.	8.4	122
Strained (Beech-Nut)	4½ oz.	7.9	137
With vegetables, junior			
(Gerber)	4½ oz.	8.8	101
With vegetables, junior			
(Heinz)	4¾ oz.	8.5	148
With vegetables, strained			
(Gerber)	4½ oz.	8.4	102
With vegetables, strained			
(Heinz)	4¾ oz.	8.4	124
Lamb:			
Junior (Beech-Nut)	3½ oz.	13.9	98
Junior (Gerber)	3½ oz.	15.3	96
Strained (Beech-Nut)	3½ oz.	13.0	92
Strained (Gerber)	3½ oz.	15.1	96
& noodles, junior (Beech-Nut)	7½ oz.	5.1	151
Lamb & lamb broth (Heinz):			
Junior	3½ oz.	13.7	97
Strained	3½ oz.	13.5	86
Liver with liver broth, strained			
(Heinz)	3½ oz.	13.6	79

Food and Description	Measure or Quantity	Protein (grams)	Calories
Macaroni:			
Alphabets & beef casserole,			
toddler (Gerber)	6⅕ oz.	11.5	148
& bacon, junior (Beech-Nut)	7½ oz.	5.5	197
& beef with vegetables,			
junior (Beech-Nut)	7½ oz.	5.3	136
With tomato, beef & bacon:			
Junior (Gerber)	7½ oz.	5.0	134
Junior (Heinz)	7½ oz.	4.1	140
Strained (Gerber)	4½ oz.	2.7	80
Strained (Heinz)	4½ oz.	3.8	89
With tomato sauce, beef &			
bacon dinner, strained			
(Beech-Nut)	4½ oz.	3.1	108
Meat sticks, junior:			
(Beech-Nut)	2½ oz.	10.5	135
(Gerber)	2½ oz.	11.0	116
(Heinz)	1 jar	7.7	106
MBF:			
Concentrate	2 T. (1.2 oz.)	2.2	43
Concentrate	13-fl.-oz. can	27.7	558
*Diluted 1 to 1	2 T.	.9	17
Modilac, regular or with iron:			
Concentrate	2 T.	1.3	40
Concentrate	13-fl.-oz. can	17.5	522
*Diluted 1 to 1	2 T.	.7	20
Noodles & beef, junior (Heinz)	7½ oz.	4.8	111
Orange-apple juice, strained:			
(Beech-Nut)	4⅕ fl. oz. (4.4 oz.)	.6	98
(Gerber)	4⅕ fl. oz. (4.6 oz.)	.4	71
Orange-apple-banana juice,			
strained:			
(Gerber)	4⅕ fl. oz. (4.6 oz.)	.3	86
(Heinz)	4½ fl. oz.	.3	89
Orange-apricot juice, strained:			
(Beech-Nut)	4⅕ fl. oz. (4.4 oz.)	.9	117
(Gerber)	4⅕ fl. oz. (4.6 oz.)	.8	80
(Heinz)	4½ fl. oz.	.4	73
Orange-banana juice, strained			
(Beech-Nut)	4⅕ fl. oz. (4.4 oz.)	.9	122
Orange juice, strained:			
(Beech-Nut)	4⅕ fl. oz. (4.4 oz.)	.7	64
(Gerber)	4⅕ fl. oz. (4.6 oz.)	.7	65

Food and Description	Measure or Quantity	Protein (grams)	Calories
(Heinz)	4½ fl. oz.	.7	69
Orange-pineapple dessert, strained (Beech-Nut)	4¾ oz.	.4	161
Orange-pineapple juice, strained:			
(Beech-Nut)	4⅕ fl. oz. (4.4 oz.)	.6	104
(Gerber)	4⅕ fl. oz. (4.6 oz.)	.7	79
(Heinz)	4½ fl. oz.	.2	72
Orange pudding, strained:			
(Gerber)	4⁷⁄₁₀ oz.	1.5	135
(Heinz)	4½ oz.	.5	119
Pea:			
Strained (Beech-Nut)	4½ oz.	5.0	86
Strained (Gerber)	4½ oz.	4.5	62
Pea, creamed (Heinz):			
Junior	7¾ oz.	6.0	158
Strained	4½ oz.	2.5	82
Pea, in butter sauce (Beech-Nut):			
Junior	7¼ oz.	7.4	152
Strained	4½ oz.	5.5	105
Peach:			
Junior (Beech-Nut)	7¾ oz.	1.5	197
Junior (Gerber)	7⁸⁄₁₀ oz.	1.3	189
Junior (Heinz)	7½ oz.	1.2	247
Strained (Beech-Nut)	4¾ oz.	.9	119
Strained (Gerber)	4⁷⁄₁₀ oz.	.8	111
Strained (Heinz)	4½ oz.	.9	160
Peach cobbler (Gerber):			
Junior	7⁸⁄₁₀ oz.	.9	195
Strained	4⁷⁄₁₀ oz.	.5	119
Peach & honey (Heinz):			
Junior	7½ oz.	1.3	155
With tapioca, strained	4½ oz.	1.0	94
Peach Melba (Beech-Nut):			
Junior	7¾ oz.	.7	247
Strained	4¾ oz.	.4	153
Peach pie (Heinz):			
Junior	7¾ oz.	1.3	212
Strained	4¾ oz.	.8	128
Pear:			
Junior (Beech-Nut)	7½ oz.	.6	157
Junior (Gerber)	7⁸⁄₁₀ oz.	.7	164

Food and Description	Measure or Quantity	Protein (grams)	Calories
Junior (Heinz)	7¾ oz.	.9	167
Strained (Beech-Nut)	4½ oz.	.4	96
Strained (Gerber)	4⁷⁄₁₀ oz.	.5	100
Strained (Heinz)	4½ oz.	.4	98
Pear & pineapple:			
Junior (Beech-Nut)	7½ oz.	.8	167
Junior (Gerber)	7⁸⁄₁₀ oz.	.8	167
Junior (Heinz)	7¾ oz.	.9	158
Strained (Beech-Nut)	4½ oz.	.4	100
Strained (Gerber)	4⁷⁄₁₀ oz.	.5	102
Strained (Heinz)	4¾ oz.	.6	104
Pineapple dessert, strained (Beech-Nut)	4¾ oz.	.7	142
Pineapple-grapefruit juice drink, strained (Gerber)	4⅕ fl. oz. (4.6 oz.)	.3	76
Pineapple juice, strained (Heinz)	4½ fl. oz.	.4	72
Pineapple-orange dessert (Heinz):			
Junior	7¾ oz.	.4	204
Strained	4½ oz.	.2	115
Pineapple pie (Heinz):			
Junior	7¾ oz.	2.4	237
Strained	4¾ oz.	1.5	145
Plum with tapioca:			
Junior (Beech-Nut)	7¾ oz.	.4	217
Junior (Gerber)	7⁸⁄₁₀ oz.	.7	221
Strained (Beech-Nut)	4¾ oz.	.3	141
Strained (Gerber)	4⁷⁄₁₀ oz.	.4	134
Strained (Heinz)	4½ oz.	.4	129
Pork:			
Junior (Beech-Nut)	3½ oz.	13.0	110
Junior (Gerber)	3½ oz.	14.5	113
Strained (Beech-Nut)	3½ oz.	12.9	111
Strained (Gerber)	3½ oz.	13.7	109
Pork with pork broth, strained (Heinz)	3½ oz.	14.4	93
Potatoes, creamed, with ham, toddler (Gerber)	6 oz.	6.2	183
Pretzel (Gerber)	1 piece (5 grams)	.7	19
Prune-orange juice:			
Strained (Beech-Nut)	4⅕ fl. oz. (4.4 oz.)	.7	95
Strained (Gerber)	4⅕ fl. oz. (4.6 oz.)	.7	101

Food and Description	Measure or Quantity	Protein (grams)	Calories
Strained (Heinz)	4½ fl. oz.	.4	82
Prune with tapioca:			
Junior (Beech-Nut)	7¾ oz.	1.3	204
Junior (Gerber)	7⁸⁄₁₀ oz.	1.3	207
Strained (Beech-Nut)	4¾ oz.	.9	129
Strained (Gerber)	4⁷⁄₁₀ oz.	.8	122
Strained (Heinz)	4¾ oz.	.8	140
Raspberry cobbler (Gerber):			
Junior	7⁸⁄₁₀ oz.	.6	182
Strained	4½ oz.	.3	103
Spaghetti & meat balls, toddler (Gerber)	6⅕ oz.	10.2	137
Spaghetti, tomato sauce & beef:			
Junior (Beech-Nut)	7½ oz.	4.5	159
Junior (Gerber)	7½ oz.	6.2	148
Junior (Heinz)	7½ oz.	5.5	168
Spaghetti, tomato sauce & meat, strained (Heinz)	4½ oz.	2.5	96
Spinach, creamed:			
Junior (Gerber)	7½ oz.	5.8	99
Strained (Gerber)	4½ oz.	3.5	56
Strained (Heinz)	4½ oz.	2.5	56
Split pea with bacon, junior (Gerber)	7½ oz.	6.6	181
Split pea, vegetables & bacon:			
Junior (Heinz)	7½ oz.	9.3	213
Strained (Heinz)	4½ oz.	5.6	120
Split pea, vegetables & ham, junior (Beech-Nut)	7½ oz.	6.6	144
Squash:			
Junior (Beech-Nut)	7½ oz.	1.7	76
Junior (Gerber)	7½ oz.	1.8	65
Strained (Beech-Nut)	4½ oz.	1.0	46
Strained (Gerber)	4½ oz.	1.1	39
Strained (Heinz)	4½ oz.	.9	50
In butter sauce, junior (Beech-Nut)	7½ oz.	1.9	108
In butter sauce, strained (Beech-Nut)	4½ oz.	.9	64
Sweet potato:			
Junior (Beech-Nut)	7¾ oz.	2.2	136
Junior (Gerber)	7⁸⁄₁₀ oz.	3.1	158
Strained (Beech-Nut)	4½ oz.	1.2	73

Food and Description	Measure or Quantity	Protein (grams)	Calories
Strained (Gerber)	4⁷⁄₁₀ oz.	1.8	95
Strained (Heinz)	4½ oz.	1.8	84
In butter sauce, junior (Beech-Nut)	7¾ oz.	2.2	155
In butter sauce, strained (Beech-Nut)	4½ oz.	1.2	91
Teething biscuit (Gerber)	1 piece (.4 oz.)	1.3	43
Teething ring, honey (Beech-Nut)	½ oz.	1.2	56
Tuna with noodles, strained (Heinz)	4½ oz.	3.2	55
Turkey:			
Junior (Beech-Nut)	3½ oz.	14.3	100
Junior (Gerber)	3½oz.	15.5	105
Strained (Beech-Nut)	3½ oz.	13.4	104
Strained (Gerber)	3½ oz.	13.7	129
Turkey dinner:			
Junior (Beech-Nut)	4½ oz.	6.5	88
Strained (Beech-Nut)	4½ oz.	6.9	106
With rice:			
Junior (Gerber)	7½ oz.	4.3	94
Strained (Beech-Nut)	4½ oz.	1.7	64
Strained (Gerber)	4½ oz.	2.6	60
& vegetables, junior (Beech-Nut)	7½ oz.	2.3	87
With vegetables:			
Junior (Gerber)	4½ oz.	7.2	102
Strained (Gerber)	4½ oz.	7.1	97
Strained (Heinz)	4¾ oz.	7.7	94
Tutti frutti dessert (Heinz):			
Junior	7¾ oz.	1.4	187
Strained	4½ oz.	.8	108
Veal:			
Junior (Beech-Nut)	3½ oz.	13.9	88
Junior (Gerber)	3½ oz.	15.4	99
Strained (Beech-Nut)	3½ oz.	13.4	109
Strained (Gerber)	3½ oz.	13.4	89
Veal dinner:			
Junior (Beech-Nut)	4½ oz.	7.2	131
Strained (Beech-Nut)	4½ oz.	6.9	120
With vegetables:			
Junior (Gerber)	4½ oz.	8.1	85
Junior (Heinz)	4¾ oz.	9.3	109

Food and Description	Measure or Quantity	Protein (grams)	Calories
Strained (Gerber)	4½ oz.	8.1	83
Strained (Heinz)	4¾ oz.	8.9	84
Veal & veal broth (Heinz):			
Junior	3½ oz.	13.7	93
Strained	3½ oz.	14.6	90
Vegetables:			
Garden, strained (Beech-Nut)	4½ oz.	2.9	59
Garden, strained (Gerber)	4½ oz.	2.9	46
Mixed, junior (Gerber)	7½ oz.	3.2	89
Mixed, junior (Heinz)	7½ oz.	3.6	100
Mixed, strained (Gerber)	4½ oz.	1.6	51
Vegetables & bacon:			
Junior (Beech-Nut)	7½ oz.	2.5	157
Junior (Gerber)	7½ oz.	4.3	139
Junior (Heinz)	7½ oz.	2.8	153
Strained (Beech-Nut)	4½ oz.	1.7	92
Strained (Gerber)	4½ oz.	2.1	96
Strained (Heinz)	4½ oz.	2.1	65
Vegetables & beef:			
Junior (Beech-Nut)	7½ oz.	4.2	136
Junior (Gerber)	7½ oz.	5.2	110
Junior (Heinz)	7½ oz.	5.1	106
Strained (Beech-Nut)	4½ oz.	2.7	88
Strained (Gerber)	4½ oz.	2.5	68
Strained (Heinz)	4½ oz.	4.3	70
Vegetables & chicken (Gerber):			
Junior	7½ oz.	4.1	105
Strained	4½ oz.	2.6	54
Vegetables, dumplings, beef & bacon (Heinz):			
Junior	7½ oz.	4.5	145
Strained	4½ oz.	2.7	79
Vegetables, egg noodles & chicken (Heinz):			
Junior	7½ oz.	4.7	134
Strained	4½ oz.	3.0	80
Vegetables, egg noodles & turkey, junior (Heinz)	7½ oz.	4.2	113
Vegetables & ham:			
Junior (Heinz)	7½ oz.	4.7	132
Strained (Beech-Nut)	4½ oz.	2.9	82
With bacon, junior (Gerber)	7½ oz.	3.6	124
With bacon, strained (Gerber)	4½ oz.	2.1	71

[23]

Food and Description	Measure or Quantity	Protein (grams)	Calories
With bacon, strained (Heinz)	4½ oz.	4.1	79
Vegetables & lamb:			
Junior (Beech-Nut)	7½ oz.	4.5	125
Junior (Gerber)	7½ oz.	4.2	108
Junior (Heinz)	7½ oz.	6.0	117
Strained (Gerber)	4½ oz.	2.6	66
Strained (Heinz)	4½ oz.	3.1	56
Vegetables & liver:			
Junior (Beech-Nut)	7½ oz.	5.1	98
Strained (Beech-Nut)	4½ oz.	2.9	56
With bacon, junior (Gerber)	7½ oz.	4.2	107
With bacon, strained (Gerber)	4½ oz.	2.9	78
Vegetables & turkey:			
Junior (Gerber)	7½ oz.	3.6	91
Strained (Gerber)	4½ oz.	2.3	58
Toddler, casserole (Gerber)	6⅕ oz.	9.7	161
Vegetable soup:			
Junior (Beech-Nut)	7½ oz.	2.3	85
Junior (Heinz)	7½ oz.	3.5	106
Strained (Beech-Nut)	4½ oz.	1.4	51

BAC ONION:

(Lawry's)	1 pkg. (3½ oz.)	17.4	372
(Lawry's)	1 tsp. (4 grams)	.6	14

BAC*OS (General Mills)

	1 T.	3.8	29

BACON, cured:

Raw:

(USDA) slab	1 lb. (weighed with rind)	35.8	2836
(USDA) sliced	1 lb.	38.1	3016
(USDA) sliced	1 oz.	2.4	189
(Hormel) Black Label	1 piece (.8 oz.)	2.1	125
(Hormel) *Range Brand*	1 piece (1.6 oz.)	5.1	275
(Wilson)	1 oz.	2.3	169

Broiled or fried, crisp, drained:

(USDA) thin slice	1 slice (5 grams)	1.5	31
(USDA) medium slice	1 slice (8 grams)	2.3	46
(USDA) thick slice	1 slice (.4 oz.)	3.6	73
(Oscar Mayer) 11-14 slices per lb., raw	1 slice (.4 oz.)	2.4	67

(USDA): United States Department of Agriculture
*Prepared as Package Directs

Food and Description	Measure or Quantity	Protein (grams)	Calories
(Oscar Mayer) 18-26 slices per lb., raw	1 slice (6 grams)	1.3	36
(Oscar Mayer) 25-30 slices per lb., raw	1 slice (4 grams)	.9	24
Canned (USDA)	3 oz.	7.2	582
BACON BITS:			
(Wilson)	1 oz.	10.7	139
Imitation:			
(Durkee)	1 tsp. (2 grams)	.7	8
(French's)	1 tsp. (2 grams)	.6	7
(McCormick)	1 tsp. (2 grams)	.8	8
BACON, CANADIAN:			
Unheated:			
(USDA)	1 oz.	5.7	61
(Oscar Mayer)	1-oz. slice	6.2	45
(Wilson)	1 oz.	5.3	42
Broiled or fried, drained			
(USDA)	1 oz.	7.8	79
BAGEL:			
Egg or water (USDA)	3" dia. (1.9 oz.)	6.0	165
(Lender's):			
Egg	2-oz. bagel	6.1	164
Garlic, onion or poppyseed	2-oz. bagel	6.0	161
Plain	2-oz. bagel	6.1	156
Pumpernickel or rye	2-oz. bagel	5.7	160
Raisin	2-oz. bagel	6.2	166
BAKING POWDER (USDA)	1 tsp.	Tr.	6
BAKON DELITES (Wise):			
Regular	½-oz. bag	8.7	72
Barbecue flavor	½-oz. bag	8.4	70
BAMBOO SHOOT, raw (USDA):			
Untrimmed	½ lb. (weighed untrimmed)	1.7	18
Trimmed	4 oz.	2.9	31

(USDA): United States Department of Agriculture

Food and Description	Measure or Quantity	Protein (grams)	Calories
BANANA:			
Common:			
Fresh:			
Whole (USDA)	1 lb. (weighed with skin)	3.4	262
Small size (USDA)	4.9-oz. banana (7¾" × 1¹¹⁄₃₂")	1.0	81
Medium size (USDA)	6.2-oz. banana (8¾" × 1¹³⁄₃₂")	1.3	101
Large size (USDA)	7-oz. banana (9¾" × 1⁷⁄₁₆")	1.5	116
Chunks (USDA)	1 cup (5 oz.)	1.6	122
Mashed (USDA)	1 cup (2 med., 7.8 oz.)	2.4	189
Sliced (USDA)	1 cup (1¼ med., 5.1 oz.)	1.6	124
Small size (Del Monte)	1 peeled banana (3.5 oz.)	1.1	84
Unpeeled (Dole)	7" banana (4.7 oz.)	1.5	114
Dehydrated (USDA):			
Flakes	½ cup (1.8 oz.)	2.2	170
Powder	1 oz.	1.2	96
Red, fresh, whole (USDA)	1 lb. (weighed with skin)	3.7	278
Red, fresh, peeled (USDA)	4 oz.	1.4	102
BANANA, BAKING (see **PLANTAIN**)			
BANANA CAKE MIX:			
*(Betty Crocker) layer	¹⁄₁₂ of cake	3.3	205
*(Duncan Hines)	¹⁄₁₂ of cake (2.7 oz.)	2.3	203
*(Pillsbury)	¹⁄₁₂ of cake	3.0	190
BANANA PIE cream or custard:			
Home recipe (USDA)	⅙ of 9" pie (5.4 oz.)	6.8	336
(Tastykake)	4-oz. pie	4.7	485
Frozen (Banquet)	2½-oz. serving	1.8	185
Frozen (Morton)	⅙ of 16-oz. pie	1.6	190
Frozen (Mrs. Smith's)	⅙ of 8" pie (2 oz.)	1.2	214
BANANA PUDDING, canned			
(Del Monte)	5-oz. container	3.3	187

(USDA): United States Department of Agriculture
*Prepared as Package Directs

Food and Description	Measure or Quantity	Protein (grams)	Calories
BANANA PUDDING & PIE MIX:			
*Instant (Jell-O)	½ cup (5.3 oz.)	4.3	178
*Instant (Royal)	½ cup (5.1 oz.)	4.4	176
*Regular (Jell-O)	½ cup (5.2 oz.)	4.3	173
*Regular (Royal)	½ cup (5.1 oz.)	4.3	163
BANANA SOFT DRINK, High Protein (Yoo-Hoo)	6 fl. oz.	6.0	100
BARBADOS CHERRY (see **ACEROLA**)			
BARBECUE DINNER MIX (Hunt's) Skillet	2-lb. 1-oz. pkg.	48.4	1404
BARBECUE SAUCE (see **SAUCE,** Barbecue)			
BARBECUE SEASONING (French's)	1 tsp. (2 grams)	.3	7
BARLEY:			
Pearled, dry:			
Light:			
(USDA)	¼ cup (1.8 oz.)	4.1	174
(Albers)	¼ cup	4.2	177
(Quaker Scotch)	¼ cup (1.7 oz.)	4.6	173
Pot or Scotch, dry (USDA)	2 oz.	5.4	197
BARRACUDA, raw, meat only (USDA)	4 oz.	23.8	128
BASS (USDA):			
Black sea:			
Raw, whole	1 lb. (weighed whole)	34.0	165
Baked, home recipe, stuffed with bacon, butter, onion, celery & bread crumbs	4 oz.	18.4	294

(USDA): United States Department of Agriculture
*Prepared as Package Directs

Food and Description	Measure or Quantity	Protein (grams)	Calories
Smallmouth & largemouth, raw:			
Whole	1 lb. (weighed whole)	26.6	146
Meat only	4 oz.	21.4	118
Striped:			
Raw, whole	1 lb. (weighed whole)	36.9	205
Raw, meat only	4 oz.	21.4	119
Oven-fried, made with milk, breadcrumbs & butter	4 oz.	24.4	222
White, raw, whole	1 lb. (weighed whole)	31.8	173
White, raw, meat only	4 oz.	20.4	111
BAVARIAN PIE FILLING			
canned (Lucky Leaf)	8 oz.	.2	306
BAVARIAN-STYLE BEANS & SPAETZLE, frozen (Birds Eye)	⅓ of 10-oz. pkg.	3.5	135
BEAN, BAKED:			
Canned with brown sugar sauce:			
(B & M) red kidney bean	1 cup (8 oz.)	20.1	360
(B & M) yellow eye bean	1 cup (8 oz.)	15.4	359
(Homemaker's) red kidney bean	1 cup (8 oz.)	15.4	337
Canned in molasses sauce:			
(Heinz)	1 cup (9¼ oz.)	15.4	283
& brown sugar sauce (Campbell)	1 cup	12.2	310
Canned in pork:			
(Campbell) Home Style	1 cup	14.4	302
(Hunt's) Snack Pack	5-oz. can	8.0	169
(Van Camp)	1 cup (7.7 oz.)	14.0	286
Canned with pork & molasses sauce:			
(USDA)	1 cup (9 oz.)	15.8	382
(B & M) Michigan Pea, New England-style	1 cup (7.9 oz.)	15.5	336
(Heinz) Boston-style	1 cup (8¾ oz.)	15.1	303

(USDA): United States Department of Agriculture

Food and Description	Measure or Quantity	Protein (grams)	Calories
(Homemaker's) Michigan Pea, New England-style	1 cup (8 oz.)	14.8	320
Canned with pork & tomato sauce:			
(USDA)	1 cup (9 oz.)	15.6	311
(Campbell)	1 cup	16.1	262
(Heinz)	1 cup (9¼ oz.)	14.8	293
(Morton House)	½ of 16-oz. can	12.6	280
Canned with tomato sauce:			
(USDA)	1 cup (9 oz.)	16.1	306
(Heinz) *Campside*	1 cup (9½ oz.)	15.1	350
(Heinz) Vegetarian	1 cup (9¼ oz.)	14.5	267
(Van Camp)	1 cup (8.3 oz.)	14.2	286
BEAN, BARBECUE (Campbell)	1 cup	13.0	287
BEAN, BAYO, dry (USDA)	4 oz.	25.3	384
BEAN 'N BEEF (Campbell)	1 cup	16.1	259
BEAN, BLACK, dry (USDA)	4 oz.	25.3	384
BEAN, BROWN, dry (USDA)	4 oz.	25.3	384
BEAN, CALICO, dry (USDA)	4 oz.	26.0	396
BEAN & FRANKFURTER, canned:			
(USDA)	1 cup (9 oz.)	19.4	367
(Campbell) in tomato & molasses sauce	1 cup	19.1	364
(Heinz)	1 can (8¾ oz.)	17.0	399
(Van Camp) *Beanie-Weenee*	1 cup (7.8 oz.)	16.6	316
BEAN & FRANKFURTER DINNER, frozen:			
(Banquet)	10¾-oz. dinner	16.8	528
(Morton)	12-oz. dinner	20.4	547
(Swanson)	11½-oz. dinner	17.9	610
BEAN, GREAT NORTHERN (see **BEAN, WHITE**)			

(USDA): United States Department of Agriculture

Food and Description	Measure or Quantity	Protein (grams)	Calories
BEAN, GREEN or SNAP:			
Fresh (USDA):			
Whole	1 lb. (weighed untrimmed)	7.6	128
1½" to 2" pieces	½ cup (1.8 oz.)	1.0	17
French-style	½ cup 1.4 oz.)	.8	13
Boiled, drained, whole (USDA)	½ cup (2.2 oz.)	1.0	16
Boiled, drained, 1½" to 2" pieces (USDA)	½ cup (2.4 oz.)	1.1	17
Canned, regular pack:			
Solids & liq. (USDA)	½ cup (4.2 oz.)	1.2	22
Drained solids, cut (USDA)	½ cup (2.5 oz.)	1.0	17
Drained liq. (USDA)	4 oz.	.5	11
Solids & liq. (Del Monte)	½ cup (4 oz.)	1.0	18
Solids & liq., any style (Green Giant)	¼ of 16-oz. can	.8	19
Solids & liq. (Stokely-Van Camp)	½ cup (3.9 oz.)	1.1	20
Drained solids (Butter Kernel)	½ cup (3.9 oz.)	1.1	20
Drained, cut or French style (Comstock)	½ cup (2.1 oz.)	.7	13
Drained, whole or cut (Del Monte)	½ cup (2.5 oz.)	.8	16
Seasoned, solids & liq. (Del Monte)	½ cup (4 oz.)	.9	19
Seasoned, drained solids (Del Monte)	½ cup (2.5 oz.)	.8	15
Seasoned, drained liq. (Del Monte)	4 oz.	.3	11
With bacon (Comstock)	4 oz.	1.4	26
With mushroom, solids & liq. (Comstock)	4 oz.	.9	15
Canned, dietetic pack:			
Solids & liq. (USDA)	4 oz.	1.2	18
Drained solids (USDA)	4 oz.	1.7	25
Drained liq. (USDA)	4 oz.	.5	9
Cut, solids & liq. (Blue Boy)	4 oz.	1.0	26
Cut (S and W) *Nutradiet* unseasoned	4 oz.	1.1	18
Solids & liq. (Diet Delight)	½ cup (4.2 oz.)	1.0	20
Solids & liq. (Tillie Lewis)	½ cup (4.2 oz.)	1.3	20

(USDA): United States Department of Agriculture

Food and Description	Measure or Quantity	Protein (grams)	Calories
Frozen:			
Whole (Birds Eye)	⅓ of 9-oz. pkg.	1.4	23
Cut, not thawed (USDA)	10-oz. pkg.	4.8	74
Cut, boiled, drained (USDA)	½ cup (2.8 oz.)	1.3	20
Cut (Birds Eye)	⅓ of 9-oz. pkg.	1.4	22
French-style, not thawed (USDA)	10-oz. pkg.	4.8	77
French-style, boiled, drained (USDA)	½ cup (2.8 oz.)	1.3	21
French-style (Birds Eye)	⅓ of 9-oz. pkg.	1.4	22
French-style with sliced mushrooms (Birds Eye)	⅓ of 9-oz. pkg.	1.4	26
French-style, with toasted almonds (Birds Eye)	½ cup (3 oz.)	2.1	52
In butter sauce, cut (Green Giant)	⅓ of 9-oz. pkg.	.8	34
In mushroom sauce, casserole (Green Giant)	⅓ of 12-oz. pkg.	1.1	52
In mushroom sauce, cut (Green Giant)	⅓ of 10-oz. pkg.	1.9	37
With onions & bacon bits in light sauce (Green Giant)	⅓ of 9-oz. pkg.	1.3	37

BEAN, ITALIAN (see **BROADBEAN**)

BEAN, KIDNEY or RED:

Dry:			
(USDA)	4 oz.	25.5	389
(USDA)	½ cup (3.3 oz.)	20.9	319
Cooked (USDA)	½ cup (3.3 oz.)	7.2	109
Canned:			
Solids & liq. (USDA)	½ cup (4.5 oz.)	7.3	115
Drained solids (Butter Kernel)	½ cup (4.2 oz.)	6.8	108
Red kidney & chili gravy (Nalley's)	4 oz.	6.4	120

BEAN, LIMA, young:

Raw, whole (USDA)	1 lb. (weighed in pod)	15.2	223
Raw, without shell (USDA)	1 lb. (weighed shelled)	38.1	558

(USDA): United States Department of Agriculture

Food and Description	Measure or Quantity	Protein (grams)	Calories
Boiled, drained (USDA)	½ cup (3 oz.)	6.5	94
Canned, regular pack:			
Solids & liq. (USDA)	½ cup (4.4 oz.)	5.1	88
Drained solids (USDA)	½ cup (3.1 oz.)	4.7	84
Drained liq. (USDA)	4 oz.	1.5	23
Drained solids (Del Monte)	½ cup (3.1 oz.)	5.2	89
Seasoned, drained solids (Del Monte)	½ cup (3.1 oz.)	6.2	88
Solids & liq. (Stokely-Van Camp)	½ cup (4.1 oz.)	4.7	82
With ham (Nalley's)	4 oz.	7.9	132
Canned, dietetic pack:			
Solids & liq., low sodium (USDA)	4 oz.	5.0	79
Drained solids, low sodium (USDA)	4 oz.	6.6	108
Solids & liq., unseasoned (Blue Boy)	4 oz.	6.6	79
Frozen:			
Baby butter beans (Birds Eye)	⅓ of 10-oz. pkg.	8.4	123
Baby limas:			
Not thawed (USDA)	4 oz.	8.6	138
Boiled, drained (USDA)	½ cup (3 oz.)	6.4	101
(Birds Eye)	½ cup (3.3 oz.)	6.9	111
In butter sauce (Green Giant)	⅓ of 10-oz. pkg.	4.7	116
Tiny (Birds Eye)	½ cup (2.5 oz.)	5.4	83
Fordhooks:			
Not thawed (USDA)	4 oz.	7.0	116
Boiled, drained (USDA)	½ cup (3 oz.)	5.0	83
(Birds Eye)	⅓ of 10-oz. pkg.	5.7	94
BEAN, LIMA, mature:			
Dry:			
Baby (USDA)	½ cup (3.4 oz.)	19.6	331
Large (USDA)	½ cup (3.1 oz.)	18.0	304
Boiled, drained (USDA)	½ cup (3.4 oz.)	7.8	131
BEAN, MUNG, dry (USDA)	½ cup (3.7 oz.)	25.4	357
BEAN, NAVY or PEA (see **BEAN, WHITE**)			

(USDA): United States Department of Agriculture

Food and Description	Measure or Quantity	Protein (grams)	Calories
BEAN, PINTO:			
Dry (USDA)	½ cup (3.4 oz.)	22.0	335
*(Uncle Ben's) including broth	¾ cup (6 oz.)	9.8	148
BEAN, RED (see **BEAN, KIDNEY** or **BEAN, RED MEXICAN**)			
BEAN, RED MEXICAN, dry (USDA)	4 oz.	26.0	396
BEAN SALAD: canned, solids & liq.:			
(Comstock)	4 oz.	2.0	81
(Hunt's) *Snack Pack*	5-oz. can	6.7	111
(Le Sueur)	¼ of 1-lb. 1-oz. can	2.4	86
BEAN, SEMI-MATURE, shelled, in brine, drained (B&M)	½ of 8¾-oz. can	11.3	155
BEAN SOUP, canned:			
*(Manischewitz)	1 cup	5.9	111
*With bacon (Campbell)	1 cup	7.7	152
With pork, condensed (USDA)	8 oz. (by wt.)	14.5	304
*With pork, prepared with equal volume water (USDA)	1 cup (8.8 oz.)	8.0	168
With smoked ham (Heinz) *Great American*	1 cup (8¾ oz.)	10.4	201
*With smoked pork (Heinz)	1 cup (8½ oz.)	6.4	157
BEAN SOUP, BLACK, canned *(Campbell)	1 cup	5.3	91
*** BEAN SOUP, LIMA,** canned (Manischewitz)	1 cup	4.5	93
BEAN SOUP MIX (Lipton) *Cup-a-Soup*	1.1-oz. pkg.	5.5	111
BEAN SOUP, NAVY, dehydrated (USDA)	1 oz.	5.0	93

(USDA): United States Department of Agriculture
*Prepared as Package Directs

[33]

Food and Description	Measure or Quantity	Protein (grams)	Calories
BEAN SPROUT:			
Mung:			
Raw (USDA)	½ lb.	8.6	80
Raw (USDA)	½ cup (1.6 oz.)	1.7	16
Boiled, drained (USDA)	½ cup (2.2 oz.)	2.0	17
Soy:			
Raw (USDA)	½ lb.	14.0	104
Raw (USDA)	½ cup (1.9 oz.)	3.3	25
Boiled, drained (USDA)	4 oz.	6.0	43
Canned (Hung's)	8 oz.	1.4	22
BEAN, WHITE, dry:			
Raw:			
Great Northern (USDA)	½ cup (3.1 oz.)	19.8	303
Navy or pea (USDA)	½ cup (3.7 oz.)	23.2	354
All other white (USDA)	4 oz.	25.3	386
Cooked:			
Great Northern (USDA)	½ cup (3 oz.)	6.6	100
Navy or pea (USDA)	½ cup (3.4 oz.)	7.5	113
All other white (USDA)	4 oz.	8.8	134
BEAN, YELLOW or WAX:			
Raw, whole (USDA)	1 lb. (weighed untrimmed)	6.8	108
Boiled, drained (USDA)	4 oz.	1.6	25
Boiled, 1″ pieces, drained (USDA)	½ cup (2.9 oz.)	1.1	18
Canned, regular pack:			
Solids & liq. (USDA)	½ cup (4.2 oz.)	1.2	23
Drained solids (USDA)	½ cup (2.2 oz.)	.9	15
Drained liq. (USDA)	4 oz.	.5	12
Drained solids (Butter Kernel)	½ cup (3.9 oz.)	1.1	20
Solids & liq. (Del Monte)	½ cup (4 oz.)	.6	18
Drained solids (Del Monte)	½ cup (2.5 oz.)	.6	14
Solids & liq., cut (Green Giant)	½ of 8.5-oz. can	1.2	19
Solids & liq. (Stokely-Van Camp)	½ cup (4.1 oz.)	1.1	22
Drained, cut or French style (Comstock)	½ cup (2.2 oz.)	1.0	16
Canned, dietetic pack:			
Solids & liq. (USDA)	4 oz.	1.0	17

(USDA): United States Department of Agriculture

[34]

Food and Description	Measure or Quantity	Protein (grams)	Calories
Drained solids (USDA)	4 oz.	1.4	24
Drained liq. (USDA)	4 oz.	.5	8
Solids & liq. (Blue Boy)	4 oz.	.9	24
Frozen:			
Cut, not thawed (USDA)	4 oz.	2.0	32
Boiled, drained (USDA)	4 oz.	1.9	31
Cut (Birds Eye)	⅓ of 9-oz. pkg.	1.4	24
BEAVER, roasted (USDA)	4 oz.	33.1	281
BEECHNUT:			
Whole (USDA)	4 oz. (weighed in shell)	13.4	393
Shelled (USDA)	1 oz.	5.5	161

BEEF. Values for beef cuts are given below for "lean and fat" and for "lean only." Beef purchased by the consumer at the retail store usually is trimmed to about one-half inch layer of fat. This is the meat described as "lean and fat." If all the fat that can be cut off with a knife is removed, the remainder is the "lean only." These cuts still contain flecks of fat known as "marbling" distributed through the meat. Cooked meats are medium done. Choice grade cuts (USDA):

Brisket:			
Raw, lean & fat	1 lb. (weighed with bone)	62.9	1284
Raw, lean & fat	1 lb. (weighed without bone)	74.8	1529
Raw, lean only	1 lb.	95.7	748
Lean & fat	4 oz.	26.0	467
Lean only	4 oz.	33.7	252
Chuck:			
Raw, lean & fat	1 lb. (weighed with bone)	71.6	984

(USDA): United States Department of Agriculture

Food and Description	Measure or Quantity	Protein (grams)	Calories
Raw, lean & fat	1 lb. (weighed without bone)	84.8	1166
Raw, lean only	1 lb.	96.6	717
Braised or pot-roasted:			
Lean & fat	4 oz.	29.5	371
Lean only	4 oz.	34.0	243
Dried (see **BEEF, CHIPPED**)			
Fat, separable, raw	1 oz.	1.9	203
Fat, separable, cooked	1 oz.	1.6	207
Filet Mignon. There are no data available on its composition. For dietary estimates, the data for sirloin steak, lean only, afford the closest approximation.			
Flank:			
Raw, 100% lean	1 lb.	98.0	653
Braised, 100% lean	4 oz.	34.6	222
Foreshank:			
Raw, lean & fat	1 lb. (weighed with bone)	50.8	531
Raw, lean only	1 lb.	98.0	603
Simmered:			
Lean & fat	4 oz.	31.6	310
Lean only	4 oz.	34.9	209
Ground:			
Regular:			
Raw	1 lb.	81.2	1216
Raw	1 cup (8 oz.)	40.5	606
Broiled	4 oz.	27.4	324
Lean:			
Raw	1 lb.	93.9	812
Raw	1 cup (8 oz.)	46.8	405
Broiled	4 oz.	31.1	248
Heel of round:			
Raw, lean & small amount of fat	1 lb.	89.8	966
Raw, lean only	1 lb.	98.0	612
Roasted:			
Lean & fat	4 oz.	30.6	296
Lean only	4 oz.	34.0	204

Food and Description	Measure or Quantity	Protein (grams)	Calories
Hindshank:			
Raw, lean & fat	1 lb. (weighed with bone)	38.1	604
Raw, lean & fat	1 lb. (weighed without bone)	82.6	1311
Raw, lean only	1 lb.	98.4	608
Simmered:			
Lean & fat	4 oz.	28.5	409
Lean only	4 oz.	34.8	209
Neck:			
Raw, lean & fat	1 lb. (weighed with bone)	70.4	820
Pot-roasted:			
Lean & fat	4 oz.	30.6	332
Lean only	4 oz.	34.5	222
Oxtail, raw	1 lb. (weighed with bone)	23.5	165
Oxtail, raw	1 lb. (weighed without bone)	94.3	662
Plate:			
Raw, lean & fat	1 lb. (weighed with bone)	59.7	1615
Raw, lean & fat	1 lb. (weighed without bone)	67.1	1814
Raw, lean only	1 lb.	95.7	744
Simmered:			
Lean & fat	4 oz.	23.4	538
Lean only	4 oz.	33.7	252
Rib roast:			
Raw, lean & fat	1 lb. (weighed with bone)	61.8	1673
Raw, lean & fat	1 lb. (weighed without bone)	67.1	1819
Raw, lean only	1 lb.	93.9	875
Roasted:			
Lean & fat	4 oz.	22.6	499
Lean only	4 oz.	32.0	273
Lean only, chopped	1 cup (4.5 oz.)	36.1	308
Lean only, diced	1 cup (5 oz.)	40.3	345
Round:			
Raw, lean & fat	1 lb. (weighed with bone)	88.5	863

Food and Description	Measure or Quantity	Protein (grams)	Calories
Raw, lean & fat	1 lb. (weighed without bone)	91.6	894
Raw, lean only	1 lb.	98.0	612
Broiled:			
Lean & fat	4 oz.	32.4	296
Lean only	4 oz.	35.5	214
Rump:			
Raw, lean & fat	1 lb. (weighed with bone)	67.0	1167
Raw, lean & fat	1 lb. (weighed without bone)	78.9	1374
Raw, lean only	1 lb.	96.2	717
Roasted:			
Lean & fat	4 oz.	26.8	393
Lean only	4 oz.	33.0	236
Steak, club:			
Raw, lean & fat	1 lb. (weighed with bone)	58.9	1443
Raw, lean & fat	1 lb. (weighed without bone)	70.3	1724
Raw, lean only	1 lb.	94.3	826
Broiled:			
Lean & fat	4 oz.	23.4	515
Lean only	4 oz.	33.6	277
One 8-oz. steak (weighed without bone before cooking) will give you:			
Lean & fat	5.9 oz.	34.2	754
Lean only	3.4 oz.	28.4	234
Steak, porterhouse:			
Raw, lean & fat	1 lb. (weighed with bone)	60.8	1603
Broiled:			
Lean & fat	4 oz.	22.3	527
Lean only	4 oz.	34.2	254
One 16-oz. steak (weighed with bone before cooking) will give you:			
Lean & fat	10.2 oz.	56.7	1339
Lean only	5.9 oz.	50.1	372
Steak, ribeye:			
Raw, lean & fat	1 lb. (weighed without bone)	67.1	1819

Food and Description	Measure or Quantity	Protein (grams)	Calories
One 10-oz. steak (weighed without bone before cooking) will give you:			
Lean & fat	7.3 oz.	41.2	911
Lean only	3.8 oz.	30.2	258
Steak, sirloin, double-bone:			
Raw, lean & fat	1 lb. (weighed with bone)	61.1	1240
Raw, lean & fat	1 lb. (weighed without bone)	74.4	1510
Raw, lean only	1 lb.	96.6	717
Broiled:			
Lean & fat	4 oz.	25.2	463
Lean only	4 oz.	34.7	245
One 16-oz. steak (weighed with bone) before cooking will give you:			
Lean & fat	8.9 oz.	55.9	1028
Lean only	5.9 oz.	50.8	359
One 12-oz. steak (weighed with bone before cooking) will give you:			
Lean & fat	6.6 oz.	41.7	767
Lean only	4.4 oz.	37.9	268
Steak, sirloin, hipbone:			
Raw, lean & fat	1 lb. (weighed with bone)	55.8	1585
Raw, lean & fat	1 lb. (weighed without bone)	65.8	1869
Raw, lean only	1 lb.	94.8	811
Broiled:			
Lean & fat	4 oz.	21.7	552
Lean only	4 oz.	33.8	272
Steak, sirloin, wedge & roundbone:			
Raw, lean & fat	1 lb. (weighed with bone)	71.1	1316
Raw, lean & fat	1 lb. (weighed without bone)	76.7	1420
Raw, lean only	1 lb.	97.5	649
Broiled:			
Lean & fat	4 oz.	26.1	439
Lean only	4 oz.	36.5	235

Food and Description	Measure or Quantity	Protein (grams)	Calories
Steak, T-bone:			
Raw, lean & fat	1 lb. (weighed with bone)	59.1	1596
Broiled:			
Lean & fat	4 oz.	22.1	536
Lean only	4 oz.	34.5	253
One 16-oz. steak (weighed with bone before cooking) will give you:			
Lean & fat	9.8 oz.	54.2	1315
Lean only	5.5 oz.	47.4	348
BEEFAMATO COCKTAIL			
(Mott's)	½ cup	.8	49
BEEFARONI, canned (Chef Boy-Ar-Dee)	⅕ of 40-oz. can	9.3	206
BEEF BOUILLON/BROTH, cubes or powder (see also **BEEF SOUP**):			
(Herb-Ox)	1 cube (4 grams)	.7	6
(Herb-Ox) instant	1 packet (4 grams)	.8	8
(Steero)	1 cube (4 grams)	.6	6
(Wyler's) instant	1 envelope (5 grams)	.9	11
(Wyler's) no salt added	1 cube (4 grams)	.4	10
BEEF & CABBAGE, casserole, frozen (Mrs. Paul's)	12-oz. pkg.	15.0	432
BEEF, CHIPPED:			
Uncooked:			
(USDA)	2 oz. (about ⅓ cup)	19.6	116
(USDA)	½ cup (2.9 oz.)	28.1	166
(Armour Star)	1 oz.	9.8	48
Cooked, creamed, home recipe (USDA)	1 cup (8.6 oz.)	20.1	377
Cooked, (Oscar Mayer) thin sliced	1 slice (5 grams)	1.1	7
Canned, creamed (Swanson)	1 cup	13.8	192
Frozen, creamed (Banquet)	5-oz. bag	13.0	126

(USDA): United States Department of Agriculture

Food and Description	Measure or Quantity	Protein (grams)	Calories
BEEF, CHOPPED or DICED,			
canned:			
(Armour Star)	12-oz. can	49.0	1042
(Hormel)	12-oz. can	55.1	867
Freeze dry (Wilson) *Campsite:*			
Dry	2 oz.	45.8	257
*Reconstituted	4 oz.	35.3	198
BEEF, CORNED (see **CORNED BEEF**)			
BEEF DINNER, frozen:			
(Banquet)	11-oz. dinner	30.3	312
(Morton)	11-oz. dinner	28.7	292
(Swanson)	11½-oz. dinner	32.1	371
(Swanson) 3-course	15-oz. dinner	35.3	567
Beef steak & carrots (Weight Watchers)	10-oz. luncheon	24.4	412
Beef steak & cauliflower (Weight Watchers)	11-oz. luncheon	25.4	431
Chopped (Banquet)	11-oz. dinner	18.1	443
Chopped (Weight Watchers)	18-oz. dinner	37.0	665
Chopped sirloin (Swanson)	10-oz. dinner	26.9	447
Pot roast, includes whole oven-browned potatoes, peas & corn (USDA)	10 oz.	37.2	301
Sliced (Morton)	11-oz. dinner	28.1	290
Sliced (Morton) 3-course	16-oz. dinner	36.3	636
BEEF & EGGPLANT, casserole, frozen (Mrs. Paul's)	12-oz. pkg.	14.6	418
BEEF GOULASH:			
Canned (Heinz)	8½-oz. can	15.4	253
Seasoning mix (Lawry's)	1.7-oz. pkg.	4.8	127
BEEF & GREEN PEPPER, casserole, frozen (Mrs. Paul's)	12-oz. pkg.	15.2	384
BEEF, GROUND, seasoning mix:			
(Durkee)	1⅛-oz. pkg.	2.4	92
With onions (French's)	1⅛-oz. pkg.	2.0	79

(USDA): United States Department of Agriculture
*Prepared as Package Directs   [41]

Food and Description	Measure or Quantity	Protein (grams)	Calories
BEEF HASH, ROAST:			
(Hormel) *Mary Kitchen*	7½ oz.	20.2	390
(Stouffer's)	11½-oz. pkg.	34.6	460
BEEF JERKY:			
(General Mills)	1 piece (¼ oz.)	3.8	25
(Lowrey's)	1 piece (¼ oz.)	2.0	21
BEEF PATTIES:			
Canned, freeze dry (Wilson)			
Campsite:			
Dry	2-oz. can	33.7	280
*Reconstituted	4 oz.	27.1	216
Frozen (Morton House):			
& Burgundy sauce	4⅙-oz. serving	9.5	154
& Italian sauce	4⅙-oz. serving	9.7	171
& Mexican sauce	4⅙-oz. serving	9.5	157
BEEF PIE:			
Baked, home recipe (USDA)	4¼″ pie (8 oz. before baking)	22.9	558
Baked, home recipe (USDA)	⅓ of 9″ pie (7.4 oz.)	21.2	517
Frozen:			
Commercial, unheated			
(USDA)	1 pie (7.6 oz.)	15.8	415
(Banquet)	8-oz. pie	16.3	409
(Banquet)	2-lb. 4-oz. pie	68.8	1311
(Morton)	8-oz. pie	15.9	368
(Stouffer's)	10-oz. pkg.	21.3	572
(Swanson)	8-oz. pie	15.6	434
(Swanson) deep dish	16-oz. pie	33.7	703
BEEF, POTTED (USDA)	1 oz.	5.0	70
BEEF PUFFS, frozen (Durkee)	1 piece (.5 oz.)	2.2	47
BEEF, ROAST, canned:			
(USDA)	4 oz.	28.0	254
(Wilson) *Tender Made*	4 oz.	26.8	134
BEEF SAUSAGE SNACK			
(Lowrey's)	1 piece (½ oz.)	3.0	88

(USDA): United States Department of Agriculture
*Prepared as Package Directs

Food and Description	Measure or Quantity	Protein (grams)	Calories
BEEF, SLICED, with barbecue sauce (Banquet)	5-oz. bag	13.8	152
BEEF SOUP, canned:			
*(Campbell)	1 cup	9.1	99
(Campbell) *Chunky*	1 cup	12.6	185
*Barley (Manischewitz)	1 cup	3.5	83
Bouillon, condensed (USDA)	8 oz. (by wt.)	9.5	59
*Bouillon, prepared with equal volume water (USDA)	1 cup (8.5 oz.)	5.0	31
Broth:			
Condensed (USDA)	8 oz. (by wt.)	9.5	59
*Prepared with equal volume water (USDA)	1 cup (8.5 oz.)	5.0	31
(Swanson)	1 cup	2.4	18
*Cabbage (Manischewitz)	1 cup	3.5	62
Consommé:			
Condensed (USDA)	8 oz. (by wt.)	9.5	59
*Prepared with equal volume water (USDA)	1 cup (8.5 oz.)	5.0	31
*(Campbell)	1 cup	5.6	33
Noodle:			
Condensed (USDA)	8 oz. (by wt.)	7.3	129
*Prepared with equal volume water (USDA)	1 cup (8.5 oz.)	3.8	67
*(Campbell)	1 cup	3.6	67
*(Heinz)	1 cup (8.5 oz.)	4.7	74
*(Manischewitz)	1 cup	4.5	64
*Curly, with chicken (Campbell)	1 cup	4.4	77
With dumplings (Heinz) *Great American*	1 cup (8¾ oz.)	6.2	109
Sirloin burger (Campbell) *Chunky*	1 cup	10.4	162
*Vegetable (Manischewitz)	1 cup	3.0	59
*With broth (Campbell)	1 cup	4.1	25
BEEF SOUP MIX:			
*Barley (Wyler's)	6 fl. oz.	1.6	54
Broth (Lipton) *Cup-a-Broth*	1 pkg. (8 grams)	.8	19
Noodle:			
(USDA)	1-oz.	3.9	110
*(USDA)	1 cup (8.1 oz.)	2.3	64

(USDA): United States Department of Agriculture
*Prepared as Package Directs

[43]

Food and Description	Measure or Quantity	Protein (grams)	Calories
(Lipton) *Cup-a-Soup*	1 pkg. (.4 oz.)	1.6	34
*With vegetable (Lipton)	1 cup	2.7	66
*(Wyler's)	6 fl. oz.	1.4	37
BEEF STEAK, freeze dry, canned (Wilson) *Campsite:*			
Dry	2-oz. can	46.9	229
*Reconstituted	4 oz.	38.1	199
BEEF STEW:			
Home recipe, made with lean beef chuck (USDA)	1 cup (8.6 oz.)	15.7	218
Canned:			
(USDA)	1 cup (8.6 oz.)	14.2	194
(Armour Star)	24-oz. can	35.4	590
(B & M)	1 cup (7.9 oz.)	12.8	152
(Dinty Moore)	8 oz.	13.4	190
(Heinz)	8½-oz. can	14.1	253
(Morton House)	⅓ of 24-oz. can	12.5	241
(Nalley's)	8 oz.	12.0	218
(Swanson)	1 cup	16.4	181
(Van Camp)	½ cup (4.6 oz.)	7.5	102
(Wilson)	15½-oz. can	25.0	343
Dietetic (Claybourne)	8-oz. can	19.3	365
Dietetic (Slim-ette)	8-oz. can	13.6	195
Meatball (Hormel)	1-lb. 8-oz. can	36.0	631
Freeze dry (Wilson) *Campsite:*			
Dry	4½-oz. can	40.7	576
*Reconstituted	16-oz.	40.4	536
Frozen, buffet (Banquet)	2-lb. pkg.	41.5	720
BEEF STEW SEASONING MIX:			
(Durkee)	1 pkg. (1¾ oz.)	.7	99
(French's)	1 pkg. (1⅞ oz.)	3.5	133
*(Kraft)	1 oz.	2.5	51
(Lawry's)	1 pkg. (1⅗ oz.)	5.7	131
BEEF STOCK BASE (French's)	1 tsp. (4 grams)	.4	9
BEEF STROGANOFF:			
Canned (Hormel)	1-lb. can	63.1	645
Mix (Chef Boy-Ar-Dee)	6⅔-oz. pkg.	9.6	232
Mix (Hunt's) *Skillet*	1-lb. 2-oz. pkg.	18.9	794

(USDA): United States Department of Agriculture
*Prepared as Package Directs

Food and Description	Measure or Quantity	Protein (grams)	Calories
Mix, dinner (Jeno's) *Add 'n Heat*	40-oz. pkg.	123.6	1928
*Mix, *Noodle-Roni*	4 oz.	6.7	126
Seasoning mix (Lawry's)	1½-oz. pkg.	4.8	118
BEEF & ZUCCHINI, casserole, frozen (Mrs. Paul's)	12-oz. pkg.	17.8	381
BEER and ALE, canned:			
Regular:			
(USDA) 4.5% alcohol	12 fl. oz. (12.7 oz.)	1.1	151
Black Horse Ale, 5% alcohol	12 fl. oz. (12.7 oz.)	1.4	162
Pfeifer, regular, 4.8% alcohol	12 fl. oz.	1.2	165
Pfeifer, 3.2 low gravity, 3.8% alcohol	12 fl. oz.	1.2	142
Schmidt, regular, 4.8% alcohol	12 fl. oz.	1.2	165
Schmidt, extra special, 4.8% alcohol	12 fl. oz.	1.2	165
Schmidt, 3.2 low gravity, 3.8% alcohol	12 fl. oz.	1.2	142
Low carbohydrate:			
Gablinger's, 4.5% alcohol	12 fl. oz.	.9	99
Meister Brau Lite, 4.6% alcohol	12 fl. oz.	.9	96
BEER, NEAR:			
Kingsbury (Heileman) 0.4% alcohol	12 fl. oz.	.6	62
Metbrew, 0.4% alcohol	12 fl. oz. (12.5 oz.)	.9	73
BEET:			
Raw (USDA)	1 lb. (weighed with skins, part tops)	3.6	96
Raw (USDA)	1 lb. (weighed with skins, without tops)	5.1	137
Raw, diced (USDA)	½ cup (2.4 oz.)	1.1	29
Boiled:			
Whole, drained (USDA)	2 beets (2" dia., 3.5 oz.)	1.1	32
Diced, drained (USDA)	½ cup (3 oz.)	.9	27

(USDA): United States Department of Agriculture
*Prepared as Package Directs

[45]

Food and Description	Measure or Quantity	Protein (grams)	Calories
Sliced, drained (USDA)	½ cup (3.6 oz.)	1.1	33
Canned, regular pack:			
Solid & liq. (USDA)	½ cup (4.3 oz.)	1.1	42
Drained solids, whole or diced (USDA)	½ cup (2.8 oz.)	.8	30
Drained solids, sliced (USDA)	½ cup (3.1 oz.)	.9	33
Drained liq. (USDA)	4 oz.	.9	29
Solids & liq. (Del Monte)	½ cup (4 oz.)	.8	30
Solids & liq. (Stokely-Van Camp)	½ cup (4.1 oz.)	1.0	39
Drained solids (Butter Kernel)	½ cup (4.1 oz.)	1.0	38
Drained solids, sliced (Comstock)	½ cup (2.6 oz.)	.9	26
Drained solids (Del Monte)	½ cup (3.1 oz.)	.7	26
Harvard, solids & liq. (Greenwood's)	½ cup (3.8 oz.)	1.1	50
Pickled, solids & liq. (Del Monte)	½ cup (4 oz.)	.6	75
Pickled, drained solids (Del Monte)	½ cup (3.1 oz.)	.6	62
Pickled, drained solids (Greenwood's)	½ cup (2.9 oz.)	1.0	40
Pickled, drained liq. (Del Monte)	4 oz.	.5	75
Pickled with onion, sliced, drained (Greenwood's)	½ cup (2.9 oz.)	1.1	45
Canned, dietetic pack:			
Solids & liq. (USDA)	4 oz.	1.0	36
Drained solids (USDA)	4 oz.	1.0	42
Drained liq. (USDA)	4 oz.	.9	28
Whole (Blue Boy)	10 small (3.5 oz.)	1.3	43
Diced, solids & liq. (Blue Boy)	4 oz.	.9	25
Diced, solids & liq. (Tillie Lewis)	½ cup (4.3 oz.)	1.1	40
Sliced (Blue Boy)	10 slices (3.5 oz.)	.9	32
Sliced (S and W) *Nutradiet*, unseasoned	4 oz.	.9	32
Frozen, sliced, in orange flavor glaze (Birds Eye)	⅓ of 10-oz. pkg.	1.1	54

(USDA): United States Department of Agriculture

Food and Description	Measure or Quantity	Protein (grams)	Calories
BEET GREENS (USDA):			
Raw, whole	1 lb. (weighed untrimmed)	5.6	61
Boiled leaves & stems, drained	½ cup (2.6 oz.)	1.2	13
BERRY PIE (Hostess)	4½-oz. pie	4.1	421
BEVERAGE (see individual listings)			
BIF (Wilson) canned luncheon meat	3 oz.	12.9	272
BIG MAC (McDonald's)	1 hamburger (6.5 oz.)	26.2	561
BIG WHEEL (Hostess)	1 cake (1.3 oz.)	1.3	185
BISCUIT:			
Baking powder, home recipe (USDA)	1-oz. biscuit (2″ dia.)	2.0	103
Egg (Stella D'oro):			
Dietetic	1 piece (.4 oz.)	1.6	42
Regular	1 piece (.4 oz.)	1.5	42
Roman	1 piece (1.1 oz.)	2.8	135
Sugared	1 piece (.5 oz.)	1.5	59
BISCUIT DOUGH:			
Frozen, commercial (USDA)	1 oz.	1.6	93
Refrigerated:			
Commercial (USDA)	1 oz.	2.1	79
(Borden):			
Big 10's, buttermilk, flaky or Southern style	1 biscuit (.9 oz.)	2.0	82
Buttered Up, flaky	1 biscuit (.9 oz.)	1.8	90
Buttermilk	1 biscuit (.8 oz.)	2.0	57
Gem, flaky	1 biscuit (.9 oz.)	2.0	82
Southern style	1 biscuit (.8 oz.)	2.0	57
(Pillsbury):			
Baking powder:			
1869 Brand	1 biscuit	1.5	80
Heat 'N Serve, *1869 Brand*	1 biscuit	1.5	100

(USDA): United States Department of Agriculture
*Prepared as Package Directs

Food and Description	Measure or Quantity	Protein (grams)	Calories
Tenderflake	1 biscuit	1.0	60
Ballard, ovenready	1 biscuit	1.5	60
Buttermelts	1 biscuit	1.0	60
Buttermilk:			
Regular	1 biscuit	1.5	60
Ballard	1 biscuit	1.5	60
1869 Brand	1 biscuit	1.5	80
Extra light	1 biscuit	1.5	55
Extra rich, *Hungry Jack*	1 biscuit	1.5	60
Flaky, *Hungry Jack*	1 biscuit	2.0	85
Heat 'N Serve, *1869 Brand*	1 biscuit	1.5	95
Butter Tastin':			
1869 Brand	1 biscuit	1.5	115
Hungry Jack	1 biscuit	1.5	90
Country style	1 biscuit	1.5	60
Flaky, *Hungry Jack*	1 biscuit	2.0	80
Fluffy, *Hungry Jack*	1 biscuit	1.5	90
BISCUIT MIX:			
Dry, with enriched flour (USDA)	1 oz.	2.2	120
*Baked from mix, with added milk (USDA)	1 oz. biscuit	2.0	92
Bisquick (Betty Crocker)	1 cup	8.7	503
BITS O'BACON (Wilson)	1 oz.	10.7	139
BLACKBERRY:			
Fresh (includes boysenberry, dewberry, youngberry) with hulls (USDA)	1 lb. (weighed untrimmed)	5.2	250
Hulled (USDA)	½ cup (2.6 oz.)	.9	42
Canned, regular, solids & liq. (USDA):			
Juice pack	4 oz.	.9	61
Light syrup	4 oz.	.9	82
Heavy syrup	½ cup (4.6 oz.)	1.0	118
Extra heavy syrup	4 oz.	.9	125
Canned, water pack, solids & liq. (USDA)	½ cup (4.3 oz.)	1.0	49

(USDA): United States Department of Agriculture
*Prepared as Package Directs

Food and Description	Measure or Quantity	Protein (grams)	Calories
Canned, low calorie, solids & liq. (S & W) *Nutradiet*	4 oz.	1.0	50
Frozen (USDA):			
Sweetened, not thawed	4 oz.	.9	109
Unsweetened, not thawed	4 oz.	1.4	54
BLACKBERRY JELLY:			
Sweetened (Smucker's)	1 T.(.7 oz.)	.1	50
Low calorie:			
(Slenderella)	1 T. (.7 oz.)	<.1	25
(Smucker's)	1 T.	<.1	4
& apple (Kraft)	1 oz.	<.1	34
BLACKBERRY JUICE, canned, unsweetened (USDA)	½ cup (4.3 oz.)	.4	45
BLACKBERRY PIE:			
Home recipe, 2 crust (USDA)	⅙ of 9″ pie (5.6 oz.)	4.1	384
(Tastykake)	4-oz. pie	3.5	386
Frozen (Banquet)	5-oz. serving	4.0	376
BLACKBERRY PIE FILLING, canned:			
(Comstock)	1 cup (10¾ oz.)	.1	438
(Lucky Leaf)	8 oz.	1.2	258
BLACKBERRY PRESERVE or JAM:			
Sweetened:			
(Bama)	1 T. (.7 oz.)	.1	54
(Smucker's)	1 T.	.3	53
Low calorie:			
(Diet Delight)	1 T. (.6 oz.)	<.1	22
(S & W) *Nutradiet*	1 T. (.5 oz.)	<.1	10
BLACK-EYED PEA, frozen (see also **COWPEA**):			
Not thawed (USDA)	10-oz. pkg.	25.5	371
Cooked, drained (USDA)	½ cup (3 oz.)	7.6	111
(Birds Eye)	½ cup (2.5 oz.)	6.4	92
BLANCMANGE (see **VANILLA PUDDING**)			

(USDA): United States Department of Agriculture

[49]

Food and Description	Measure or Quantity	Protein (grams)	Calories
BLOOD PUDDING or SAUSAGE (USDA)	1 oz.	4.0	112
BLOODY MARY MIX (Sacramento)	½ cup (4.3 oz.)	1.0	28
BLUEBERRY:			
Fresh, whole (USDA)	1 lb. (weighed untrimmed	2.9	259
Fresh, trimmed (USDA)	½ cup (2.6 oz.)	.5	45
Canned, solids & liq. (USDA):			
Syrup pack, extra heavy	½ cup (4.4 oz.)	.5	126
Water pack	½ cup (4.3 oz.)	.6	47
Frozen:			
Sweetened, solids & liq. (USDA)	½ cup (4 oz.)	.7	120
Unsweetened, solids & liq. (USDA)	½ cup (2.9 oz.)	.6	45
Quick thaw (Birds Eye)	½ cup (5 oz.)	.7	114
BLUEBERRY PIE:			
Home recipe, 2 crust (USDA)	⅙ of 9″ pie (5.6 oz.)	3.8	382
(Hostess)	4½-oz. pie	3.7	421
(Tastykake)	4-oz. pie	3.4	376
Frozen:			
(Banquet)	5-oz. serving	3.2	366
(Morton)	⅙ of 24-oz. pie	2.7	289
(Mrs. Smith's)	⅙ of 8″ pie (4.2 oz.)	2.0	294
(Mrs. Smith's)	⅛ of 10″ pie (5.6 oz.)	2.6	389
(Mrs. Smith's) natural juice	⅙ of 8″ pie (4.2 oz.)	2.7	347
Tart (Pepperidge Farm)	1 pie tart (3 oz.)	2.2	277
BLUEBERRY PIE FILLING, canned:			
(Comstock)	1 cup (10¾ oz.)	1.0	332
(Lucky Leaf)	8 oz.	1.0	256
(Wilderness)	21-oz. can	2.4	714
BLUEBERRY PRESERVE or JAM, sweetened:			
(Bama)	1 T. (.7 oz.)	.1	54
(Smucker's)	1 T.	Tr.	52

(USDA): United States Department of Agriculture

Food and Description	Measure or Quantity	Protein (grams)	Calories
BLUEBERRY TURNOVER:			
(Pepperidge Farm)	1 turnover (3.3 oz.)	2.8	321
(Pillsbury)	1 turnover	2.0	150
BLUEFISH (USDA):			
Raw, whole	1 lb. (weighed whole)	47.4	271
Raw, meat only	4 oz.	23.2	133
Baked or broiled with butter	4.4-oz. piece (3½″ × 3″ × ½″)	32.8	199
Fried, made with egg, milk or water & bread crumbs	5.3-oz. piece (3½″ × 3″ × ½″)	34.0	308
BOCKWURST (USDA)	1 oz.	3.2	75
BOLOGNA:			
All meat (USDA)	1 oz.	3.8	79
All meat, very thin slice (USDA)	½-oz. slice (3″ × ⅛″)	1.7	36
With cereal (USDA)	1 oz.	4.0	74
(Eckrich):			
Beef	1-oz. slice	3.5	92
Lunch or garlic	1. oz.	3.0	94
Thick sliced	1.7-oz. slice	5.0	150
All meat (Armour Star)	1 oz.	2.9	99
All meat (Hormel)	1 oz.	3.3	85
All meat (Oscar Mayer):			
8–10 slices per ¾ lb.	1 slice (1.3 oz.)	4.2	120
8 slices per ½ lb.	1 slice (1 oz.)	3.1	89
10 slices per ½ lb.	1 slice (.8 oz.)	2.5	73
Garlic	1-oz. slice	2.5	73
Wisconsin made, coarse	1 oz.	3.4	82
Wisconsin made, fine	1 oz.	3.1	88
Coarse ground (Hormel)	1 oz.	4.1	75
Fine ground (Hormel)	1 oz.	3.5	80
German brand (Oscar Mayer)	.8-oz. slice	3.4	55
Lebanon, pure beef (Oscar Mayer)	.8-oz. slice	4.8	46
Pure beef (Oscar Mayer):			
8 slices per ¾ lb.	1 slice (1.3 oz.)	4.2	120
10 slices per ½ lb.	1 slice (.8 oz.)	2.5	72
(Wilson)	1 oz.	3.5	87

(USDA): United States Department of Agriculture

[51]

Food and Description	Measure or Quantity	Protein (grams)	Calories
BONITO, raw (USDA):			
Whole	1 lb. (weighed whole)	63.1	442
Meat only	4 oz.	27.2	191
*****BORSCHT** (Manischewitz)	1 cup	.5	72
BOSCO (Best Foods)	1 T. (.7 oz.)	.6	57
BOSTON BROWN BREAD (see **BREAD**)			
BOSTON CREAM PIE:			
Home recipe (USDA)	1/12 of 8" pie (2.4 oz.)	3.4	208
Frozen (Mrs. Smith's)	1/6 of 8" pie (3.3 oz.)	2.5	329
*****Mix (Betty Crocker)	1/8 of pie	4.0	265
BOUILLON CUBE (see also individual flavors) (USDA):			
Flavor not indicated	1 cube (approx. ½", 4 grams)	.8	5
BOYSENBERRY:			
Fresh (see **BLACKBERRY,** fresh)			
Canned, unsweetened or low calorie:			
Water pack, solids & liq. (USDA)	4 oz.	.8	41
Solids & liq. (S and W) *Nutradiet*	4 oz.	.9	36
Frozen, not thawed (USDA):			
Sweetened	4 oz.	.9	109
Unsweetened	4 oz.	1.4	54
BOYSENBERRY JELLY			
Sweetened (Smucker's)	1 T. (.7 oz.)	.1	50
BOYSENBERRY PIE, frozen:			
(Banquet)	5-oz. serving	4.0	374
(Morton)	1/6 of 20-oz. pie	2.2	249

(USDA): United States Department of Agriculture
*Prepared as Package Directs

Food and Description	Measure or Quantity	Protein (grams)	Calories
BOYSENBERRY PIE FILLING			
(Comstock)	1 cup (10¾ oz.)	1.6	370
BOYSENBERRY PRESERVE or JAM			
Sweetened (Smucker's)	1 T. (.7 oz.)	.2	52
Low calorie:			
(S and W) *Nutradiet*	1 T. (.5 oz.)	.1	11
(Slenderella)	1 T. (.7 oz.)	<.1	25
(Tillie Lewis)	1 T. (.5 oz.)	Tr.	10
BRAINS, all animals, raw (USDA)	4 oz.	11.8	142
BRAN (USDA):			
With added sugar & defatted wheat germ	1 oz.	3.1	67
With added sugar & malt extract	1 oz.	3.6	68
BRAN BREAKFAST CEREAL:			
Plain:			
All-Bran (Kellogg's)	½ cup (1 oz.)	3.4	64
Bran Buds (Kellogg's)	⅓ cup (1 oz.)	3.3	73
40% bran flakes (USDA)	½ cup (6 oz.)	1.8	53
40% bran flakes (Kellogg's)	¾ cup (1 oz.)	3.0	70
40% bran flakes (Post)	⅔ cup (1 oz.)	2.5	97
100% bran (Nabisco)	½ cup (1 oz.)	3.1	97
Raisin bran flakes:			
(USDA)	½ cup (.9 oz.)	2.1	72
(Kellogg's)	¾ cup (1⅜ oz.)	3.4	101
(Post)	½ cup (1 oz.)	2.0	92
Cinnamon (Post)	½ cup (1 oz.)	2.0	92
BRATWURST (Oscar Mayer)	1 oz.	4.0	94
BRAUNSCHWEIGER:			
(USDA)	2 slices (2″ × ¼″, .7 oz.)	3.0	64
(Oscar Mayer)	1 oz.	3.7	107
(Wilson)	1. oz.	4.2	90
Liver cheese (Oscar Mayer)	1 slice (6 per ½ lb.)	6.1	102

(USDA): United States Department of Agriculture

Food and Description	Measure or Quantity	Protein (grams)	Calories
BRAZIL NUT (USDA):			
Whole	1 lb. (weighed in shell)	31.1	1424
Whole	1 cup (14 nuts, 4.3 oz. with shell)	8.4	383
Shelled	½ cup (2.5 oz.)	10.0	458
Shelled	4 nuts (.6 oz.)	2.5	114
BREAD (listed by type or brand name; toasting does not affect these nutritive values, only weight):			
Boston brown (USDA)	1.7-oz. slice (3" × ¾")	2.6	101
Cheese, party (Pepperidge Farm)	1 slice (6 grams)	.8	18
Cornbread (see **CORNBREAD**)			
Corn & Molasses (Pepperidge Farm)	.9-oz. slice	1.8	71
Cracked-wheat:			
(USDA)	.8-oz. slice	2.0	60
(USDA) 20 slices to 1 lb.	.9-oz. slice	2.2	66
(Pepperidge Farm)	.9-oz. slice	2.2	69
Honey (Wonder)	.8-oz. slice	1.9	61
Daffodil Farm (Wonder)	.8-oz. slice	2.2	58
Date-nut loaf (Thomas')	1.1-oz. slice	1.7	94
Finn Crisp	1 piece (6 grams)	.7	22
French:			
(USDA) 20 slices to 1 lb.	.8-oz. slice	2.1	67
(Pepperidge Farm)	1" slice (1.1 oz.)	2.7	84
(Wonder)	1-oz. slice	2.5	75
Glutogen Gluten (Thomas')	.5-oz. slice	2.1	35
Honey Wheatberry (Pepperidge Farm)	1.1-oz. slice	3.0	78
Italian:			
(USDA) 20 slices to 1 lb.	.8-oz. slice	2.1	63
(Pepperidge Farm)	1" slice (1.2 oz.)	2.9	88
Natural Health (Arnold)	.9-oz. slice	2.8	73
Oatmeal:			
(Arnold)	.8-oz. slice	2.0	64
(Pepperidge Farm)	.9-oz. slice	2.2	68
Profile, dark (Wonder)	.8-oz. slice	2.2	59
Profile, light (Wonder)	.8-oz. slice	2.1	59
Protogen Protein (Thomas')	.7-oz. slice	2.5	41

(USDA): United States Department of Agriculture

Food and Description	Measure or Quantity	Protein (grams)	Calories
Pumpernickel:			
(USDA) 20 slices to 1 lb.	.8-oz. slice	2.1	57
(Arnold) Jewish	1.4-oz. slice	3.6	104
(Pepperidge Farm) family	1.2-oz. slice	2.8	79
(Pepperidge Farm) party	1 slice (8 grams)	.7	20
(Wonder)	.8-oz. slice	2.0	54
Raisin:			
(USDA) 18 slices to 1 lb.	.9-oz. slice	1.7	66
Cinnamon (Pepperidge Farm)	.9-oz. slice	1.6	74
Cinnamon (Thomas')	.8-oz. slice	1.8	60
Orange (Arnold)	.9-oz. slice	2.0	76
Tea (Arnold)	.9-oz. slice	2.2	76
Rite Diet (Thomas')	.7-oz. slice	2.4	50
Roman Meal	.8-oz. slice	2.3	63
Rye:			
Light, 18 slices to 1 lb. (USDA)	.9-oz. slice	2.3	61
(Arnold) Melba thin, Jewish	.6-oz. slice	1.4	43
(Arnold) Jewish, seeded or unseeded	1.2-oz. slice	3.3	94
(Pepperidge Farm) family	1.2-oz. slice	2.7	82
(Pepperidge Farm) party	slice (6 grams)	.5	16
(Pepperidge Farm) seedless	1.2-oz. slice	2.7	82
Ry-King (Wasa):			
Brown	1 piece (.4 oz.)	1.0	40
Golden	1 piece (10 grams)	.9	34
Lite	1 piece (8 grams)	.7	28
Seasoned	1 piece (9 grams)	.8	33
(Wonder)	.8-oz. slice	2.3	55
(Wonder) *Beefsteak*	.8-oz. slice	2.0	59
Salt-rising (USDA)	.9-oz. slice	2.0	67
Seven grain (Mannafood)	.9-oz. slice	7.4	66
Soy & Wheat:			
(Mannafood)	.9-oz. slice	9.1	65
Salt free (Mannafood)	.9-oz. slice	8.9	60
Sprouted wheat (Mannafood)	.9-oz. slice	9.5	58
Toaster cake (see **TOASTER CAKE**)			
Vienna, 20 slices to 1 lb. (USDA)	.8-oz. slice	2.1	67
Wheat (Wonder):			
Golden	.9-oz. slice	2.1	60

(USDA): United States Department of Agriculture

Food and Description	Measure or Quantity	Protein (grams)	Calories
Home Pride	.8-oz. slice	2.1	59
Wheat berry (Wonder)	.8-oz. slice	2.2	56
Wheat germ (Pepperidge Farm)	.9-oz. slice	2.8	68
White, enriched or unenriched:			
Prepared with 1-2% nonfat dry milk (USDA)	.8-oz. slice	2.0	62
Prepared with 3-4% nonfat dry milk (USDA)	.8-oz. slice	2.0	62
Prepared with 5–6% nonfat dry milk (USDA)	.8-oz. slice	2.1	63
(Arnold) Melba thin, diet slice	.5-oz. slice	1.4	43
(Arnold) sandwich, soft, 1½-lb. loaf	.9-oz. slice	2.2	70
(Arnold) small family	.8-oz. slice	2.1	68
(Arnold) toasting	1.1-oz. slice	3.4	88
(Pepperidge Farm) large loaf	1-oz. slice	2.1	75
(Pepperidge Farm) large loaf, Calif. only	.8-oz. slice	1.9	65
(Pepperidge Farm) sandwich	.8-oz. slice	1.8	70
(Pepperidge Farm) sliced	.9-oz. slice	2.3	74
(Pepperidge Farm) toasting	1.2-oz. slice	2.8	84
(Pepperidge Farm) very thin slice, East	.5-oz. slice	1.5	41
(Pepperidge Farm) very thin slice, Mid-west	.6-oz. slice	1.6	44
(Thomas')	.9-oz. slice	2.3	69
(Wonder) 32 slices to 24-oz. loaf	¾-oz. slice	1.8	55
(Wonder) 30 slices to 24-oz. loaf	.8-oz. slice	1.9	59
(Wonder) 26 slices to 24-oz. loaf	.9-oz. slice	2.2	68
(Wonder) 24 slices to 24-oz. loaf	1-oz. slice	2.4	74
(Wonder) 22 slices to 24-oz. loaf	1.1-oz. slice	2.6	80
Brick Oven (Arnold) 1-lb. loaf	.8-oz. slice	2.1	68
Brick Oven (Arnold) 30-oz. loaf	1.1-oz. slice	2.7	85
Brick Oven, golden (Arnold) 2-lb. loaf	1-oz. slice	2.6	82

(USDA): United States Department of Agriculture

Food and Description	Measure or Quantity	Protein (grams)	Calories
English Tea Loaf			
(Pepperidge Farm)	.9-oz. slice	2.2	72
Hearthstone (Arnold) 1-lb.			
loaf	.9-oz. slice	2.2	71
Hearthstone (Arnold) 2-lb.			
loaf	1.1-oz. slice	2.6	84
Home Pride (Wonder)	.9-oz. slice	1.9	59
Whole grain wheat			
(Mannafood)	1-oz. slice	9.4	62
Whole-wheat:			
Prepared with 2% nonfat dry			
milk (USDA)	.9-oz. slice	2.6	61
Prepared with water (USDA)	.8-oz. slice	2.1	55
Prepared with water (USDA)	.9-oz. slice	2.3	60
(Arnold) Melba thin, diet			
slice	.6-oz. slice	1.6	44
(Arnold) small family	.8-oz. slice	2.3	65
(Pepperidge Farm)	.9-oz. slice	2.3	62
(Thomas') 100%	.9-oz. slice	2.6	64
(Wonder)	.8-oz. slice	2.2	59
Brick Oven (Arnold) 1-lb. loaf	.8-oz. slice	2.3	65
Brick Oven (Arnold) 2-lb. loaf	1.1-oz. slice	3.1	84
BREAD, CANNED:			
Banana nut (Dromedary)	½″ slice (1 oz.)	1.8	75
Brown, plain (B&M)	½″ slice (1. oz.)	2.4	83
Brown with raisins (B&M)	½″ slice (1.6 oz.)	2.6	83
Chocolate nut (Dromedary)	½″ slice (1 oz.)	1.6	86
Date & nut (Dromedary)	½″ slice (1 oz.)	1.5	74
Orange nut (Dromedary)	½″ slice (1 oz.)	1.5	78
BREAD CRUMBS:			
Dry, grated (USDA)	1 cup (3.5 oz.)	12.6	392
Dry, grated (USDA)	1 T. (6 grams)	.8	25
(Buitoni)	4 oz.	18.4	415
(Old London)	1 cup (4.5 oz.)	18.9	468
Seasoned (Contadina)	1 cup (4.1 oz.)	16.7	397
BREAD DOUGH, frozen			
(Morton)	1 oz.	2.2	82
BREAD MIX:			
*Apricot nut (Pillsbury)	¹⁄₁₂ of loaf	3.0	180

(USDA): United States Department of Agriculture
*Prepared as Package Directs

[57]

Food and Description	Measure or Quantity	Protein (grams)	Calories
*Banana or Blueberry (Pillsbury)	½₂ of loaf	2.0	150
*Cranberry (Pillsbury)	½₂ of loaf	3.0	170
*Date (Pillsbury)	½₂ of loaf	3.0	170
*Nut (Pillsbury)	½₂ of loaf	3.0	190
BREAD PUDDING, with raisins, home recipe, made with milk, butter, eggs & breadcrumbs (USDA)	1 cup (9.3 oz.)	14.8	496
BREAD STICK:			
Cheese (Keebler)	1 piece (3 grams)	.4	10
Dietetic (Stella D'oro)	1 piece (9 grams)	1.2	39
Garlic (Keebler)	1 piece (3 grams)	.4	11
Onion (Keebler)	1 piece (3 grams)	.4	10
Onion (Stella D'oro)	1 piece (.4 oz.)	1.2	42
Regular (Stella D'oro)	1 piece (10 grams)	1.1	40
Salt:			
(USDA)	1 piece (3 grams)	.4	12
Vienna type (USDA)	1 piece (3 grams)	.3	9
(Keebler)	1 piece (3 grams)	.4	10
Sesame (Keebler)	1 piece (3 grams)	.4	11
Sesame (Stella D'oro)	1 piece (9 grams)	1.3	42
BREAD STUFFING MIX:			
Dry (USDA)	1 cup (2.5 oz.)	9.2	263
*Crumb type, prepared with water & fat (USDA)	1 cup (5 oz.)	9.2	505
*Moist type, prepared with water, egg & fat (USDA)	1 cup (7.2 oz.)	8.9	422
Corn bread (Pepperidge Farm)	8-oz. pkg.	24.4	836
Cube (Pepperidge Farm)	7-oz. pkg.	23.4	756
Herb seasoned (Pepperidge Farm)	8-oz. pkg.	25.5	836
Seasoned (Uncle Ben's) *Stuff 'n Such*	6-oz. pkg.	21.9	615
*Seasoned (Uncle Ben's) *Stuff 'n Such,* no added butter	½ cup (2.9 oz.)	4.2	118
*(Uncle Ben's) *Stuff 'n Such,* with added butter	½ cup (3.3 oz.)	4.3	191

(USDA): United States Department of Agriculture
*Prepared as Package Directs

Food and Description	Measure or Quantity	Protein (grams)	Calories
BREADFRUIT, fresh (USDA):			
Whole	1 lb. (weighed untrimmed)	5.9	360
Peeled & trimmed	4 oz.	1.9	117
BRIGHT & EARLY	6 fl. oz. (6.6 oz.)	.1	100
BROADBEAN:			
Raw:			
Immature seed (USDA)	1 lb. (weighed in pod)	13.0	162
Immature seed (USDA)	1 oz. (without pod)	2.4	30
Mature seed, dry (USDA)	1 oz.	7.1	96
Canned, regular pack, drained (Del Monte)	½ cup (2.5 oz.)	1.9	24
Frozen, Italian bean (Birds Eye)	⅓ of 9-oz. pkg.	1.8	23
Frozen, Italian bean in tomato sauce (Green Giant)	⅓ of 10-oz. pkg.	1.9	50
BROCCOLI:			
Raw, whole (USDA)	1 lb. (weighed untrimmed)	10.0	89
Raw, large leaves removed (USDA)	1 lb. (weighed partially trimmed)	12.7	113
Boiled, drained (USDA):			
Whole stalk	1 stalk (6.3 oz.)	5.6	47
½" pieces	½ cup (2.8 oz.)	2.4	20
Frozen:			
Chopped or cut:			
Not thawed (USDA)	10-oz. pkg.	9.1	82
Boiled, drained (USDA)	1⅜ cups (10-oz. pkg.)	7.8	65
(Birds Eye)	⅓ of 10-oz. pkg.	2.9	27
In cheese sauce (Green Giant)	⅓ of 10-oz. pkg.	3.8	60
Spears:			
Not thawed (USDA)	10-oz. pkg.	9.4	79
Boiled, drained (USDA)	½ cup (3.3 oz.)	2.9	24
(Birds Eye)	⅓ of 10-oz. pkg.	3.0	26
& noodle casserole (Green Giant)	⅓ of 10-oz. pkg.	3.5	95
Baby spears (Birds Eye)	⅓ of 10-oz. pkg.	3.1	26

(USDA): United States Department of Agriculture

Food and Description	Measure or Quantity	Protein (grams)	Calories
In butter sauce (Green Giant)	⅓ of 10-oz. pkg.	2.1	48
In Hollandaise sauce (Birds Eye)	⅓ of 10-oz. pkg.	3.2	100
BROTH & SEASONING (see also individual kinds):			
Golden (George Washington)	1 packet (4 grams)	<.1	5
Rich brown (George Washington)	1 packet (4 grams)	<.1	5
BROTWURST (Oscar Mayer)	3-oz. link	11.9	275
BROWNIE (See **COOKIE**)			
BRUSSEL SPROUT:			
Raw trimmed (USDA)	1 lb.	20.4	188
Boiled, 1¼-1½" dia., drained (USDA)	1 cup (7–8 sprouts, 5.5 oz.)	6.5	56
Frozen:			
Not thawed (USDA)	10-oz. pkg.	9.4	102
Boiled, drained (USDA)	4 oz.	3.6	37
Au gratin, casserole (Green Giant)	⅓ of 10-oz. pkg.	4.2	71
Baby sprouts (Birds Eye)	½ cup (3.3 oz.)	3.1	34
In butter sauce (Green Giant)	⅓ of 10-oz. pkg.	2.8	58
BUCKWHEAT:			
Flour (see **FLOUR**)			
Groats:			
Whole-grain (USDA)	1 oz.	3.3	95
Cracked (Pocono)	1 oz.	2.6	104
Whole, brown (Pocono)	1 oz.	4.3	104
Whole, white (Pocono)	1 oz.	3.5	102
Wolff's Kasha (Birkett)	1 oz.	2.6	108
BUC WHEATS, cereal (General Mills)	1 cup	2.9	102
BUFFALOFISH, raw (USDA):			
Whole	1 lb. (weighed whole)	25.4	164

(USDA): United States Department of Agriculture

Food and Description	Measure or Quantity	Protein (grams)	Calories
Meat only	4 oz.	19.8	128
BULGUR, from hard red winter wheat (USDA):			
Dry	1 lb.	50.8	1605
Canned, seasoned	1 cup (4.8 oz.)	8.4	246
Canned, unseasoned	4 oz.	7.0	191
BULLHEAD, raw (USDA):			
Whole	1 lb. (weighed whole)	14.1	72
Meat only	4 oz.	18.5	95
BULLOCK'S-HEART (see **CUSTARD APPLE**)			
BUN (see **ROLL**)			
BURBOT, raw (USDA):			
Whole	1 lb. (weighed whole)	11.8	56
Meat only	4 oz.	19.7	93
BUTTER:			
(USDA)	¼ lb. (1 stick, ½ cup)	.7	812
(USDA)	1 T. (⅛ stick, .5 oz.)	.1	100
(USDA)	1 pat (1″ × 1″ × ⅓″, 5 grams)	<.1	72
(Breakstone)	1 T. (.5 oz.)	<.1	100
(Sealtest)	1 T. (.5 oz.)	<.1	110
Whipped (USDA)	2.7 oz. (1 stick, ½ cup)	.5	544
Whipped (USDA)	1 T. (⅛ stick, 9 grams)	.1	64
Whipped (USDA)	1 pat (1¼″ × 1¼″ × ⅓″, 4 grams)	<.1	29
Whipped (Breakstone)	1 T. (9 grams)	<.1	67
Whipped (Sealtest)	1 T. (9 grams)	.1	68
BUTTER BEAN (see **BEAN, LIMA**)			
*****BUTTER BRICKLE LAYER CAKE MIX** (Betty Crocker)	¹⁄₁₂ of cake	2.9	203

(USDA): United States Department of Agriculture
*Prepared as Package Directs

Food and Description	Measure or Quantity	Protein (grams)	Calories
BUTTERFISH, raw (USDA):			
Gulf:			
Whole	1 lb. (weighed whole)	37.5	220
Meat only	4 oz.	18.4	108
Northern:			
Whole	1 lb. (weighed whole)	41.9	391
Meat only	4 oz.	20.5	192
BUTTERMILK (see **MILK**)			
BUTTERNUT (USDA):			
Whole	1 lb. (weighed in shell)	15.1	399
Shelled	4 oz.	26.9	713
BUTTER OIL, or dehydrated butter (USDA)	1 cup (7.2 oz.)	.6	1787
***BUTTERSCOTCH CAKE MIX** (Pillsbury) *Bundt*	1/12 of cake	4.0	310
BUTTERSCOTCH PIE:			
Home recipe (USDA)	1/6 of 9″ pie (5.4 oz.)	6.7	406
Frozen, cream (Banquet)	2½-oz. serving	2.0	187
BUTTERSCOTCH PIE FILLING MIX (see **BUTTERSCOTCH PUDDING MIX**)			
BUTTERSCOTCH PUDDING:			
Chilled (Breakstone)	5-oz. container	2.0	252
Chilled (Sanna)	4¼ oz. container	3.1	167
Chilled (Sealtest)	4 oz.	3.0	124
Canned:			
(Betty Crocker)	½ cup	2.7	171
(Del Monte)	5-oz. can	3.3	191
(Hunt's)	5-oz. can	2.2	238
(Thank You)	½ cup (4.5 oz.)	2.5	169

(USDA): United States Department of Agriculture
*Prepared as Package Directs

Food and Description	Measure or Quantity	Protein (grams)	Calories
BUTTERSCOTCH PUDDING or			
PIE MIX:			
Sweetened:			
*Instant (Jell-O)	½ cup (5.3 oz.)	4.3	178
*Instant (Royal)	½ cup (5.1 oz.)	4.3	176
*Regular (Jell-O)	½ cup (5 oz.)	4.3	173
*Regular (Royal)	½ cup (5.1 oz.)	4.3	190
*Low calorie (D-Zerta)	½ cup (4.6 oz.)	4.8	107
B-V (Wilson)	1 tsp. (7 grams)	2.0	11

C

CABBAGE:			
White (USDA):			
Raw:			
Whole	1 lb. (weighed untrimmed)	4.7	86
Coarsely shredded or sliced	1 cup (2.5 oz.)	.9	17
Finely shredded or chopped	1 cup (3.2 oz.)	1.2	22
Wedge	3½″ × 4½″ wedge (3.5 oz.)	1.3	24
Boiled until tender:			
Shredded, small amount of water, drained	½ cup (2.6 oz.)	.8	15
Wedges, in large amount of water, drained	½ cup (3.2 oz.)	.9	17
Dehydrated	1 oz.	3.5	87
Red, raw (USDA):			
Whole	1 lb. (weighed untrimmed)	7.2	111
Coarsely shredded	1 cup (2.5 oz.)	1.4	22
Red, canned (Greenwood's)	½ cup (4.8 oz.)	1.6	77
Savoy, raw (USDA):			
Whole	1 lb. (weighed untrimmed)	8.6	86
Coarsely shredded	1 cup (2.5 oz.)	1.7	17
CABBAGE, CHINESE or			
CELERY, raw (USDA):			
Whole	1 lb. (weighed untrimmed)	5.3	62

(USDA): United States Department of Agriculture
*Prepared as Package Directs

Food and Description	Measure or Quantity	Protein (grams)	Calories
1" pieces, leaves with stalk	½ cup (1.3 oz.)	.5	5
CABBAGE, SPOON or WHITE MUSTARD or PAKCHOY (USDA):			
Raw	1 lb. (weighed untrimmed)	6.9	69
Boiled, drained	½ cup (3 oz.)	1.2	12
CAKE. Most cakes are listed elsewhere by kind of cake such as **ANGEL FOOD** or **CHOCOLATE** or brand name such as *YANKEE DOODLES.* (USDA):			
Plain, home recipe:			
Without icing	⅑ of 9" sq. cake (3" × 3" × 1", 3 oz.)	3.9	313
With boiled white icing	⅑ of 9" sq. cake (4 oz.)	4.3	401
With chocolate icing	1/16 of 10" layer cake (3.5 oz.)	4.2	368
With uncooked white icing	1/16 of 10" layer cake (3.5 oz.)	3.4	367
White, home recipe:			
Without icing	⅑ of 9" sq. cake (3" × 3" × 1", 3 oz.)	4.0	322
With coconut icing	1/16 of 10" layer cake (3.5 oz.)	3.7	371
With uncooked white icing	1/16 of 10" layer cake (3.5 oz.)	3.3	375
Yellow, home recipe:			
Without icing	⅑ of 9" sq. cake (3" × 3" × 1", 3 oz.)	3.9	312
With caramel icing	1/16 of 10" layer cake (3.5 oz.)	4.0	362
With chocolate icing, 2-layer	1/16 of 9" cake (2.6 oz.)	3.2	274
CAKE DECORATOR (Pillsbury)	1 T.	0.	70
CAKE FROSTING (see **CAKE ICING & CAKE ICING MIX**)			

(USDA): United States Department of Agriculture

Food and Description	Measure or Quantity	Protein (grams)	Calories
CAKE ICING:			
Butterscotch (Betty Crocker)	½2 of 16.5-oz. can	.2	164
Caramel, home recipe (USDA)	4 oz.	1.5	408
Cherry (Betty Crocker)	½2 of 16.5-oz. can	.2	167
Chocolate, home recipe (USDA)	1 cup (9.7 oz.)	8.8	1034
Chocolate (Betty Crocker)	½2 of 16.5-oz. can	.6	162
Coconut, home recipe (USDA)	1 cup (5.8 oz.)	3.2	604
Dark Dutch fudge (Betty Crocker)	½2 of 16.5-oz. can	.8	153
Lemon (Betty Crocker)	½2 of 16.5-oz. can	.2	166
Milk chocolate (Betty Crocker)	½2 of 16.5-oz. can	.4	164
Vanilla (Betty Crocker)	½2 of 16.5-oz. can	.2	166
White, boiled, home recipe (USDA)	1 cup (3.3 oz.)	1.3	297
White, uncooked, home recipe (USDA)	4 oz.	.6	426
CAKE ICING MIX:			
*Banana (Betty Crocker)	½2 of cake's icing	.1	139
*Butter Brickle (Betty Crocker)	½2 of cake's icing	Tr.	140
*Caramel (Betty Crocker)	½2 of cake's icing	.2	139
*Caramel apple (Betty Crocker)	½2 of cake's icing	.1	130
*Cherry, creamy (Betty Crocker)	½2 of cake's icing	.2	139
*Cherry fluff (Betty Crocker)	½2 of cake's icing	.4	59
*Cherry fudge (Betty Crocker)	½2 of cake's icing	.8	132
*Chocolate, fluffy (Betty Crocker)	½2 of cake's icing	.2	70
Chocolate fudge (USDA)	1 oz.	.7	116
*Chocolate fudge, prepared with water & fat (USDA)	8 oz.	5.0	857
*Chocolate fudge (Betty Crocker)	½2 of cake's icing	.8	134
*Chocolate malt (Betty Crocker)	½2 of cake's icing	.6	136
*Chocolate walnut (Betty Crocker)	½2 of cake's icing	.9	131
*Coconut-pecan (Betty Crocker)	½2 of cake's icing	.5	102
*Coconut, toasted (Betty Crocker)	½2 of cake's icing	.3	141
*Dark chocolate fudge (Betty Crocker)	½2 of cake's icing	.6	130

(USDA): United States Department of Agriculture
*Prepared as Package Directs

[65]

Food and Description	Measure or Quantity	Protein (grams)	Calories
Fudge, creamy, contains nonfat dry milk (USDA):			
Dry	1 oz.	.9	109
Prepared with water (USDA)	1 cup (8.6 oz.)	6.9	831
Prepared with water & fat (USDA)	8 oz.	5.9	869
*Fudge nugget (Betty Crocker)	1/12 of cake's icing	.7	133
*Lemon, creamy (Betty Crocker)	1/12 of cake's icing	Tr.	134
*Lemon fluff (Betty Crocker)	1/12 of cake's icing	.4	58
*Milk chocolate (Betty Crocker)	1/12 of cake's icing	.5	130
*Orange (Betty Crocker)	1/13 of cake's icing	.2	134
*Pineapple (Betty Crocker)	1/12 of cake's icing	Tr.	134
*Sour cream, chocolate fudge (Betty Crocker)	1/12 of cake's icing	.6	129
*Sour cream, white (Betty Crocker)	1/12 of cake's icing	.1	130
*Spice, creamy (Betty Crocker)	1/12 of cake's icing	.2	139
*White, creamy (Betty Crocker)	1/12 of cake's icing	.1	140
*White, fluffy (Betty Crocker)	1/12 of cake's icing	.4	58
CAKE MIX. Most cake mixes are listed by kind of cake, such as **ANGEL FOOD CAKE MIX, CHOCOLATE CAKE MIX,** etc.			
White:			
(USDA)	1 oz.	1.2	123
*Made with egg whites & water, with chocolate icing, 2-layers (USDA)	1/16 of 9" cake (2.5 oz.)	2.8	249
*(Betty Crocker) layer	1/12 of cake	2.7	190
*(Duncan Hines)	1/12 of cake (2.6 oz.)	2.2	190
*(Pillsbury)	1/12 of cake	3.0	200
*Sour cream (Betty Crocker) layer	1/12 of cake	2.8	191
*Whipping cream (Pillsbury)	1/12 of cake	4.0	230
Yellow:			
(USDA)	1 oz.	1.1	124
*Made with eggs & water, with chocolate icing (USDA)	1/16 of 9" cake (2.6 oz.)	3.1	253
*(Betty Crocker) layer	1/12 of cake	2.9	202

(USDA): United States Department of Agriculture
*Prepared as Package Directs

Food and Description	Measure or Quantity	Protein (grams)	Calories
*Butter recipe (Betty Crocker)	¹⁄₁₂ of cake	3.1	278
*(Duncan Hines)	¹⁄₁₂ of cake (2.7 oz.)	3.0	202
*Golden butter (Duncan Hines)	¹⁄₁₂ of cake (3.3 oz.)	3.7	283
*Butter (Pillsbury)	¹⁄₁₂ of cake	3.0	210
*(Swans Down)	¹⁄₁₂ of cake (2.5 oz.)	3.1	186

CANDIED FRUIT (see individual kinds)

CANDY. The following values of candies from the U.S. Department of Agriculture are representative of the types sold commercially. These values may be useful when individual brands or sizes are not known:

Almond:			
Chocolate-coated	1 oz.	3.5	161
Chocolate-coated	1 cup (6.3 oz.)	22.1	1024
Sugar-coated or Jordan	1 oz.	2.2	129
Butterscotch	1 oz.	Tr.	113
Candy corn	1 oz.	Tr.	103
Caramel:			
Plain	1 oz.	1.1	113
Plain with nuts	1 oz.	1.3	121
Chocolate	1 oz.	1.1	113
Chocolate with nuts	1 oz.	1.3	121
Chocolate-flavored roll	1 oz.	.6	112
Chocolate:			
Bittersweet	1 oz.	2.2	135
Milk:			
Plain	1 oz.	2.2	147
With almonds	1 oz.	2.6	151
With peanuts	1 oz.	4.0	154
Semisweet	1 oz.	1.2	144
Sweet	1 oz.	1.2	150
Chocolate discs, sugar-coated	1 oz.	1.5	132
Coconut center, chocolate-coated	1 oz.	.8	124
Fondant, plain	1 oz.	< .1	103

*Prepared as Package Directs

Food and Description	Measure or Quantity	Protein (grams)	Calories
Fondant, chocolate-covered	1 oz.	.5	116
Fudge:			
Chocolate fudge	1 oz.	.8	113
Chocolate fudge,			
chocolate-coated	1 oz.	1.1	122
Chocolate fudge with nuts	1 oz.	1.1	121
Chocolate fudge with nuts,			
chocolate-coated	1 oz.	1.4	128
Vanilla fudge	1 oz.	.9	113
Vanilla fudge with nuts	1 oz.	1.2	120
With peanuts & caramel			
chocolate-coated	1 oz.	2.7	130
Gumdrops	1 oz.	<.1	98
Hard	1 oz.	0.	109
Honeycombed hard candy,			
with peanut butter,			
chocolate-covered	1 oz.	1.9	131
Jelly beans	1 oz.	Tr.	104
Marshmallows	1 oz.	.6	90
Mints, uncoated	1 oz.	<.1	103
Nougat & caramel,			
chocolate-covered	1 oz.	1.1	118
Peanut bar	1 oz.	5.0	146
Peanut brittle	1 oz.	1.6	119
Peanuts, chocolate-covered	1 oz.	4.6	159
Raisins, chocolate-covered	1 oz.	1.5	120
Vanilla creams,			
chocolate-covered	1 oz.	1.1	123
CANDY, COMMERCIAL (see also **CANDY, DIETETIC**):			
Almonds, chocolate-covered:			
Candy-coated (Hershey's)	1 oz.	2.4	142
(Kraft)	1 piece (3 grams)	.3	14
Almond Cluster (Kraft)	1 piece (.4 oz.)	1.4	63
Almond Toffee Bar (Kraft)	1 oz.	1.2	142
Brazil nuts, chocolate-covered			
(Kraft)	1 piece (6 grams)	.6	32
Bridge mix:			
Almond (Kraft)	1 piece (4 grams)	.5	22
Caramelette (Kraft)	1 piece (3 grams)	.2	12
Jelly (Kraft)	1 piece (3 grams)	.1	12
Malted milk ball (Kraft)	1 piece (3 grams)	.2	11

Food and Description	Measure or Quantity	Protein (grams)	Calories
Mintette (Kraft)	1 piece (3 grams)	<.1	12
Peanut (Kraft)	1 piece (1 gram)	.3	8
Peanut crunch (Kraft)	1 piece (5 grams)	.4	23
Raisin (Kraft)	1 piece (1 gram)	<.1	5
(Nabisco)	1 piece (2 grams)	.1	8
Butternut (Hollywood)	1¼-oz. bar	4.0	168
Butterscotch Skimmers			
(Nabisco)	1 piece (6 grams)	Tr.	25
Caramel:			
Caramelette (Kraft)	1 piece (3 grams)	.2	12
Chocolate (Kraft)	1 piece (8 grams)	.5	33
Chocolate, bar (Kraft)	1 piece (6 grams)	.4	26
Coconut (Kraft)	1 piece (8 grams)	.4	32
Vanilla (Kraft)	1 piece (8 grams)	.5	33
Vanilla, bar (Kraft)	1 piece (6 grams)	.4	26
Vanilla, chocolate-covered			
(Kraft)	1 piece (9 grams)	.6	39
Vanilla, *Twisteroo* (Kraft)	1 piece (6 grams)	.3	25
Cashew cluster (Kraft)	1 piece (.4 oz.)	1.3	58
Charleston Chew:			
5¢ size	1 piece (¾ oz.)	.7	90
10¢ size	1 bar (1⅛ oz.)	1.1	149
Bite-size	1 piece (7 grams)	.2	30
Cherry, chocolate-covered:			
Dark (Nabisco)	1 piece (.6 oz.)	.2	67
Dark (Nabisco) *Welch's*	1 piece (.6 oz.)	.2	67
Milk (Nabisco)	1 piece (.6 oz.)	.2	66
Milk (Nabisco) *Welch's*	1 piece (.6 oz.)	.2	66
Chocolate bar:			
Milk chocolate:			
(Ghirardelli)	1.1-oz. bar	2.0	169
(Hershey's)	1¼-oz. bar	3.1	218
(Hershey's)	¼-oz. miniature	.5	39
Mint chocolate (Ghirardelli)	1.1-oz. bar	2.0	171
Semisweet (Ghirardelli) *Eagle*	1 sq. (1 oz.)	1.0	151
Special Dark (Hershey's)	1 oz.	1.1	145
Special Dark (Hershey's)	¼-oz. miniature	.3	37
Chocolate bar with almonds:			
(Ghirardelli)	1.1-oz. bar	2.3	173
(Hershey's)	1.3-oz.	3.2	205
Chocolate block, milk:			
(Ghirardelli)	1 sq. (1 oz.)	1.5	147
(Hershey's)	1 oz.	1.3	145

Food and Description	Measure or Quantity	Protein (grams)	Calories
Chocolate Crisp Bar (Ghirardelli)	1-oz. bar	1.9	161
Chocolate Crisp Bar (Kraft)	1 oz.	2.2	132
Cluster:			
Crispy (Nabisco)	1 piece (.6 oz.)	.4	65
Peanut, chocolate-covered (Kraft)	1 piece (.4 oz.)	1.8	59
Royal Clusters (Nabisco)	1 piece (.6 oz.)	1.5	78
Coco-Mello (Nabisco)	1 piece (.7 oz.)	.8	91
Coconut:			
Bar (Nabisco) Welch's	1 piece (1.1oz.)	1.0	132
Cream egg (Hershey's)	1 oz.	.8	142
Squares (Nabisco)	1 piece (.5 oz.)	.6	64
Fiddle Faddle	1½-oz. packet	2.0	177
Frappe (Nabisco) Welch's	1 piece (1.1 oz.)	.7	132
Fruit Roll (Sahadi):			
Apple	1 piece (2 oz.)	1.0	195
Apricot	1 piece (2 oz.)	1.9	192
Cherry	1 piece (2 oz.)	1.2	196
Grape	1 piece (2 oz.)	.8	195
Plum	1 piece (2 oz.)	.8	195
Prune	1 piece (2 oz.)	.8	195
Raspberry	1 piece (2 oz.)	1.0	195
Strawberry	1 piece (2 oz.)	.9	196
Fudge:			
Bar (Nabisco) Welch's	1 piece (1.1 oz.)	.9	144
Bar (Tom Houston)	1 bar (1.5 oz.)	1.6	179
Fudgies, bar (Kraft)	1 piece (7 grams)	.2	27
Fudgies, regular (Kraft)	1 piece (8 grams)	.2	33
Fudgies, Twisteroo (Kraft)	1 piece (6 grams)	.2	26
Home Style (Nabisco)	1 piece (.7 oz.)	.6	90
Nut, bars or squares (Nabisco)	1 piece (.5 oz.)	.2	71
Good & Fruity	1 oz.	.2	106
Good & Plenty	1 oz.	<.1	100
Hard candy, Sherbit (F&F)	1 piece	0.	9
Hershey-Ets, candy-coated	1.1 oz. pkg.	1.0	154
Hollywood	1½-oz. bar	3.2	183
Jelly (see individual flavors and brand names in this section)			
Kisses, milk chocolate (Hershey's)	1 piece (5 grams)	.4	26
Krackel Bar (Hershey's)	1.4-oz. bar	2.6	212
Krackel Bar (Hershey's)	¼-oz. miniature	.5	38

Food and Description	Measure or Quantity	Protein (grams)	Calories
Licorice:			
Twist (American Licorice Co.):			
Black	1 piece (10 grams)	.3	27
Red	1 piece (9 grams)	.3	33
Life Savers (Beech-Nut):			
Drop	1 piece (3 grams)	0.	10
Mint	1 piece (2 grams)	0.	7
Malted Milk Crunch (Nabisco)	1 piece (2 grams)	.1	9
Marshmallow:			
(Campfire)	1 oz.	.7	111
Chocolate (Kraft)	1 piece (7 grams)	.2	24
Coconut (Kraft)	1 piece (.4 oz.)	.3	40
Flavored miniature (Kraft)	1 piece (< 1 gram)	< .1	2
Flavored, regular (Kraft)	1 piece (7 grams)	.1	23
White, miniature (Kraft)	1 piece (< 1 gram)	< .1	2
White, regular (Kraft)	1 piece (7 grams)	.1	23
Milk Shake (Hollywood)	1¼-oz. bar	1.3	148
Mint or peppermint:			
Buttermint (Kraft)	1 piece (2 grams)	0.	8
Encore (Kraft)	1 piece (2 grams)	0.	6
Mini-mint (Kraft)	1 piece (3 grams)	< .1	12
Party (Kraft)	1 piece (2 grams)	0.	8
Pattie, chocolate-covered:			
Junior Mint Pattie (Nabisco)	1 piece (2 grams)	Tr.	10
Peppermint pattie (Nabisco)	1 piece (.5 oz.)	.2	64
Sherbit, pressed mints (F & F)	1 piece	0.	7
Thin (Nabisco)	1 piece (.4 oz.)	.2	42
Wafers (Nabisco)	1 piece (2 grams)	< .1	10
Mr. Goodbar (Hershey's)	1.8-oz. bar	7.3	283
Mr. Goodbar (Hershey's)	¼-oz. miniature	1.0	39
North Pole (F & F)	1⅜-oz. bar	1.5	150
Nougat centers (Nabisco)			
Chuckles	1 piece (4 grams)	Tr.	17
Nutty Crunch (Nabisco)	1 piece (½ oz.)	.5	71
Payday (Hollywood)	1¼-oz. bar	3.6	154
Peanut, chocolate-covered:			
(Hershey's) candy-coated	1 oz.	2.6	139
(Kraft)	1 piece (2 grams)	.3	12
(Nabisco)	1 piece (4 grams)	.7	24

Food and Description	Measure or Quantity	Protein (grams)	Calories
(Tom Houston)	1 oz.	4.6	157
Peanut Brittle:			
(Kraft)	1 oz.	2.6	126
Coconut (Kraft)	1 oz.	1.0	125
Jumbo Peanut Block Bar			
(Planters)	1 oz.	4.4	139
Peanut Butter Cup (Reese's)	.9-oz. cup	3.4	133
Peanut Butter Egg (Reese's)	1 oz.	3.2	133
Peanut Plank (Tom Houston)	1 bar (1.5 oz.)	7.4	216
Pom Poms (Nabisco)	1 piece (3 grams)	.2	14
Raisin, chocolate-covered:			
(Ghirardelli)	1 bar (1.1 oz.)	1.8	160
(Nabisco)	1 piece (< 1 gram)	<.1	4
Rally (Hershey's)	1.8-oz. bar	5.4	260
Screaming Yellow Zonkers	1 oz.	.9	120
Spearmint leaves (Quaker City)	1 oz.	.1	107
Sprigs, sweet chocolate			
(Hershey's)	1 oz.	1.4	136
Stars, chocolate:			
(Kraft)	1 piece (3 grams)	.2	13
(Nabisco)	1 piece (3 grams)	.2	15
Sugar Babies (Nabisco)	1 piece (2 grams)	<.1	6
Sugar Daddy (Nabisco):			
Caramel sucker	1 piece (1.1 oz.)	.6	121
Giant sucker, caramel	1 piece (1 lb.)	10.0	1809
Junior	1 piece (.4 oz.)	.3	50
Junior sucker, choco-flavored	1 piece (.4 oz.)	.7	51
Nugget	1 piece (.4 oz.)	.3	48
Nugget	1 piece (7 grams)	.1	27
Sugar Mama (Nabisco)	1 piece (.8 oz.)	.8	101
Sugar Wafer (F & F)	1¼-oz. pkg.	1.0	180
Toffee:			
Chocolate (Kraft)	1 piece (7 grams)	.4	27
Coffee, rum butter or vanilla			
(Kraft)	1 piece (7 grams)	.3	28
Tootsie Roll:			
Regular:			
1¢ size or midgee	1 piece (7 grams)	.2	26
2¢ size	1 piece (.4 oz.)	.3	43
5¢ size	1 piece (¾ oz.)	.7	87
10¢ size	1 piece (1½ oz.)	1.4	174
Twin pak	10¢ size (1¼ oz.)	1.2	137
Twin pak	15¢ size (2 oz.)	1.8	219

Food and Description	Measure or Quantity	Protein (grams)	Calories
Vending-machine size	1 piece (5 grams)	.1	21
Pop, 2 for 5¢	1 piece (.5 oz.)	.1	55
Pop, 5¢ size	1 piece (1 oz.)	.2	110
Pop-drop	1 piece (5 grams)	<.1	18
Twizzlers (Y & S):			
Licorice, bars	1¾-oz. pkg.	1.7	183
Strawberry, bars	1 oz.	.9	102
Walnut Hill (F & F)	1⅜-oz. bar	1.8	177
Whirligigs (Nabisco)	1 piece (6 grams)	.1	26
CANDY DIETETIC:			
Chocolate, assorted:			
Milk (Estee)	1 piece (8 grams)	.8	49
Slimtreats	1 piece (2 grams)	.4	13
Chocolate bar, almonds:			
(Estee)	¾-oz. bar	1.9	125
(Estee)	3-oz. bar	7.6	497
(Estee) *ST*	⅝-oz. bar	2.1	106
Chocolate bar, bittersweet:			
(Estee)	¾-oz. bar	1.5	124
(Estee)	3-oz. bar	6.0	496
Chocolate bar, coconut			
(Estee)	¾-oz. bar	1.7	129
Chocolate bar, crunch:			
(Estee)	⅝-oz. bar	1.4	101
(Estee)	2½-oz. bar	5.7	41
Chocolate bar, fruit-nut			
(Estee)	3-oz. bar	7.6	493
Chocolate bar, milk:			
(Estee)	¾-oz. bar	1.9	128
(Estee)	3-oz. bar	7.6	509
(Estee) *ST*	⅝-oz. bar	1.9	100
Chocolate bar, peppermint			
(Estee)	¾-oz.bar	1.7	126
Chocolate bar, white (Estee)	3-oz. bar	9.4	473
Creams (Estee) assorted or peppermint	1 piece (8 grams)	.8	49
Gum Drops (Estee):			
Fruit	1 piece (2 grams)	Tr.	3
Licorice	1 piece (2 grams)	Tr.	2
Hard candy:			
Assorted (Estee)	1 piece (3 grams)	Tr.	12

Food and Description	Measure or Quantity	Protein (grams)	Calories
Slimtreats	1 piece (3 grams)	0.	9
Mint, any flavor (Estee)	1 piece (1 gram)	0.	4
Nut, chocolate-covered (Estee)	1 piece (8 grams)	.9	47
Peanut, chocolate-covered (Estee)	1 piece (1 gram)	.2	7
Peanut butter cup (Estee)	1 piece (8 grams)	1.3	45
Raisin, chocolate-covered (Estee)	1 piece (1 gram)	<.1	5
TV mix (Estee)	1 piece (2 grams)	.2	9
CANE SYRUP (USDA)	1 T. (.7 oz.)	0.	55
CANTALOUPE, fresh:			
Whole (USDA)	1 lb. (weighed whole)	1.6	68
Whole, medium (USDA)	5″ dia. melon (1⅔ lbs., weighed with skin & cavity contents)	2.7	115
Cubed (USDA)	½ cup (2.9 oz.)	.6	24
CAPE GOOSEBERRY (see **GROUND-CHERRY**)			
CAPICOLA or CAPACOLA SAUSAGE (USDA)	1 oz.	5.7	141
CAP'N CRUNCH:			
(Quaker)	¾ cup (1 oz.)	1.3	123
Crunchberries (Quaker)	¾ cup (1 oz.)	1.3	119
Peanut butter (Quaker)	¾ cup (1 oz.)	2.1	128
Vanilly (Quaker)	¾ cup (1 oz.)	1.3	116
CARAMBOLA, raw (USDA):			
Whole	1 lb. (weighed whole)	3.0	149
Flesh only	4 oz.	.8	40
CARAMEL CAKE, home recipe (USDA):			
Without icing	⅑ of 9″ sq. cake (3 oz.)	3.9	331
With caramel icing	3 oz.	3.1	322

(USDA): United States Department of Agriculture

Food and Description	Measure or Quantity	Protein (grams)	Calories
CARAMEL CAKE MIX:			
*(Duncan Hines)	¹⁄₁₂ of cake (2.7 oz.)	2.3	202
*Apple (Betty Crocker) layer	¹⁄₁₂ of cake	2.9	203
*Pudding (Betty Crocker)	⅙ of cake	2.5	225
*CARAMEL PUDDING MIX,			
nut (Royal)	½ cup (5.1 oz.)	4.8	194
CARAWAY SEED (information supplied by General Mills Laboratory)	1 oz.	5.4	72
CARISSA or NATAL PLUM, raw:			
Whole (USDA)	1 lb. (weighed whole)	2.0	273
Flesh only (USDA)	4 oz.	.6	79
CARNATION INSTANT BREAKFAST:			
Regular, any flavor	1 pkg.	7.2	128
Special Morning, chocolate or chocolate malt	1 pkg. (1 oz.)	12.7	188
Special Morning, strawberry or vanilla	1 pkg. (1 oz.)	12.8	188
CAROB FLOUR (see **FLOUR**)			
CARP, raw (USDA):			
Whole	1 lb. (weighed whole)	24.5	156
Meat only	4 oz.	20.4	130
CARROT:			
Raw (USDA):			
Whole	1 lb. (weighed with full tops)	2.8	112
Partially trimmed	1 lb. (weighed without tops, with skins)	4.1	156
Trimmed	5½" × 1" carrot (1.8 oz.)	.6	21

(USDA): United States Department of Agriculture
*Prepared as Package Directs

[75]

Food and Description	Measure or Quantity	Protein (grams)	Calories
Trimmed	25 thin strips (1.8 oz.)	.6	21
Chunks	½ cup (2.4 oz.)	.8	29
Diced	½ cup (2.5 oz.)	.8	30
Grated or shredded	½ cup (1.9 oz.)	.6	23
Slices	½ cup (2.3 oz.)	.7	27
Strips	½ cup (2 oz.)	.6	24
Boiled (USDA):			
Chunks, drained	½ cup (2.9 oz.)	.7	25
Diced, drained	½ cup (2.5 oz.)	.6	22
Slices, drained	½ cup (2.7 oz.)	.7	24
Canned, regular pack:			
Diced, solids & liq. (USDA)	½ cup (4.3 oz.)	.7	34
Diced, drained solids (USDA)	½ cup (2.8 oz.)	.6	24
Drained liq. (USDA)	4 oz.	.5	25
Drained solids (Butter Kernel)	½ cup (4.1 oz.)	.7	29
Drained solids (Del Monte)	½ cup (2.8 oz.)	.6	26
Sliced, solids & liq. (Comstock)	½ cup (2.6 oz.)	.6	22
Small, solid & liq. (Le Sueur)	¼ of 15-oz. can	.4	23
Solids & liq. (Del Monte)	½ cup (4 oz.)	.8	32
Solids & liq. (Stokely-Van Camp)	½ cup (4 oz.)	.7	32
Canned, dietetic pack:			
Low sodium, solids & liq. (USDA)	4 oz. (by wt.)	.8	25
Low sodium, drained solids (USDA)	½ cup (2.8 oz.)	.6	20
Diced, solids & liq. (Blue Boy)	4 oz.	.9	48
Diced, solids & liq. (Tillie Lewis)	½ cup (4.3 oz.)	.9	26
Sliced, solids & liq. (Blue Boy)	4 oz.	.9	27
Sliced (S and W) *Nutradiet*, unseasoned	4 oz.	.6	25
Dehydrated (USDA)	1 oz.	1.9	97
Frozen:			
Nuggets in butter sauce (Green Giant)	⅓ of 10-oz. pkg.	.6	52
Sliced, honey glazed (Green Giant)	⅓ of 10-oz. pkg.	.7	77

(USDA): United States Department of Agriculture

Food and Description	Measure or Quantity	Protein (grams)	Calories
With brown sugar glaze (Birds Eye)	½ cup (3.3 oz.)	.8	87
CASABA MELON, fresh (USDA):			
Whole	1 lb. (weighed whole)	2.7	61
Flesh only	4 oz.	1.4	31
CASHEW NUT:			
(USDA)	1 oz.	4.9	159
(USDA)	½ cup (2.5 oz.)	12.0	393
(USDA)	5 large or 8 med., .4 oz.)	1.8	60
Freshnut	1 oz.	4.6	169
(Tom Houston)	15 nuts (1.1 oz.)	5.2	168
Dry roasted (Flavor House)	1 oz.	5.6	172
Dry roasted (Planters)	1 oz.	5.0	171
Dry roasted (Skippy)	1 oz.	5.6	164
Oil roasted (Planters)	15¢ bag (.9 oz.)	4.5	159
Oil roasted (Skippy)	1 oz.	5.0	177
CATFISH, freshwater, raw, fillet			
(USDA)	4 oz.	20.0	117
CATSUP:			
Regular pack:			
(USDA)	½ cup (5 oz.)	2.8	149
(USDA)	1 T. (.6 oz.)	.4	19
(Bama)	1 T. (.6 oz.)	.3	19
(Del Monte)	1 T. (.7 oz.)	.2	20
(Heinz)	1 T.	.2	16
(Hunt's)	½ cup (4.8 oz.)	2.3	146
(Hunt's)	1 T. (.6 oz.)	.3	18
(Nalley's)	1 oz.	.6	31
(Smucker's)	1 T. (.6 oz.)	.3	19
(Stokely-Van Camp)	1 T. (.6 oz.)	.4	19
Dietetic pack:			
Low sodium (USDA)	½ cup (5 oz.)	2.8	149
Low sodium (USDA)	1 T. (.6 oz.)	.4	19
(Tillie Lewis)	1 T. (.7 oz.)	.3	8

(USDA): United States Department of Agriculture

Food and Description	Measure or Quantity	Protein (grams)	Calories
CAULIFLOWER.			
Raw (USDA):			
Whole	1 lb. (weighed untrimmed)	4.8	48
Flowerbuds	1 lb. (weighed trimmed)	12.2	122
Buds	½ cup (1.8 oz.)	1.4	14
Slices	½ cup (1.5 oz.)	1.1	11
Boiled, drained (USDA)	½ cup (2.2 oz.)	1.4	14
Frozen:			
Not thawed (USDA)	10-oz. pkg.	5.7	62
Boiled, drained (USDA)	½ cup (3.2 oz.)	1.7	16
(Birds Eye)	⅓ of 10-oz. pkg.	2.0	21
Au gratin (Stouffer's)	⅓ of 10-oz. pkg.	4.2	113
Cut, in butter sauce (Green Giant)	⅓ of 10-oz. pkg.	1.3	44
Hungarian, with sour cream sauce, casserole (Green Giant)	⅓ of 10-oz. pkg.	2.4	70
In cheese sauce (Green Giant)	⅓ of 10-oz. pkg.	3.5	63
CAULIFLOWER, SWEET, PICKLED (Smucker's)	1 bud (.5 oz.)	4.0	24
CAVIAR, STURGEON (USDA):			
Pressed	1 oz.	9.8	90
Whole eggs	1 oz.	7.6	74
Whole eggs	1 T. (.6 oz.)	4.3	42
CELERIAC ROOT, raw (USDA):			
Whole	1 lb. (weighed unpared)	7.0	156
Pared	4 oz.	2.0	45
CELERY, all varieties:			
Raw (USDA):			
Whole	1 lb. (weighed untrimmed)	3.1	58
1 large outer stalk	8″ × 1½″ at root end (1.4 oz.)	.4	7
3 small inner stalks	5″ x ¾″ (1.8 oz.)	.4	8

(USDA): United States Department of Agriculture

Food and Description	Measure or Quantity	Protein (grams)	Calories
Diced, chopped or cut in chunks	½ cup (2.1 oz.)	.5	10
Slices	½ cup (1.9 oz.)	.5	9
Boiled, drained:			
Diced or cut in chunks	½ cup (2.7 oz.)	.6	11
Slices	½ cup (3 oz.)	.7	12

CELERY CABBAGE (see **CABBAGE, CHINESE**)

CELERY SEASONING

(French's)	1 tsp. (5 grams)	.2	2

CELERY SOUP, Cream of:

Condensed (USDA)	8 oz. (by wt.)	3.2	163
*Prepared with equal volume water (USDA)	1 cup (8.5 oz.)	1.7	86
*Prepared with equal volume milk (USDA)	1 cup (8.6 oz.)	6.4	169
*(Campbell)	1 cup	1.6	75
*(Heinz)	1 cup (8½ oz.)	1.7	101

CEREAL BREAKFAST FOODS (see kind of cereal such as **CORN FLAKES** or brand name such as **KIX**)

CERVELAT (USDA):

Dry	1 oz.	7.0	128
Soft	1 oz.	5.3	87

CHARD, swiss (USDA):

Raw, whole	1 lb. (weighed untrimmed)	10.0	104
Raw, trimmed	4 oz.	2.7	28
Boiled, drained	½ cup (3.4 oz.)	1.7	17

CHARLOTTE RUSSE, with lady-fingers, whipped cream filling, home recipe (USDA)

	4 oz.	6.7	324

(USDA): United States Department of Agriculture
*Prepared as Package Directs

Food and Description	Measure or Quantity	Protein (grams)	Calories
CHAYOTE, raw (USDA):			
Whole	1 lb. (weighed unpared)	2.3	108
Pared	4 oz.	.7	32
***CHEDDAR CHEESE SOUP,** canned (Campbell)	1 cup	4.5	141
CHEERIOS, cereal (General Mills)	1¼ cups (1 oz.)	3.8	112
CHEESE:			
American or cheddar:			
Natural:			
(USDA)	1 oz.	7.1	113
(USDA)	1″ cube (.6 oz.)	4.2	68
Diced (USDA)	1 cup (4.6 oz.)	32.8	521
Grated or shredded (USDA)	1 cup (3.9 oz.)	27.8	442
Grated or shredded (USDA)	1 T. (7 grams)	1.7	27
(Kraft)	1 oz.	7.1	113
Cheddar (Sealtest)	1 oz.	7.1	115
Grated (Kraft)	1 oz.	8.6	129
Shredded (Kraft)	1 oz.	7.1	113
Process:			
(USDA)	1 oz.	6.6	105
(USDA)	1″ cube (.6 oz.)	4.2	67
(USDA)	3½″ × 3⅜″ × ⅛″ slice	6.6	105
(Borden)	¾-oz. slice	4.5	83
American (Borden) *Miracle Melt*	1 T. (.5 oz.)	2.1	38
Cheddar (Borden) *Miracle Melt*	1 T. (.5 oz.)	2.2	39
(Kraft) loaf or slice	1 oz.	6.3	105
(Sealtest)	1 oz.	6.3	105
Dried, sharp cheddar (Data from General Mills Laboratory)	1 oz.	9.2	171
American blue (Borden) *Miracle Melt*	1 T. (.5 oz.)	2.1	38
Asiago (Frigo)	1 oz.	7.1	113

(USDA): United States Department of Agriculture
*Prepared as Package Directs

Food and Description	Measure or Quantity	Protein (grams)	Calories
Bleu or Blue:			
Natural (USDA)	1 oz.	6.1	104
Natural (USDA)	1″ cube (.6 oz.)	3.7	63
Natural, crumbled (USDA)	1 cup (4.8 oz.)	29.0	497
(Frigo)	1 oz.	5.8	99
(Kraft) natural	1 oz.	5.8	99
(Stella)	1 oz.	6.9	112
Bondost (Kraft) natural	1 oz.	6.5	103
Brick:			
(USDA) natural	1 oz.	6.3	105
(Kraft) natural	1 oz.	6.7	103
(Kraft) process, slices	1 oz.	6.2	101
Camembert, domestic:			
Natural (USDA)	1 oz.	5.0	85
Natural (USDA)	2¼″ × 2⅛″ × 1⅛″ wedge (3 to a 4-oz. pkg.).	6.6	114
(Borden)	1 oz.	4.9	86
(Kraft) natural	1 oz.	5.0	85
Caraway (Kraft) natural	1 oz.	7.0	111
Casino Swiss (Kraft) natural	1 oz.	7.9	104
Chantelle, natural (Kraft)	1 oz.	5.7	90
Cheddar (See American)			
Cheez-ola, process (Fisher)	1 oz.	7.1	90
Colby, natural (Kraft)	1 oz.	7.0	111
Cottage, large or small curd:			
Creamed, unflavored:			
(USDA)	1 oz.	3.9	30
(USDA)	8-oz. pkg.	30.8	240
(USDA)	1 packed cup (8.6 oz.)	33.3	260
(USDA)	1 T. (.5 oz.)	2.0	16
(Axelrod)	8-oz. container	28.8	218
(Borden)	8-oz. container	30.8	240
(Dean)	8-oz. container	28.1	218
(Kraft)	1 oz.	3.4	27
(Sealtest)	1 cup (7.9 oz.)	27.6	213
California (Breakstone)	8-oz. container	27.9	216
California (Breakstone)	1 T. (.6 oz.)	1.9	15
Light n' Lively (Sealtest)	1 cup (7.9 oz.)	28.0	155
Lite Line (Borden)	1 cup	30.0	189
Low fat (Breakstone)	8-oz. container	27.2	184
Low fat (Breakstone)	1 T. (.6 oz.)	1.9	13

(USDA): United States Department of Agriculture

[81]

Food and Description	Measure or Quantity	Protein (grams)	Calories
Low fat, 2% fat (Sealtest)	1 cup (7.9 oz.)	30.7	193
Tangy small curd (Breakstone)	8-oz. container	27.9	216
Tangy small curd (Breakstone)	1 T. (.6 oz.)	1.9	15
Tiny soft curd (Breakstone)	8-oz. container	27.9	216
Tiny soft curd (Breakstone)	1 T. (.6 oz.)	1.9	15
Creamed, flavored:			
Chive (Breakstone)	8-oz. container	27.9	216
Chive (Breakstone)	1 T. (.6 oz.)	1.9	15
Chive (Sealtest)	1 cup (7.9 oz.)	27.3	211
Chive-pepper (Sealtest)	1 cup (7.9 oz.)	26.0	206
Peach, low fat (Breakstone)	8-oz. container	23.1	232
Peach, low fat (Breakstone)	1 T. (.6 oz.)	1.7	17
Peach-pineapple (Sealtest)	1 cup (7.9 oz.)	22.2	228
Pineapple, low fat (Breakstone)	8-oz. container	22.9	268
Pineapple, low fat (Breakstone)	1 T. (.6 oz.)	1.6	19
Pineapple (Sealtest)	1 cup (7.9 oz.)	22.2	222
Spring Garden Salad (Sealtest)	1 cup (7.9 oz.)	24.4	208
Uncreamed:			
(USDA)	8-oz. pkg.	38.6	195
(USDA)	1 oz.	4.8	24
(USDA)	1 packed cup (7 oz.)	34.0	172
(Dean)	8-oz. container	42.2	191
(Dean) 2% fat	8-oz. container	27.2	193
(Sealtest)	1 cup (7.9 oz.)	41.7	179
Pot, unsalted (Borden)	8-oz. pkg.	38.8	195
Pot style (Breakstone)	8-oz. container	37.9	172
Pot style (Breakstone)	1 T. (.6 oz.)	2.6	12
Skim milk (Breakstone)	8-oz. container	42.2	182
Skim milk (Breakstone)	1 T. (.6 oz.)	3.0	13
Country Charm, natural (Fisher)	1 oz.	6.9	89
Cream cheese:			
Plain, unwhipped:			
(USDA)	1 oz.	2.3	106
(USDA)	3-oz. pkg. (2⅞″ × ⅞″)	6.8	318
(USDA)	8-oz. pkg.	18.2	849
(USDA)	½ cup (4.1 oz.)	9.2	430

(USDA): United States Department of Agriculture

Food and Description	Measure or Quantity	Protein (grams)	Calories
(USDA)	1" cube (.6 oz.)	1.3	60
(USDA)	1 T. (.5 oz.)	1.1	52
(Borden)	1 oz.	2.0	101
(Breakstone)	1 oz.	2.4	98
(Breakstone)	1 T. (.5 oz.)	1.2	49
(Kraft)	1 oz.	2.2	104
(Sealtest)	1 oz.	2.4	98
Hostess (Kraft)	1 oz.	2.4	98
Neufchâtel (Borden)	1 oz.	2.8	74
Philadelphia (Kraft)	1 oz.	2.2	104
Philadelphia, imitation (Kraft)	1 oz.	3.5	52
Plain, whipped (Breakstone):			
Temp-Tee	1 oz.	2.4	98
Temp-Tee	1 T. (9 grams)	.8	32
Flavored, unwhipped:			
Chive (Kraft) *Hostess*	1 oz.	2.1	84
Chive (Kraft) *Philadelphia*	1 oz.	3.5	84
Olive-pimento (Kraft) *Hostess*	1 oz.	1.7	84
Pimento (Kraft) *Hostess*	1 oz.	2.0	85
Pimento (Kraft) *Philadelphia*	1 oz.	2.0	85
Pineapple (Kraft) *Hostess*	1 oz.	1.6	86
Roquefort (Kraft) *Hostess*	1 oz.	2.5	80
Flavored, whipped (Kraft):			
Bacon & horseradish	1 oz.	2.2	96
Blue	1 oz.	2.8	97
Catalina	1 oz.	2.1	94
Chive	1 oz.	2.1	92
Onion	1 oz.	2.0	93
Pimento	1 oz.	2.1	91
Salami	1 oz.	2.4	88
Smoked salmon	1 oz.	2.4	90
Edam:			
(House of Gold)	1 oz.	7.9	105
Natural (Kraft)	1 oz.	7.9	104
Farmer, midget (Breakstone)	1 oz.	4.3	40
Farmer, midget (Breakstone)	1 T. (.5 oz.)	2.0	19
Farmer (Dean)	1 oz.	4.4	46
Fontina, natural (Kraft)	1 oz.	7.2	113
Fontina (Stella)	1 oz.	6.9	112
Frankenmuth, natural (Kraft)	1 oz.	7.1	113

(USDA): United States Department of Agriculture

Food and Description	Measure or Quantity	Protein (grams)	Calories
Gjetost, natural (Kraft)	1 oz.	2.0	134
Gorgonzola, natural (Kraft)	1 oz.	7.1	111
Gouda, natural (Kraft)	1 oz.	7.2	107
Gruyere, process (Borden)	1 oz.	6.2	93
Gruyere, natural (Kraft)	1 oz.	8.0	110
Jack-dry, natural (Kraft)	1 oz.	6.5	101
Jack-fresh, natural (Kraft)	1 oz.	6.2	95
Kisses, mild (Borden)	1 kiss (6 grams)	1.1	18
Kisses, tangy (Borden)	1 kiss (6 grams)	1.1	19
Lager-Kase, natural (Kraft)	1 oz.	6.7	107
Leyden, natural (Kraft)	1 oz.	10.7	80
Liederkranz (Borden)	1 oz.	4.8	86
Limburger:			
Natural (USDA)	1 oz.	6.0	98
Natural (Kraft)	1 oz.	6.0	98
Monterey Jack (Frigo)	1 oz.	6.5	103
Monterey Jack, natural (Kraft)	1 oz.	6.5	102
Mozzarella:			
(Frigo)	1 oz.	8.9	79
Low moisture, part-skim, natural (Kraft)	1 oz.	8.0	84
Low moisture, part-skim, pizza, natural (Kraft)	1 oz.	8.8	79
Shredded (Kraft)	1 oz.	8.8	79
Muenster:			
Natural (Kraft)	1 oz.	6.5	100
Process, slices (Kraft)	1 oz.	6.2	102
Neufchâtel:			
Natural (Kraft) Calorie-Wise	1 oz.	3.2	70
Process (Borden)	1 oz.	2.7	73
Nuworld, natural (Kraft)	1 oz.	6.2	103
Old English, process, loaf or slices (Kraft)	1 oz.	6.4	105
Parmesan:			
Natural:			
(USDA)	1 oz.	10.2	111
(Frigo)	1 oz.	10.9	107
(Kraft)	1 oz.	10.9	107
(Stella)	1 oz.	10.0	103
Grated:			
Natural (USDA)	1 oz.	12.1	132
Natural (USDA)	1 cup loosely packed (3.7 oz.)	45.3	494

(USDA): United States Department of Agriculture

Food and Description	Measure or Quantity	Protein (grams)	Calories
Natural (USDA)	1 cup pressed down (4.9 oz.)	59.8	652
Natural (USDA)	1 T. loosely packed (7 grams)	2.8	31
Natural (USDA)	1 T. pressed down (9 grams)	3.8	42
(Borden)	1 oz.	10.8	143
(Buitoni)	1 oz.	12.1	118
(Frigo)	1 T. (6 grams)	2.8	27
(Kraft)	1 oz.	13.0	127
Shredded (Kraft)	1 oz.	11.7	114
Parmesan & Romano, grated:			
(Borden) natural	1 oz.	12.0	135
(Kraft)	1 oz.	12.6	130
Pepato (Frigo)	1 oz.	9.4	110
Pimento American, process:			
(USDA)	1 oz.	6.5	105
Loaf or slices (Kraft)	1 oz.	6.4	103
Pinconning, natural (Kraft)	1 oz.	7.1	113
Pizza:			
(Frigo)	1 oz.	9.4	73
Low fat, part skim, shredded (Kraft)	1 oz.	8.6	86
Port du Salut, natural (Kraft)	1 oz.	6.7	100
Primost, natural (Kraft)	1 oz.	2.0	134
Provolone:			
(Frigo)	1 oz.	7.1	99
Natural (Kraft)	1 oz.	7.1	99
Ricotta:			
(Breakstone)	1 oz.	2.8	45
(Breakstone)	1 T. (.6 oz.)	1.6	25
Natural (Kraft)	1 oz.	2.9	47
(Sierra)	1 oz.	3.2	50
Romano:			
(Frigo)	1 oz.	9.4	110
Natural (Stella)	1 oz.	9.4	106
Grated:			
(Borden)	1 oz.	8.0	136
(Buitoni)	1 oz.	11.4	123
(Frigo)	1 T. (.6 oz.)	2.5	29
(Kraft)	1 oz.	11.5	134
Shredded (Kraft)	1 oz.	10.3	121

(USDA): United States Department of Agriculture

Food and Description	Measure or Quantity	Protein (grams)	Calories
Romano & Parmesan:			
Plain (Kraft)	1 oz.	11.9	133
Flavored (Kraft):			
Bacon smoke	1 oz.	11.7	130
Garlic or onion	1 oz.	11.9	134
Roquefort, natural:			
(USDA)	1 oz.	6.1	104
(USDA)	1″ cube (.6 oz.)	3.7	63
(Kraft)	1 oz.	6.0	105
Sage, natural (Kraft)	1 oz.	7.1	113
Sap Sago, natural (Kraft)	1 oz.	11.5	76
Sardo Romano, natural (Kraft)	1 oz.	9.2	109
Scamorze:			
(Frigo)	1 oz.	8.9	79
Natural (Kraft)	1 oz.	7.3	100
Special cure, process, slices (Kraft)	1 oz.	6.5	105
Supercure, process, slices (Kraft)	1 oz.	6.5	105
Swiss, domestic:			
Natural:			
(USDA)	1 oz.	7.8	105
(USDA)	1″ cube (.5 oz.)	4.1	56
(USDA)	1¼-oz. slice (7½″ × 4″ × ¹⁄₁₆″)	9.6	130
(Kraft)	1 oz.	7.8	104
(Sealtest)	1 oz.	7.8	105
Process:			
(USDA)	1-oz. slice (3½″ × 3⅜″ × ⅛″)	7.5	101
(USDA)	1″ cube (.6 oz.)	4.8	64
(Borden) 1-lb. pkg.	.7-oz. slice	4.5	64
(Borden) 6- or 12-oz. pkg.	¾-oz. slice	5.1	72
(Borden) ½ or 5-lb. pkg.	.8-oz. slice	5.5	78
(Kraft) loaf	1 oz.	6.8	92
(Kraft) slices	1 oz.	7.1	95
With Muenster (Kraft) loaf	1 oz.	6.4	98
Washed curd, natural (Kraft)	1 oz.	6.7	107
CHEESE CAKE, frozen (Mrs. Smith's)	⅙ of 8″ cake (4 oz.)	10.2	214

(USDA): United States Department of Agriculture

Food and Description	Measure or Quantity	Protein (grams)	Calories
CHEESE CAKE MIX:			
*(Jell-O)	⅛ of cake including crust (3.3 oz.)	5.9	255
*(Royal) *No-Bake*	⅛ of 9" cake including crust (3.2 oz.)	9.1	278
CHEESE DIP (see **DIP**)			
CHEESE FONDUE:			
Home recipe (USDA)	4 oz.	16.8	301
Process (Borden)	6-oz. serving	24.6	354
CHEESE FOOD, process:			
American:			
(USDA)	1-oz. slice (3½" × 3⅜" × ⅛")	5.6	92
(USDA)	1" cube (.6 oz.)	3.6	58
(USDA)	1 T. (.5 oz.)	2.8	45
(Borden)	1" × 1" × 1" piece (.8 oz.)	3.9	71
Grated (Borden)	1 oz.	8.4	129
Grated, used in *Kraft Dinner*	1 oz.	8.6	129
Slices (Kraft)	1 oz.	5.8	94
Cheez'n bacon, slices (Kraft)	¾-oz. slice	4.6	76
Links (Kraft) *Handi-Snack:*			
Bacon	1 oz.	5.5	93
Garlic	1 oz.	5.6	92
Jalapeño	1 oz.	5.6	92
Nippy	1 oz.	5.6	92
Smokelle	1 oz.	5.4	93
Swiss	1 oz.	6.2	90
Loaf:			
Munst-ett (Kraft)	1 oz.	5.3	100
Pizzalone, loaf (Kraft)	1 oz.	7.1	90
Super blend (Kraft)	1 oz.	5.8	92
Pimento, slices (Kraft)	1 oz.	5.8	94
Salami, slices (Kraft)	1 oz.	5.8	94
Swiss (Borden)	.7-oz. slice	4.5	68
Swiss, cold pack (Borden)	.7-oz. slice	4.1	62
Swiss, slices (Kraft)	1 oz.	6.0	92

(USDA): United States Department of Agriculture
*Prepared as Package Directs

Food and Description	Measure or Quantity	Protein (grams)	Calories
CHEESE PIE:			
(Tastykake)	4-oz. pie	4.3	357
Frozen, pineapple (Mrs. Smith's)	⅙ of 8" pie (4 oz.)	6.2	273
Frozen, pineapple (Mrs. Smith's)	⅛ of 10" pie (5.4 oz.)	7.8	341
CHEESE PUFF, frozen (Durkee)	1 piece (.5 oz.)	2.7	59
CHEESE SOUFFLE:			
Home recipe (USDA)	¼ of 7" souffle (3.9 oz.)	10.9	240
Frozen (Stouffer's)	⅓ of 12-oz. pkg.	11.7	243
CHEESE SPREAD:			
American, process:			
(USDA)	1 oz. (2¾" × 2¼" × ¼")	4.5	82
(USDA)	1 T. (.5 oz.)	2.2	40
(USDA) shredded	1 packed cup (4 oz.)	18.1	325
(Borden)	.7-oz. slice	3.6	63
(Kraft) *Swankyswig*	1 oz.	4.4	77
(Nabisco) *Snack Mate*	1 tsp. (5 grams)	.8	15
Bacon, process:			
(Borden)	1 oz.	3.5	72
(Kraft) *Squeez-A-Snak*	1 oz.	4.8	83
(Kraft) *Swankyswig*	1 oz.	5.4	92
Cheddar, process (Nabisco) *Snack Mate*	1 tsp. (5 grams)	.8	15
Cheddar, seasoned, process (Nabisco) *Snack Mate*	1 tps. (5 grams)	.8	15
Cheddar, sharp, process (Borden) *Country Store*	1 oz.	5.0	85
Cheez Whiz, process (Kraft)	1 oz.	4.4	76
Count Down (Fisher)	1 oz.	6.8	43
Garlic, process:			
(Borden)	1 oz.	3.5	72
(Kraft) *Squeez-A-Snak*	1 oz.	4.5	84
(Kraft) *Swankyswig*	1 oz.	4.5	86
Hickory smoke, process (Nabisco) *Snack Mate*	1 tsp. (5 grams)	.8	14
Imitation, process (Kraft) *Calorie-Wise* or Tasty-loaf	1 oz.	4.6	48

(USDA): United States Department of Agriculture

Food and Description	Measure or Quantity	Protein (grams)	Calories
Imitation (Fisher) *Chef's Delight*	1 oz.	4.8	41
Imitation (Fisher) *Mellow Age*	1 oz.	4.8	41
Jalapeño, process (Kraft) *Cheez Whiz*	1 oz.	4.3	76
Limburger (Kraft)	1 oz.	4.0	69
Neufchâtel:			
Bacon & horseradish (Kraft) *Party Snacks*	1 oz.	2.6	76
Chipped beef (Kraft) *Party Snacks*	1 oz.	2.1	67
Chive (Kraft) *Party Snacks*	1 oz.	2.5	69
Clam (Kraft) *Party Snacks*	1 oz.	2.8	67
Olive-pimento (Kraft) *Swankyswig*	1 oz.	1.7	70
Onion (Kraft) *Party Snacks*	1 oz.	1.8	66
Pimento (Borden)	1 T. (.5 oz.)	.9	36
Pimento (Kraft) *Party Snacks*	1 oz.	2.3	67
Pimento (Kraft) *Swankyswig*	1 oz.	2.3	67
Pineapple (Borden)	1 T. (.5 oz.)	.9	36
Pineapple (Kraft) *Swankyswig*	1 oz.	1.8	70
Relish (Kraft) *Swankyswig*	1 oz.	1.5	71
Roka (Kraft) *Swankyswig*	1 oz.	2.5	80
Old English, process (Kraft) *Swankyswig*	1 oz.	5.4	96
Onion, French, process (Nabisco) *Snack Mate*	1 tsp. (5 grams)	.8	15
Pimento:			
(Borden) *Country Store*	1 T. (.5 oz.)	1.4	36
(Kraft) *Cheez Whiz*	1 oz.	4.4	76
Process (Kraft) *Squeez-A-Snak*	1 oz.	4.5	86
Process (Nabisco) *Snack Mate*	1 tsp. (5 grams)	.8	15
(Sealtest)	1 oz.	4.4	77
Velveeta, process (Kraft)	1 oz.	5.2	84
Sharp, process (Kraft) *Squeez-A-Snak*	1 oz.	4.6	85
Sharpie, process (Kraft)	1 oz.	5.5	90
Smoke, process (Kraft) *Squeez-A-Snak*	1 oz.	4.8	83
Smokelle, process (Kraft) *Swankyswig*	1 oz.	5.6	90

Food and Description	Measure or Quantity	Protein (grams)	Calories
Smokey, process (Borden)	1 oz.	3.5	72
Velva Kreme, process (Borden)	1 oz.	1.3	94
Velveeta, process (Kraft)	1 oz.	5.2	84
CHEESE STRAW:			
(USDA)	1 oz.	3.2	128
(USDA)	5″ × ⅜″ × ⅜″ piece (6 grams)	.7	27
Frozen (Durkee)	1 piece (8 grams)	1.2	29
CHERIMOYA, raw (USDA):			
Whole	1 lb. (weighed with skin & seeds)	3.4	247
Flesh only	4 oz.	1.5	107
CHERRY:			
Sour:			
Fresh (USDA):			
Whole	1 lb. (weighed with stems)	4.4	213
Whole	1 lb. (weighed without stems)	5.0	242
Pitted	½ cup (2.7 oz.)	.9	45
Canned, syrup pack, pitted (USDA):			
Light syrup	4 oz. (with liq.)	.9	84
Heavy syrup	4 oz. (with liq.)	.9	101
Heavy syrup	½ cup (4.6 oz.)	1.0	116
Extra heavy syrup	4 oz. (with liq.)	.9	127
Canned, water pack pitted, solids & liq.:			
(USDA)	½ cup (4.3 oz.)	1.0	52
(Stokely-Van Camp)	½ cup (4 oz.)	.9	49
Frozen, pitted (USDA):			
Sweetened	½ cup (4.6 oz.)	1.3	146
Unsweetened	4 oz.	1.1	62
Sweet:			
Fresh (USDA):			
Whole, with stems	1 lb. (weighed with stems)	5.3	286
Whole, with stems	½ cup (2.3 oz.)	.8	41
Pitted	½ cup (2.9 oz.)	1.1	57

(USDA): United States Department of Agriculture

Food and Description	Measure or Quantity	Protein (grams)	Calories
Canned, syrup pack with pits, solids & liq.:			
Light syrup (USDA)	4 oz.	1.0	70
Heavy syrup (USDA)	4 oz.	1.0	87
Heavy syrup, dark (Del Monte)	½ cup (4.3 oz.)	.8	92
Heavy syrup, Royal Anne (Del Monte)	½ cup (with liq., 4.6 oz.)	1.0	109
Heavy syrup, light or dark (Stokely-Van Camp)	½ cup (4.2 oz.)	1.1	97
Extra heavy syrup (USDA)	4 oz.	.8	108
Canned, syrup pack, pitted, solids & liq.:			
Light syrup (USDA)	4 oz.	1.0	74
Heavy syrup (USDA)	4 oz.	1.0	92
Heavy syrup (USDA)	½ cup (4.2 oz.)	1.1	96
Heavy syrup (Del Monte)	½ cup (4.3 oz.)	2.1	92
Extra heavy syrup (USDA)	4 oz.	.9	113
Canned, water pack, with pits, solids & liq.:			
(USDA)	4 oz.	1.0	52
(Tillie Lewis) unpitted	½ of 8-oz. can	1.0	54
Canned, water or dietetic pack, pitted, solids & liq.:			
(USDA)	4 oz.	1.0	54
(Blue Boy)	4 oz.	.6	52
Royal Anne:			
(Diet Delight)	½ cup (4.4 oz.)	1.1	65
(S and W) *Nutradiet,* low calorie	14 whole cherries (3.5 oz.)	.8	49
(S and W) *Nutradiet,* unsweetened	14 whole cherries (3.5 oz.)	.9	47
Dark (S and W) *Nutradiet,* low calorie	4 oz.	.6	60
Frozen, quick-thaw (Birds Eye)	½ cup (5 oz.)	1.4	122
CHERRY CAKE:			
Frozen, shortcake (Mrs. Smith's)	⅙ of 9″ cake	2.2	394

(USDA): United States Department of Agriculture

Food and Description	Measure or Quantity	Protein (grams)	Calories
*Mix (Duncan Hines)	¹⁄₁₂ of cake (2.6 oz.)	2.2	193
*Mix, chip, layer (Betty Crocker)	¹⁄₁₂ of cake	2.7	198
*Mix, upside down (Betty Crocker)	⅑ of cake	2.0	282
CHERRY, CANDIED (USDA)	1 oz.	.1	96
CHERRY DRINK (Hi-C)	6 fl. oz. (6.3 oz.)	.6	90
CHERRY JELLY (Smucker's)	1 T. (.7 oz.)	.1	50
CHERRY, MARASCHINO (USDA)	1 oz. (with liq)	<.1	33
CHERRY PIE:			
Home recipe, 2 crust (USDA)	⅙ of 9″ pie (5.6 oz.)	4.1	412
(Drake's)	2-oz. pie	1.9	203
(Hostess)	4½-oz. pie	3.7	427
Cherry-apple (Tastykake)	4-oz. pie	3.4	373
Frozen:			
Unbaked (USDA)	5 oz.	2.7	364
Baked (USDA)	5 oz.	3.1	413
(Banquet)	5 oz.	4.0	352
(Morton)	⅙ of 24-oz. pie	4.5	342
(Mrs. Smith's)	⅙ of 8″ pie (4.2 oz.)	2.3	309
(Mrs. Smith's)	⅛ of 10″ pie (5.6 oz.)	3.0	411
(Mrs. Smith's) natural juice	⅙ of 8″ pie (4.2 oz.)	2.7	344
Tart (Pepperidge Farm)	1 3-oz. pie tart	2.4	277
CHERRY PIE FILLING:			
(Comstock)	1 cup (10¾ oz.)	1.8	334
(Lucky Leaf)	8 oz.	1.4	242
(Wilderness)	21-oz. can	3.1	720
CHERRY PRESERVE or JAM:			
Sweetened:			
(Smucker's)	1 T. (.7 oz.)	<.1	52
(Bama)	1 T. (.7 oz.)	.1	54
Dietetic or low calorie (S and W) *Nutradiet*	1 T. (.5 oz.)	.2	11

(USDA): United States Department of Agriculture
*Prepared as Package Directs

Food and Description	Measure or Quantity	Protein (grams)	Calories
CHERRY SOFT DRINK			
(Yoo-Hoo) High Protein	6 fl. oz. (6.4 oz.)	6.0	100
CHERRY TURNOVER, frozen:			
(Pepperidge Farm)	3.3-oz. turnover	3.0	342
(Pillsbury)	1 turnover	2.0	150
CHERVIL, raw (USDA)	1 oz.	1.0	16
CHESTNUT (USDA):			
Fresh, in shell	1 lb. (weighed in shell)	10.7	713
Fresh, shelled	4 oz.	3.3	220
Dried, in shell	1 lb. (weighed in shell)	24.9	1402
Dried, shelled	4 oz.	7.6	428
CHESTNUT FLOUR (see **FLOUR, CHESTNUT**)			
CHEWING GUM:			
Regular:			
Beech-Nut	1 stick (3 grams)	0.	10
Doublemint	1 stick (3 grams)	<.1	8
Juicy Fruit	1 stick (3 grams)	<.1	9
Spearmint (Wrigley's)	1 stick (3 grams)	<.1	8
Dietetic:			
All flavors (Estee)	1 section (1 gram)	0.	4
*Care*Free* (Beech-Nut)	1 stick (3 grams)	0.	7
CHICKEN (see also **CHICKEN, CANNED**) (USDA):			
Broiler, cooked, meat only	4 oz.	27.0	154
Capon, raw, ready-to-cook	1 lb. (weighed with bones)	70.9	937
Capon, raw, meat with skin	4 oz.	24.5	330
Fryer:			
Raw:			
Ready-to-cook	1 lb. (weighed with bones)	57.4	382
Meat and skin	1 lb.	85.3	572
Meat only	1 lb.	87.5	485
Dark meat with skin	1 lb.	80.3	599

(USDA): United States Department of Agriculture

Food and Description	Measure or Quantity	Protein (grams)	Calories
Light meat with skin	1 lb.	90.3	544
Dark meat without skin	1 lb.	82.1	508
Light meat without skin	1 lb.	93.0	458
Skin	4 oz.	18.3	253
Back	1 lb. (weighed with bones)	40.4	385
Breast	1 lb. (weighed with bones)	74.5	394
Leg or Drumstick	1 lb. (weighed with bones)	51.2	313
Neck	1 lb. (weighed with bones)	33.7	329
Rib	1 lb. (weighed with bones)	40.9	287
Thigh	1 lb. (weighed with bones)	61.6	435
Wing	1 lb. (weighed with bones)	41.1	325
Fried. A 2½-pound chicken (weighed before cooking with bone) will give you:			
Back	1 back (2.2 oz.)	12.0	139
Breast	½ breast (3.3 oz.)	24.7	154
Leg or drumstick	1 leg (2 oz.)	12.1	87
Neck	1 neck (2.1 oz.)	11.2	121
Rib	1 rib (.7 oz.)	4.4	42
Thigh	1 thigh (2.3 oz.)	14.6	118
Wing	1 wing (1¾ oz.)	8.4	78
Fried, meat, skin & giblets	4 oz.	34.8	282
Fried, meat, & skin	4 oz.	34.7	284
Fried, meat only	4 oz.	35.4	237
Fried, dark meat with skin	4 oz.	33.9	298
Fried, light meat with skin	4 oz.	35.7	265
Fried, dark meat without skin	4 oz.	34.5	249
Fried, light meat without skin	4 oz.	36.4	223
Fried skin	1 oz.	8.0	119
Hen & cock:			
Raw:			
Ready-to-cook	1 lb. (weighed without bones)	57.6	987
Meat, skin & giblets	1 lb.	86.2	1116

Food and Description	Measure or Quantity	Protein (grams)	Calories
Meat & skin	1 lb.	86.2	1139
Meat only	1 lb.	98.0	703
Dark meat without skin	1 lb.	91.6	699
White meat without skin	1 lb.	106.1	603
Stewed:			
Meat, skin & giblets	4 oz.	29.7	354
Meat & skin	4 oz.	29.6	359
Dark meat, without skin	4 oz.	32.3	235
Light meat, without skin	4 oz.	36.5	204
Chopped	½ cup (2.5 oz.)	21.6	150
Diced	½ cup (2.4 oz.)	20.1	139
Ground	½ cup (2 oz.)	16.8	116
Roaster:			
Raw:			
Ready-to-cook	1 lb. (weighed with bones)	60.3	791
Meat, skin & giblets	1 lb.	88.9	866
Meat & skin	1 lb.	88.5	894
Meat only	1 lb.	95.7	594
Dark meat without skin	1 lb.	95.3	599
White meat without skin	1 lb.	105.7	581
Roasted:			
Total edible portion	1 lb.	114.3	1315
Meat, skin & giblets	4 oz.	30.8	274
Meat & skin	4 oz.	30.7	281
Meat only	4 oz.	33.5	208
Dark meat without skin	4 oz.	33.2	209
Light meat without skin	4 oz.	36.6	206
CHICKEN A LA KING:			
Home recipe (USDA)	1 cup (8.6 oz.)	27.4	468
Canned (Richardson & Robbins)	1 cup (7.9 oz.)	20.0	272
Canned (Swanson)	1 cup	16.0	260
Frozen (Banquet)	5-oz. bag	14.8	140
CHICKEN BOUILLON/BROTH, cube or powder (see also **CHICKEN SOUP**):			
(Herb-Ox)	1 cube (4 grams)	.7	6
(Herb-Ox) instant	1 packet (5 grams)	.7	12
(Steero)	1 cube (4 grams)	.6	6
(Wyler's) regular	1 cube (4 grams)	.4	6
(Wyler's) no salt added	1 cube (4 grams)	.4	11

(USDA): United States Department of Agriculture

Food and Description	Measure or Quantity	Protein (grams)	Calories
(Wyler's) instant	1 envelope (4 grams)	.5	8
CHICKEN CACCIATORE, canned (Hormel)	1-lb. can	48.6	386
CHICKEN, CANNED:			
Boned:			
(USDA)	4 oz.	24.6	225
(USDA)	½ cup (3 oz.)	18.4	168
(Lynden Farms) solids & liq.	5-oz. jar	27.0	229
(Lynden Farms) with broth	11-oz. jar	56.2	490
(Lynden Farms) with broth	29-oz. can	172.6	2302
(Swanson) with broth	5-oz. can	31.0	223
Fat, rendered, with onion (Lynden Farms)	¼ of 12.5-oz. jar	0.	797
Whole (Lynden Farms) solids & liq.	¼ of 52-oz. can	95.9	1170
CHICKEN, CREAMED, frozen (Stouffer's)	11½-oz. pkg.	37.2	613
CHICKEN DINNER or LUNCH:			
Canned:			
Noodle (Heinz)	8½-oz. can	10.2	186
Noodle (Lynden Farms)	14-oz. jar	17.9	413
Noodle with vegetables (Lynden Farms)	15-oz. can	17.0	434
(Weight Watchers)	10-oz. luncheon	12.6	284
Frozen:			
Boneless chicken (Swanson) *Hungry Man*	19-oz. dinner	47.1	746
Chicken & dumplings:			
Buffet (Banquet)	2-lb. pkg.	87.2	1306
(Morton)	11-oz. dinner	18.7	356
(Morton) 3-course	16-oz. dinner	27.2	699
Chicken livers & onion (Weight Watchers)	11½-oz. luncheon	9.5	234
Chicken & noodles (Banquet)	12-oz. dinner	18.7	374
Chicken & noodles (Morton)	10¼-oz. dinner	14.5	392
Creole (Weight Watchers)	12-oz. luncheon	27.2	211

(USDA): United States Department of Agriculture

Food and Description	Measure or Quantity	Protein (grams)	Calories
Fried:			
With mashed potato, carrots, peas, corn & bean (USDA)	12 oz.	43.5	588
(Banquet)	12-oz. dinner	27.8	530
(Morton)	11-oz. dinner	28.1	449
(Morton) 3-course	15½-oz. dinner	35.2	722
(Swanson)	11½-oz. dinner	37.4	600
(Swanson) 3-course	15-oz. dinner	35.7	639
With shoestring potato (Swanson)	25-oz. dinner	91.5	1879
*Mix, noodle (Jeno's) Add 'n Heat	50-oz. dinner	171.5	2155

CHICKEN & DUMPLINGS (see CHICKEN DINNER)

CHICKEN, FREEZE DRY

(Wilson) *Campsite:*			
Dry	2 oz.	48.7	246
*Reconstituted	4 oz.	39.1	199

CHICKEN FRICASSEE:

Home recipe (USDA)	1 cup (8.5 oz.)	36.7	386
Canned (Lynden Farms)	14.5-oz. can	32.9	506
Canned (Richardson & Robbins)	1 cup (7.9 oz.)	20.0	256

CHICKEN, FRIED, frozen:

(Banquet) whole chicken	2 lb.	169.2	2195
(Banquet) ½ chicken	14 oz.	74.0	960
(Swanson) halves	2 pieces (17¼ oz.)	83.4	1103
(Swanson) quarters	4 pieces (17¼ oz.)	72.3	1098
(Swanson)	16-oz. pkg.	84.1	1200
(Swanson)	32-oz. pkg.	155.9	2371
With whipped potato (Swanson)	7-oz. pkg.	24.0	412

CHICKEN GIBLETS:

Capon, raw	2 oz.	11.6	125
Fryer, raw	2 oz.	9.9	58
Fryer, fried from a 2½-lb. chicken	1 heart, gizzard & liver (2.1 oz.)	18.5	151
Hen & cock, raw	2 oz.	10.5	108

(USDA): United States Department of Agriculture
*Prepared as Package Directs

[97]

Food and Description	Measure or Quantity	Protein (grams)	Calories
Roaster, raw	2 oz.	11.2	77
CHICKEN GIZZARD (USDA):			
Raw	2 oz.	11.4	64
Simmered	2 oz.	15.3	84
CHICKEN LIVER PUFF, frozen			
(Durkee)	1 piece (.5 oz.)	2.2	48
CHICKEN & NOODLES:			
Home recipe (USDA)	1 cup (6.5 oz.)	22.3	367
Frozen (Banquet) buffet	2 lbs.	74.8	735
Frozen, escalloped (Stouffer's)	11½-oz. pkg.	26.0	589
CHICKEN PIE:			
Baked, home recipe (USDA)	4¼" pie (8 oz.)	22.9	533
Baked, home recipe (USDA)	⅓ of 9" pie (8.2 oz.)	23.4	545
Frozen:			
Commercial, unheated			
(USDA)	8-oz. pie	15.2	497
(Banquet)	8-oz. pie	15.4	427
(Banquet)	2-lb. 4-oz. pie	79.0	1408
(Morton)	8-oz. pie	18.2	445
(Stouffer's)	10-oz. pkg.	22.4	722
(Swanson)	8-oz. pie	16.3	445
(Swanson) deep dish	16-oz. pie	34.2	708
CHICKEN, POTTED (USDA)	1 oz.	5.0	70
CHICKEN PUFF, frozen			
(Durkee)	1 piece (.5 oz.)	2.2	49
CHICKEN RAVIOLI, in sauce			
(Lynden Farms)	14.5-oz. can	16.4	452
CHICKEN SOUP, canned:			
(Campbell) *Chunky*	1 cup	10.8	155
*Barley (Manischewitz)	1 cup	3.5	83
*(Campbell)	1 cup	8.4	53
(Lynden Farms)	1 cup (8 oz.)	2.3	14
(Richardson & Robbins)	1 cup (8.1 oz.)	3.7	32
(Swanson)	1 cup	4.0	31
*Dietetic (Claybourne)	8 oz. (by wt.)	.2	9

(USDA): United States Department of Agriculture
*Prepared as Package Directs

Food and Description	Measure or Quantity	Protein (grams)	Calories
*Diet (Slim-ette)	8 oz. (by wt.)	.2	7
With rice (Richardson & Robbins)	1 cup (8.1 oz.)	4.1	48
Consommé:			
Condensed (USDA)	8 oz. (by wt.)	6.4	41
*Prepared with equal volume water (USDA)	1 cup (8.5 oz.)	3.4	22
Cream of:			
Condensed (USDA)	8 oz. (by wt.)	5.4	179
*Prepared with equal volume milk (USDA)	1 cup (8.6 oz.)	7.4	179
*Prepared with equal volume water (USDA)	1 cup (8 oz.)	2.9	94
*(Campbell)	1 cup	3.2	87
*(Heinz)	1 cup (8.5 oz.)	3.1	93
(Heinz) Great American	1 cup (8.5 oz.)	5.6	108
*& Dumplings (Campbell)	1 cup	6.6	95
Gumbo:			
Condensed (USDA)	8 oz. (by wt.)	5.9	104
*Prepared with equal volume water (USDA)	1 cup (8.5 oz.)	3.1	55
*(Campbell)	1 cup	2.5	55
Creole (Heinz) Great American	1 cup (8¾ oz.)	4.5	96
*& Kasha (Manischewitz)	1 cup	1.8	41
& Noodle:			
Condensed (USDA)	8 oz. (by wt.)	6.4	120
*Prepared with equal volume water (USDA)	1 cup (8.8 oz.)	3.5	65
*(Campbell)	1 cup	3.4	62
*(Campbell) Noodle-O's	1 cup	3.4	67
*(Heinz)	1 cup (8.5 oz.)	3.5	75
*(Manischewitz)	1 cup	4.7	46
Dietetic (Tillie Lewis)	1 cup (8 oz.)	3.4	53
With dumplings (Heinz) Great American	1 cup (8.5 oz.)	5.2	89
*With stars (Campbell)	1 cup	3.9	57
*With stars (Heinz)	1 cup (8.5 oz.)	3.4	66
& Rice:			
Condensed (USDA)	8 oz. (by wt.)	5.9	89
*Prepared with equal volume water (USDA)	1 cup (8.5 oz.)	3.1	48
*(Campbell)	1 cup	3.2	49

(USDA): United States Department of Agriculture
*Prepared as Package Directs

[99]

Food and Description	Measure or Quantity	Protein (grams)	Calories
*(Heinz)	1 cup (8.5 oz.)	2.5	61
*(Manischewitz)	1 cup	3.9	47
With mushrooms (Heinz)			
Great American	1 cup (8.5 oz.)	4.2	96
Vegetable:			
Condensed (USDA)	8 oz. (by wt.)	7.7	141
*Prepared with equal volume			
water (USDA)	1 cup (8.6 oz.)	4.2	76
*(Campbell)	1 cup	3.9	68
*(Heinz)	1 cup (8.5 oz.)	3.5	85
*(Manischewitz)	1 cup	2.7	55
CHICKEN SOUP MIX:			
Cream of (Lipton) *Cup-a-Soup*	1 pkg. (.8 oz.)	2.5	95
Cream of (Wyler's)	1 pkg. (.8 oz.)	2.7	93
& Noodle:			
(USDA)	2-oz. pkg.	8.3	218
*(USDA)	1 cup (8.1 oz.)	1.8	51
*(Lipton)	1 cup	2.0	53
(Lipton) *Cup-a-Soup*	1 pkg. (.4 oz.)	1.6	38
*(Lipton) *Giggle*	1 cup	2.8	76
*(Wyler's)	6 fl. oz.	1.6	33
Ring-O-Noodle (Lipton)	1 cup	2.1	57
Ring noodle (Lipton)			
Cup-a-Soup	1 pkg. (.6 oz.)	2.1	55
*With diced chicken (Lipton)	1 cup	3.3	68
With meat (Lipton)			
Cup-a-Soup	.4-oz. pkg.	2.6	42
& Rice:			
(USDA)	1 oz.	2.6	100
*(USDA)	1 cup (8 oz.)	1.1	46
*(Lipton)	1 cup (8 oz.)	2.6	62
Rice-A-Roni	1 cup	2.5	63
*(Wyler's)	6 fl. oz.	.8	37
*Vegetable (Lipton)	1 cup	3.4	74
Vegetable (Lipton) *Cup-A-Soup*	1 pkg. (.5 oz.)	2.0	41
CHICKEN SPREAD:			
(Swanson)	5-oz. can	21.5	283
(Underwood)	4-¾ oz. can	21.2	301
(Underwood)	1 T. (.5 oz.)	2.2	31

(USDA): United States Department of Agriculture
*Prepared as Package Directs

Food and Description	Measure or Quantity	Protein (grams)	Calories
CHICKEN STEW:			
Canned:			
(B & M)	1 cup (7.9 oz.)	10.6	128
(Swanson)	1 cup	12.0	166
With dumplings (Heinz)	8.5-oz. can	9.6	202
Freeze dry, canned (Wilson)			
Campsite:			
Dry	4½-oz. can	53.3	548
*Reconstituted	1 lb.	41.7	426
CHICKEN STOCK BASE			
(French's)	1 tsp. (3 grams)	.3	8
*CHICKEN SUPREME mix,			
without added fat (Lipton)	11½-oz. serving (½ dry pkg.)	14.8	324
CHICKEN TAMALE PIE,			
canned (Lynden Farms)	½ tamale pie with sauce (3.8 oz.)	3.3	143
CHICK PEA or GARBANZO			
(USDA):			
Dry	1 lb.	93.0	1633
Dry	1 cup (7.1 oz.)	41.0	720
CHICORY GREENS, raw			
(USDA):			
Untrimmed	½ lb. (weighed untrimmed)	3.4	37
Trimmed	4 oz.	2.0	23
CHICORY, WITLOOF, Belgian			
or French endive, raw,			
bleached head (USDA):			
Untrimmed	½ lb. (weighed untrimmed)	2.0	30
Trimmed, cut	½ cup (.9 oz.)	.3	4
CHILI or CHILI CON CARNE:			
Canned, with beans:			
(USDA)	1 cup (8.8 oz.)	18.8	332
(Armour Star)	15½-oz. can	37.8	692

(USDA): United States Department of Agriculture
*Prepared as Package Directs

Food and Description	Measure or Quantity	Protein (grams)	Calories
(Chef Boy-Ar-Dee)	¼ of 30-oz. can	17.8	307
(Heinz)	8¾-oz. can	17.7	352
(Hormel)	7½ oz.	16.4	320
(Morton House)	1 cup (8 oz.)	16.8	367
(Nalley's) mild or hot	8 oz.	19.2	345
(Swanson)	1 cup	15.9	270
(Van Camp)	1 cup (8 oz.)	17.2	304
(Wilson)	½ of 15½-oz. can	15.4	315
Canned without beans:			
(USDA) not less than 60% meat, nor more than 8% cereal & seasonings	1 cup (9 oz.)	26.3	510
(Armour Star)	15½-oz. can	34.2	835
(Chef Boy-Ar-Dee)	½ of 15¼-oz. can	17.3	328
(Hormel)	7½ oz.	21.1	340
(Nalley's)	8 oz.	18.7	311
(Van Camp)	1 cup (8.1 oz.)	23.6	460
(Wilson)	½ of 15½-oz. can	12.7	420
Frozen, with beans (Banquet)	8-oz. bag	14.5	310
CHILI BEEF SOUP:			
*(Campbell)	1 cup	7.0	149
*(Heinz)	1 cup (8¾ oz.)	6.7	161
(Heinz) *Great American*	1 cup (8¾ oz.)	8.7	179
CHILI CON CARNE MIX (Durkee):			
*With meat & beans	2½ cups (2¼-oz. pkg.)	52.0	1720
*Without meat & beans	1¼ cups (2¼-oz. pkg.)	4.7	196
CHILI CON CARNE SPREAD (Oscar Mayer):			
With beans	1 oz.	3.1	62
Without beans	1 oz.	4.3	78
***CHILI DOG SAUCE MIX** (McCormick)	1 serving (.9 oz.)	.9	18
CHILI POWDER:			
With added seasoning (USDA)	1 T. (.5 oz.)	2.1	51
(Chili Products)	½ oz.	1.9	47

(USDA): United States Department of Agriculture
*Prepared as Package Directs

Food and Description	Measure or Quantity	Protein (grams)	Calories
CHILI SAUCE:			
(USDA)	½ cup (4.4 oz.)	3.1	129
(USDA)	1 T. (.5 oz.)	.4	16
(Del Monte)	1 T.	.3	18
(Heinz)	1 T.	.2	17
(Hunt's)	½ cup (4.8 oz.)	2.6	153
(Hunt's)	1 T. (.6 oz.)	.3	19
(Stokely-Van Camp)	1 T. (.5 oz.)	.4	15
CHILI SEASONING MIX:			
(Durkee)	1.7-oz. pkg.	2.1	148
(French's) *Chili-O*	1¾-oz. pkg.	4.4	123
*(Kraft)	1 oz.	3.4	44
(Lawry's)	1.6-oz. pkg.	4.7	137
*(Wyler's)	6 fl. oz.	.8	34
CHINESE DATE (see **JUJUBE**)			
CHINESE DINNER, frozen:			
Chicken chow mein (Banquet)	11-oz. dinner	17.2	291
(Swanson)	11-oz. dinner	19.7	356
CHINESE VEGETABLES (see **VEGETABLES, MIXED**)			
CHIPS (see **CRACKERS** for corn chips and **POTATO CHIPS**)			
CHITTERLINGS, canned (Hormel)	1-lb. 2-oz. can	35.2	832
CHIVES, raw (USDA)	1 oz.	.5	8
CHOCOLATE, BAKING:			
Bitter or unsweetened:			
(USDA)	1 oz.	3.0	143
Grated (USDA)	½ cup (2.3 oz.)	7.1	333
(Baker's)	1-oz. sq.	3.1	136
(Hershey's)	1 oz.	3.0	183
Sweetened:			
Bittersweet (USDA)	1 oz.	2.2	135
Chips, milk (Hershey's)	¼ cup (1.5 oz.)	2.7	234
Chips, semisweet (Baker's)	¼ cup (1.5 oz.)	1.7	191

(USDA): United States Department of Agriculture
*Prepared as Package Directs

Food and Description	Measure or Quantity	Protein (grams)	Calories
Chips, semisweet (Ghirardelli)	⅓ cup (2 oz.)	2.5	299
Chips, semisweet (Hershey's) regular & mini	¼ cup (1.5 oz.)	1.7	227
German's, sweet (Baker's)	4½ sq. (1 oz.)	1.1	141
Morsels, milk (Nestlé's)	1 oz.	1.4	152
Semisweet, small pieces (USDA)	½ cup (3 oz.)	3.6	431
Semisweet (Baker's)	1-oz. sq.	1.6	132
CHOCOLATE CAKE:			
Home recipe (USDA):			
Without icing	3 oz.	4.1	311
With chocolate icing, 2-layer	⅟₁₆ of 10″ cake (4.2 oz.)	5.4	443
With chocolate icing, 2-layer	⅟₁₆ of 9″ cake (2.6 oz.)	3.4	277
With uncooked white icing	⅟₁₆ of 10″ cake (4.2 oz.)	4.2	443
Fudge, frozen (Pepperidge Farm)	⅙ of cake (3 oz.)	3.6	315
German chocolate, frozen (Morton)	2.2-oz. serving	2.9	230
Golden, frozen (Pepperidge Farm)	⅙ of cake (3 oz.)	3.0	320
CHOCOLATE CAKE MIX (see also **FUDGE CAKE MIX**):			
Chocolate malt (USDA)	1 oz.	1.1	117
*Chocolate malt, uncooked white icing (USDA)	4 oz.	3.9	392
*Chocolate malt layer (Betty Crocker)	⅟₁₂ of cake	2.9	200
*Chocolate pudding (Betty Crocker)	⅙ of cake	2.7	221
*Deep chocolate (Duncan Hines)	⅟₁₂ of cake (2.7 oz.)	3.0	201
*Double Dutch (Pillsbury)	⅟₁₂ of cake	4.0	210
*German chocolate layer (Betty Crocker)	⅟₁₂ of cake	2.9	200
*German chocolate (Pillsbury)	⅟₁₂ of cake	3.0	210
*German chocolate, streusel (Pillsbury)	⅟₁₂ of cake	4.0	350
*Macaroon (Pillsbury) *Bundt*	⅟₁₂ of cake	4.0	360

(USDA): United States Department of Agriculture
*Prepared as Package Directs

Food and Description	Measure or Quantity	Protein (grams)	Calories
*Milk chocolate layer (Betty Crocker)	¹⁄₁₂ of cake	3.2	199
*Streusel (Pillsbury)	¹⁄₁₂ of cake	4.0	340
*Swiss chocolate (Duncan Hines)	¹⁄₁₂ of cake (2.7 oz.)	3.0	201
CHOCOLATE CANDY (see **CANDY**)			
CHOCOLATE DRINK:			
Canned (Borden)	9½-fl.-oz. can	11.5	232
Mix:			
Hot (USDA)	1 cup (4.9 oz.)	13.1	545
Hot (USDA)	1 oz.	2.7	111
Dutch, instant (Borden)	2 heaping tsps. (¾ oz.)	1.0	87
Instant (Ghirardelli)	1 T. (.4 oz.)	.4	48
CHOCOLATE, GROUND (Ghirardelli)	¼ cup (1.3 oz.)	1.9	163
CHOCOLATE, HOT, home recipe (USDA)	1 cup (8.8 oz.)	8.2	238
CHOCOLATE ICE CREAM (see also individual brands):			
(Borden) 9.5% fat	¼ pt. (2.3 oz.)	2.5	126
(Prestige) French	¼ pt. (2.6 oz.)	2.9	182
(Sealtest)	¼ pt. (2.3 oz.)	2.4	136
CHOCOLATE PIE:			
Chiffon, home recipe (USDA)	⅙ of 9″ pie (4.9 oz.)	9.5	459
Meringue, home recipe (USDA)	⅙ of 9″ pie (4.9 oz.)	6.7	353
Nut (Tastykake)	4½-oz. pie	7.1	451
Frozen:			
Cream:			
(Banquet)	2½-oz. serving	2.5	202
(Mrs. Smith's)	⅙ of 8″ pie (2.8 oz.)	1.5	247
Meringue (Mrs. Smith's)	⅛ of 10″ pie (5.2 oz.)	2.6	514
Tart (Pepperidge Farm)	1 pie tart (3 oz.)	2.9	306
Velvet nut (Kraft)	⅙ of 16¾-oz. pie	5.1	303

(USDA): United States Department of Agriculture
*Prepared as Package Directs

Food and Description	Measure or Quantity	Protein (grams)	Calories
CHOCOLATE PIE FILLING (see **CHOCOLATE PUDDING MIX**)			
CHOCOLATE PUDDING, sweetened:			
Home recipe with starch base (USDA)	½ cup (4.6 oz.)	4.0	192
Chilled:			
Dark chocolate (Breakstone)	5-oz. container	2.8	256
Light chocolate (Breakstone)	5-oz. container	2.4	254
Dark chocolate (Sanna)	4¼-oz. container	3.2	191
Light chocolate (Sanna)	4½-oz. container	4.1	198
(Sealtest)	4 oz.	3.3	136
Canned:			
(Betty Crocker)	½ cup	3.1	175
(Hunt's)	5-oz. can	2.4	239
(Thank You)	½ cup (4.5 oz.)	3.2	175
Dark 'N' Sweet (Royal) *Creamerino*	5-oz. can	3.3	250
Fudge (Betty Crocker)	½ cup	3.1	175
Fudge (Del Monte)	5-oz. can	4.2	198
Fudge (Hunt's)	5-oz. can	2.2	229
Milk chocolate (Del Monte)	5-oz. can	4.2	202
Milk chocolate (Royal) *Creamerino*	5-oz. can	3.1	244
CHOCOLATE PUDDING or PIE FILLING MIX: Sweetened:			
Regular:			
Dry (USDA)	1 oz.	8.5	102
*Prepared with milk (USDA)	½ cup (4.6 oz.)	4.4	161
*(Jell-O)	½ cup (5.2 oz.)	5.0	174
*(Royal)	½ cup (5.1 oz.)	5.1	196
*(Royal) *Dark 'N' Sweet*	½ cup (5.1 oz.)	5.3	195
*Fudge (Jell-O)	½ cup (5.2 oz.)	5.0	174
*Milk chocolate (Jell-O)	½ cup (5.2 oz.)	5.0	174
Instant:			
Dry (USDA)	1 oz.	.9	101
*Prepared with milk, without cooking (USDA)	4 oz.	3.4	142

(USDA): United States Department of Agriculture
*Prepared as Package Directs

Food and Description	Measure or Quantity	Protein (grams)	Calories
*(Jell-O)	½ cup (5.4 oz.)	5.2	190
*(Royal)	½ cup (5.1 oz.)	5.0	194
*(Royal) *Dark 'N' Sweet*	½ cup (5.1 oz.)	5.0	194
*Fudge (Jell-O)	½ cup (5.4 oz.)	5.2	190
*Low calories or dietetic (D-Zerta)	½ cup (4.6 oz.)	5.0	102
CHOCOLATE RENNET MIX:			
Powder:			
Dry (Junket)	1 oz.	1.5	116
*(Junket)	4 oz.	4.2	113
Tablet:			
Dry (Junket)	1 tablet (<1 gram)	Tr.	1
*& sugar (Junket)	4 oz.	3.7	101
CHOCOLATE SOFT DRINK			
(Yoo-Hoo) High Protein	6 fl. oz.	6.0	100
CHOCOLATE SYRUP:			
Sweetened:			
Fudge (USDA)	1 fl. oz. (1.3 oz.)	1.9	125
Fudge (USDA)	1 T. (.7 oz.)	1.0	63
Thin type (USDA)	1 fl. oz. (1.3 oz.)	.9	93
Thin type (USDA)	1 T. (.7 oz.)	.4	47
(Hershey's)	1 T. (.6 oz.)	.4	44
(Smucker's)	1 T. (.6 oz.)	.5	51
Low calorie:			
(Slim-ette) *Chocotop*	1 T. (.5 oz.)	.2	9
(Tillie Lewis)	1 T. (.5 oz.)	.2	4
CHOCO-NUT-SUNDAE CONE			
(Sealtest)	2½ fl. oz. (2.1 oz.)	2.6	186
CHOP SUEY:			
Home recipe, with meat (USDA)	1 cup (8.8 oz.)	26.0	300
Canned:			
Meat (USDA)	1 cup (8.8 oz.)	11.0	155
Chicken (Hung's)	8 oz.	5.3	120
Meatless (Hung's)	8 oz.	4.2	112
Frozen:			
Beef (Banquet) buffet	2 lbs.	51.5	554
Beef (Banquet) cooking bag	7-oz. bag	11.2	121

(USDA): United States Department of Agriculture
*Prepared as Package Directs

Food and Description	Measure or Quantity	Protein (grams)	Calories
Beef dinner (Banquet)	12-oz. dinner	13.6	282
Mix (Durkee)	1⅝-oz. pkg.	1.8	128
*Mix, with meat & vegatables (Durkee)	3½ cups	107.6	1113

CHOP SUEY VEGETABLES
(see **VEGETABLES, MIXED**)

CHOW CHOW:

Sour (USDA)	1 oz.	.4	8
Sweet (USDA)	1 oz.	.4	33

CHOW MEIN (see also
CHINESE DINNER):

Home recipe, chicken, without noodles (USDA)	4 oz.	14.1	116
Canned:			
Chicken:			
(USDA) without noodles	4 oz.	2.9	43
(Hung's)	8 oz.	4.9	104
Meatless (Hung's)	8 oz.	4.0	96
Frozen:			
Chicken:			
(Banquet)	7-oz. bag	12.2	123
(Banquet) buffet	2-lb. pkg.	56.0	563
With rice (Swanson)	8½-oz. pkg.	12.2	188
Mix (Durkee)	1⅝-oz. pkg.	1.8	128
*Mix, with meat and vegetables (Durkee)	3½ cups	42.2	1113

CHOW MEIN NOODLES (see
NOODLES, CHOW MEIN)

CHOW MEIN VEGETABLES
(see **VEGETABLES, MIXED**)

CHUB, raw (USDA):

Whole	1 lb. (weighed whole)	22.9	217
Meat only	4 oz.	17.4	164

CIDER (see **APPLE CIDER**)

(USDA): United States Department of Agriculture
*Prepared as Package Directs

Food and Description	Measure or Quantity	Protein (grams)	Calories
CINNAMON:			
Ground (Information supplied by General Mills Laboratory)	1 oz.	1.0	114
With sugar (French's)	1 tsp. (4 grams)	Tr.	15
*CINNAMON CAKE MIX,			
streusel (Pillsbury)	½₂ of cake	4.0	350
CINNAMON STICKS, frozen			
(Aunt Jemima)	3 pieces (1¾ oz.)	3.2	145
CISCO (see **LAKE HERRING**)			
CITRON, CANDIED (USDA)	1 oz.	<.1	89
CITRUS COOLER (Hi-C)	6 fl. oz. (6.3 oz.)	.7	90
CLACKERS, cereal (General Mills)	1 cup (1 oz.)	1.9	111
CLAM:			
Raw, all kinds, meat & liq. (USDA)	4 oz.	9.2	60
Raw, all kinds, meat only (USDA)	4 med. clams (3 oz.)	10.7	65
Raw, hard or round (USDA):			
Meat & liq.	1 lb. (weighed in shell)	9.4	71
Meat only	1 cup (7 round chowders, 8 oz.)	25.2	182
Raw, soft (USDA):			
Meat & liq.	1 lb. (weighed in shell)	22.6	142
Meat only	1 cup (19 large, 8 oz.)	31.8	186
Canned, all kinds:			
Solids & liq. (USDA)	4 oz.	9.0	59
Meat only (USDA)	½ cup (2.8 oz.)	12.6	78
Chopped (Snow)	4 oz.	14.4	68
Creamed, with mushrooms (Snow)	4 oz.	9.3	127
Minced (Snow)	4 oz.	28.8	137
Solids & liq. (Bumble Bee)	4½-oz. can	10.0	66
Frozen, breaded (Mrs. Paul's)	5-oz. pkg.	19.1	506

(USDA): United States Department of Agriculture
*Prepared as Package Directs

Food and Description	Measure or Quantity	Protein (grams)	Calories
CLAM CAKE, frozen, thins (Mrs. Paul's)	2½-oz. cake	7.5	147
CLAM CHOWDER:			
Manhattan, canned:			
Condensed (USDA)	8 oz. (by wt.)	4.1	150
*Prepared with equal volume water (USDA)	1 cup (8.6 oz.)	2.2	81
*(Campbell)	1 cup	2.2	72
(Campbell) *Chunky*	1 cup	5.9	120
*(Doxsee)	1 cup (8.6 oz.)	3.9	112
(Heinz) *Great American*	1 cup (8½ oz.)	4.1	111
(Snow)	8 oz.	8.4	85
New England:			
Canned:			
*(Campbell)	1 cup	8.3	157
*(Doxsee)	1 cup (8.6 oz.)	10.2	214
*(Snow)	⅓ of 15-oz. can	7.1	87
Frozen:			
Condensed (USDA)	8 oz. (by wt.)	8.4	243
*Prepared with equal volume water (USDA)	1 cup (8.5 oz.)	4.3	130
*Prepared with equal volume milk (USDA)	1 cup (8.6 oz.)	9.1	211
Mix (Lipton) *Cup-a-Soup*	1 pkg. (.8 0z.)	2.1	102
Mix, cream of (Wyler's)	1 pkg. (.8 oz.)	2.6	112
CLAM FRITTERS, home recipe, made with flour, baking powder, butter & eggs (USDA)	1 fritter (2″ × 1¾″, 1.4 oz.)	4.6	124
CLAM JUICE/LIQUOR, canned:			
(USDA)	1 cup (8.3 oz.)	5.4	45
(Snow)	8 oz.	5.7	31
CLAM STEW, New England (Snow)	8 oz.	13.3	198
CLAM STICK, frozen, breaded (Mrs. Paul's)	1 piece (.8 oz.)	2.1	46

(USDA): United States Department of Agriculture
*Prepared as Package Directs

Food and Description	Measure or Quantity	Protein (grams)	Calories
CLAMATO COCKTAIL (Mott's)	4 oz.	.8	43
COCOA, dry:			
Plain (USDA):			
Medium-low fat	½ cup (1.5 oz.)	8.3	95
Medium-low fat	1 T. (5 grams)	1.0	12
Medium-high fat	½ cup (1.5 oz.)	7.4	114
Medium-high fat	1 T. (5 grams)	.9	14
Processed with alkali (USDA):			
Medium-low fat	½ cup (1.5 oz.)	8.3	92
Medium-low fat	1 T. (5 grams)	1.0	12
Medium-high fat	½ cup (1.5 oz.)	7.4	112
Medium-high fat	1 T. (5 grams)	.9	14
(Hershey's)	½ cup (1.5 oz.)	8.4	185
(Hershey's)	1-oz. packet	5.6	122
(Hershey's)	1 T. (5 grams)	1.0	22
COCOA, HOME RECIPE (USDA)	1 cup (8.8 oz.)	9.5	242
COCOA KRISPIES, cereal (Kellogg's)	1 cup (1 oz.)	1.4	111
COCOA MIX:			
With nonfat dry milk (USDA)	1 oz.	5.3	102
Without nonfat dry milk (USDA)	1 oz.	1.1	98
(Kraft)	1 oz.	4.5	105
*(Kraft)	1 cup	5.7	129
Hot (Hershey's)	1-oz. packet	3.1	116
Instant:			
(Hershey's)	¾-oz.-packet	.8	84
Swiss Miss	1 oz. (6 fl. oz. serving	5.0	103
Chocolate marshmallow (Carnation)	1 pkg. (1 oz.)	3.5	112
Milk chocolate (Carnation)	1 pkg. (1 oz.)	3.5	120
Mini marshmallow, *Swiss Miss*	1.1-oz. dry (6 fl. oz.)	3.3	117
Rich chocolate (Carnation)	1 pkg. (1 oz.)	3.5	109
COCOA PEBBLES, cereal (Post)	⅞ cup (1 oz.)	.9	111

(USDA): United States Department of Agriculture
*Prepared as Package Directs   [111]

Food and Description	Measure or Quantity	Protein (grams)	Calories
COCOA PUFFS, cereal (General Mills)	1 cup (1 oz.)	1.4	109
COCONUT:			
Fresh (USDA):			
Whole	1 lb. (weighed in shell)	8.3	816
Meat only	4 oz.	4.0	392
Meat only	2" × 2" × ½" piece (1.6 oz.)	1.6	156
Grated or shredded	1 firmly packed cup (4.6 oz.)	4.6	450
Grated	1 lightly packed cup (2.9 oz.)	2.8	277
Cream, liq. expressed from grated coconut	4 oz.	5.0	379
Milk, liq. expressed from mixture of grated coconut & water	4 oz.	3.6	286
Water, liq. from coconut	1 cup (8.5 oz.)	.7	53
Dried, canned or packaged:			
Sweetened, shredded (USDA)	½ lightly packed cup (1.6 oz.)	1.7	252
Unsweetened (USDA)	½ lightly packed cup (1.6 oz.)	3.3	305
Angel Flake (Baker's)	½ cup (1.3 oz.)	1.2	178
Chocolate (Durkee)	½ cup (1.3 oz.)	1.9	176
Cookie (Baker's)	½ cup (2 oz.)	2.0	280
Crunchies (Baker's)	½ cup (2.1 oz.)	3.4	352
Lemon (Durkee)	½ cup (1.3 oz.)	1.5	181
Orange (Durkee)	½ cup (1.3 oz.)	1.5	181
Peppermint (Durkee)	½ cup (1.3 oz.)	1.5	182
Premium shred (Baker's)	½ cup (1.5 oz.)	1.4	210
Southern style (Baker's)	½ cup (1.5 oz.)	1.4	170
COCONUT CAKE, frozen (Pepperidge Farm)	⅙ of cake (3 oz.)	2.7	323
***COCONUT CAKE MIX** (Duncan Hines)	¹⁄₁₂ of cake (2.7 oz.)	3.0	200

(USDA): United States Department of Agriculture
*Prepared as Package Directs

Food and Description	Measure or Quantity	Protein (grams)	Calories
COCONUT PIE:			
Cream:			
(Tastykake)	4-oz. pie	6.4	467
Frozen:			
(Banquet)	2½-oz. serving	2.0	209
(Morton)	⅙ of 16-oz. pie	1.6	206
(Mrs. Smith's)	⅙ of 8″ pie (2.8 oz.)	1.2	233
Tart (Pepperidge Farm)	3-oz. pie tart	4.1	310
Custard:			
Home recipe (USDA)	⅙ of 9″ pie (5.4 oz.)	9.1	357
Frozen:			
Baked (USDA)	5 oz.	8.5	354
Unbaked (USDA)	5 oz.	7.4	291
(Banquet)	5-oz. serving	6.8	294
(Morton)	⅙ of 22-oz. pie	3.2	224
(Mrs. Smith's)	⅙ of 8″ pie (4 oz.)	7.7	265
(Mrs. Smith's)	⅛ of 10″ pie (5.5 oz.)	9.5	333
Meringue, frozen (Mrs. Smith's)	⅛ of 10″ pie (5.2 oz.)	1.6	438
COCONUT PIE FILLING MIX (see also **COCONUT PUDDING MIX**):			
Custard & pie crust, dry (USDA)	1 oz.	.9	133
*Custard made with egg yolk & milk (USDA)	5 oz. (including crust)	6.1	288
COCONUT PUDDING MIX:			
*Cream, regular (Jell-O)	½ cup (5.2 oz.)	4.5	175
*Cream, instant (Jell-O)	½ cup (5.3 oz.)	4.5	188
*Toasted, instant (Royal)	½ cup (5.1 oz.)	4.5	184
COCONUT SOFT DRINK (Yoo-Hoo) High Protein	6 fl. oz. (6.4 oz.)	6.0	100
COCO WHEATS, cereal	3 T. (1.3 oz.)	4.1	132
COD:			
Raw, whole (USDA)	1 lb. (weighed whole)	24.7	110
Raw, meat only (USDA)	4 oz.	20.0	88

(USDA): United States Department of Agriculture
*Prepared as Package Directs

[113]

Food and Description	Measure or Quantity	Protein (grams)	Calories
Broiled (USDA)	4 oz.	32.3	143
Canned (USDA)	4 oz.	21.8	96
Dehydrated (USDA)	4 oz.	92.8	425
Dried, salted (USDA)	4 oz.	32.9	147
Dried, salted (USDA)	5½" × 1½" × ½" (2.8 oz.)	23.2	104
Frozen (Gorton)	⅓ of 1-lb. pkg.	27.0	117

CODFISH CAKE (see **FISH CAKE**)

COFFEE:
Regular:
*Max Pax	¾ cup	Tr.	2
*(Maxwell House)	¾ cup	Tr.	2
*Yuban	¾ cup	Tr.	2

Instant:
Dry (USDA)	1 oz.	Tr.	37
Dry (USDA)	1 rounded tsp. (2 grams)	Tr.	3
(Borden)	1 rounded tsp. (2 grams)	.3	5
*(Chase & Sanborn)	¾ cup	Tr.	1
Kava (Borden)	1 tsp. (1 gram)	.2	3
*(Maxwell House)	¾ cup	Tr.	4
*(Yuban)	¾ cup	Tr.	4

Decaffeinated:
*Sanka regular	¾ cup	Tr.	2
*Sanka instant	¾ cup	Tr.	4
*Siesta	¾ cup	Tr.	1

Freeze-dried:
*Maxim	¾ cup	Tr.	4
*Sanka	¾ cup	Tr.	4

COFFEE CAKE:
(Drake's) junior	1 cake (1.1 oz.)	1.6	126
(Drake's) large	¼ of 11-oz. cake	4.0	308
(Drake's) small	1 small cake	3.9	280
Almond Danish with icing (Pillsbury)	1 roll	2.5	140
Butterfly (Mrs. Smith's)	1 piece (2¾ oz.)	3.0	299
Cherry (Mrs. Smith's)	1 piece (2¾ oz.)	3.0	238
Cinnamon (Pillsbury) country	2" × 4" piece	1.5	105

(USDA): United States Department of Agriculture
*Prepared as Package Directs

Food and Description	Measure or Quantity	Protein (grams)	Calories
Cinnamon with icing (Pillsbury)	1 roll	1.5	115
Cinnamon with icing (Pillsbury) *Ballard*	1 roll	2.0	105
Cinnamon with icing (Pillsbury) *Hungry Jack*	1 roll	2.0	145
Cinnamon-raisin (Mrs. Smith's)	1 piece (2¾ oz.)	3.1	295
Cinnamon twist (Pepperidge Farm)	⅙ cake (1.8 oz.)	2.7	156
Danish without fruit or nuts:			
Individual round (USDA)	1 piece (2.3 oz.)	4.8	273
Packaged ring (USDA)	12-oz. cake	25.2	1435
Danish (Mrs. Smith's) tray pack	1¾-oz. piece	2.0	190
Danish, apple (Morton)	1 cake (13.5 oz.)	13.1	1130
Danish, caramel (Pillsbury)	1 roll	2.0	155
Danish, cinnamon & raisin (Pillsbury)	1 roll	2.0	135
Danish, orange (Pillsbury)	1 roll	2.0	130
Danish pecan twist (Morton)	12-oz. cake	16.7	1369
Melt-A-Way (Morton)	13-oz. cake	19.2	1511
Meltaway (Mrs. Smith's)	1 piece (2¾ oz.)	2.0	351
Pecan roll (Mrs. Smith's)	1 piece (2½ oz.)	6.0	396
COFFEE CAKE MIX:			
Dry (USDA)	1 oz.	1.7	122
*Prepared with egg & milk (USDA)	2 oz.	3.6	183
*(Aunt Jemima)	⅛ of cake (1.9 oz.)	3.0	186
COLESLAW, not drained (USDA):			
Prepared with commercial French dressing	4 oz.	1.4	108
Prepared with homemade French dressing	4 oz.	1.2	146
Prepared with mayonnaise	4 oz.	1.5	163
Prepared with mayonnaise-type salad dressing	4 oz.	1.4	112
COLLARDS:			
Raw (USDA):			
Leaves including stems	1 lb.	16.3	181
Leaves only	½ lb.	7.4	70

(USDA): United States Department of Agriculture
*Prepared as Package Directs

Food and Description	Measure or Quantity	Protein (grams)	Calories
Boiled, drained (USDA):			
Leaves, cooked in large amount of water	½ cup (3.4 oz.)	3.2	29
Leaves & stems, cooked in small amount of water	4 oz.	3.1	33
Leaves, cooked in small amount water	½ cup (3.4 oz.)	3.4	31
Frozen:			
Not thawed (USDA)	10-oz. pkg.	8.8	91
Boiled, chopped, drained (USDA)	½ cup (3 oz.)	2.5	26
Chopped (Birds Eye)	⅓ pkg. (3.3 oz.)	2.9	29
CONCENTRATE, cereal (Kellogg's)	⅓ cup (1 oz.)	11.7	108
CONSOMME, canned, dietetic pack (Slim-ette)	8 oz. (by wt.)	6.3	60
COOKIE, COMMERCIAL. The following are listed by type or brand name:			
Almond crescent (Nabisco)	1 piece (7 grams)	.4	34
Almond toast, Mandel (Stella D'oro)	1 piece (.5 oz.)	1.0	49
Angelica Goodies (Stella D'oro)	1 piece (.8 oz.)	1.7	100
Anginetti (Stella D'oro)	1 piece (5 grams)	.8	28
Animal Cracker:			
(USDA)	1 oz.	1.9	122
(Nabisco) Barnum's	1 piece (3 grams)	.2	12
(Sunshine) regular	1 piece (2 grams)	.2	10
(Sunshine) iced	1 piece (5 grams)	.1	26
Anisette sponge (Stella D'oro)	1 piece (.5 oz.)	1.1	39
Anisette toast (Stella D'oro)	1 piece (.4 oz.)	.8	39
Applesauce (Sunshine) regular or iced	1 piece (.6 oz.)	.9	86
Arrowroot (Sunshine)	1 piece (4 grams)	.2	16
Assortment:			
(USDA)	1 oz.	1.4	136
(Stella D'oro) Lady Stella	1 piece (8 grams)	.4	37
(Sunshine) Lady Joan	1 piece (9 grams)	.6	42
(Sunshine) Lady Joan, iced	1 piece (.4 oz.)	.7	47
Aunt Sally, iced (Sunshine)	1 piece (.8 oz.)	.9	96

(USDA): United States Department of Agriculture

Food and Description	Measure or Quantity	Protein (grams)	Calories
Bana-Bee (Nabisco)	6 pieces (1¾-oz. pkg.)	4.0	253
Big Treat (Sunshine)	1 piece (1.3 oz.)	1.2	153
Bordeaux (Pepperidge Farm)	1 piece (8 grams)	.4	36
Breakfast Treats (Stella D'oro)	1 piece (.8 oz.)	1.5	99
Brown edge wafers (Nabisco)	1 piece (6 grams)	.3	28
Brownie:			
(Hostess) 2 to pkg.	1 piece (.9 oz.)	1.5	100
(Tastykake)	1 pkg. (2¼ oz.)	3.3	242
Chocolate nut (Pepperidge Farm)	1 piece (.4 oz.)	.7	54
Peanut butter (Tastykake)	1 pkg. (1¾ oz.)	4.8	239
Pecan fudge (Keebler)	1 piece (.9 oz.)	1.9	115
Frozen, with nuts & chocolate icing (USDA)	1 oz.	1.4	119
Brussels (Pepperidge Farm)	1 piece (8 grams)	.5	42
Butter:			
Thin, rich (USDA)	1 oz.	1.7	130
(Nabisco)	1 piece (5 grams)	.3	23
(Sunshine)	1 piece (5 grams)	.3	23
Buttercup (Keebler)	1 piece (5 grams)	.3	24
Butterscotch Fudgies (Tastykake)	1 pkg. (1¾ oz.)	4.4	251
Capri (Pepperidge Farm)	1 piece (.6 oz.)	.9	82
Cardiff (Pepperidge Farm)	1 piece (4 grams)	.2	18
Cherry Coolers (Sunshine)	1 piece (6 grams)	.3	29
Chinese almond (Stella D'oro)	1 piece (1.2 oz.)	2.2	178
Chocolate & chocolate-covered:			
(USDA)	1 oz.	2.0	126
Como (Stella D'oro)	1 piece (.7 oz.)	2.0	155
Creme (Wise)	1 piece (1.1 oz.)	.6	32
Peanut bars (Nabisco) *Ideal*	1 piece (.6 oz.)	1.2	94
Pinwheels (Nabisco)	1 piece (1.1 oz.)	.8	139
Snaps (Nabisco)	1 piece (4 grams)	.2	18
Snaps (Sunshine)	1 piece (3 grams	.2	14
Wafers (Nabisco) *Famous*	1 piece (6 grams)	.6	28
Chocolate chip:			
(USDA)	1 oz.	1.5	134
(Drake's)	1 piece (.5 oz.)	.8	74
(Keebler) old fashioned	1 piece (.6 oz.)	.7	80
(Nabisco)	1 piece (7 grams)	.3	33
(Nabisco) *Chips Ahoy*	1 piece (.4 oz.)	.5	51
(Nabisco) *Family Favorites*	1 piece (7 grams)	.3	33

(USDA): United States Department of Agriculture

Food and Description	Measure or Quantity	Protein (grams)	Calories
(Nabisco) Snaps	1 piece (4 grams)	.2	21
(Pepperidge Farm)	1 piece (.4 oz.)	.6	52
(Sunshine) *Chip-A-Roos*	1 piece (.4 oz.)	.6	63
(Tastykake) *Choc-O-Chip*	4 pieces (1¾-oz. pkg.)	3.6	283
Cinnamon:			
Crisp (Keebler)	1 piece (4 grams)	.2	17
Spice, vanilla, sandwich (Nabisco) *Crinkles*	6 pieces (1⅝-oz. pkg.)	2.0	228
Sugar (Pepperidge Farm)	1 piece (.4 oz.)	.6	52
Toast (Sunshine)	1 piece (3 grams)	.2	13
Coconut:			
Bar (USDA)	1 oz.	1.8	140
Bar (Nabisco)	1 piece (9 grams)	.5	45
Bar (Sunshine)	1 piece (.4 oz.)	.5	47
(Nabisco) *Family Favorites*	1 piece (3 grams)	.2	16
Chocolate chip (Nabisco)	1 piece (.5 oz.)	.8	77
Chocolate chip (Sunshine)	1 piece (.6 oz.)	.8	80
Chocolate drop (Keebler)	1 piece (.5 oz.)	.7	75
Coconut Kiss (Tastykake)	4 pieces (1¾-oz. pkg.)	3.2	318
Commodore (Keebler)	1 piece (.5 oz.)	.8	65
Como Delight (Stella D'oro)	1 piece (1.1 oz.)	2.3	153
Cowboys and Indians (Nabisco)	1 piece (2 grams)	.1	10
Cream Lunch (Sunshine)	1 piece (.4 oz.)	.7	45
Creme Wafer Stick (Nabisco)	1 piece (9 grams)	.4	50
Cup Custard (Sunshine):			
Chocolate	1 piece (.5 oz.)	.8	70
Vanilla	1 piece (.5 oz.)	.8	71
Devil's Food Cake (Nab)	2 pieces (1¼-oz. pkg.)	1.9	135
Devil's Food Cake (Nabisco)	1 piece (.5 oz.)	.7	49
Dixie Vanilla (Sunshine)	1 piece (.5 oz.)	.9	60
Dresden (Pepperidge Farm)	1 piece (.6 oz.)	.8	83
Egg Jumbo (Stella D'oro)	1 piece (.4 oz.)	.8	40
Fig bar:			
(USDA)	1 oz.	1.1	101
(Keebler)	1 piece (.7 oz.)	.6	71
(Nab) *Fig Newtons*	1 piece (1-oz. pkg.)	.9	104
(Nab) *Fig Newtons*	2 pieces (2-oz. pkg.)	1.8	208

(USDA): United States Department of Agriculture

Food and Description	Measure or Quantity	Protein (grams)	Calories
(Nabisco) *Fig Newtons*	1 piece (.6 oz.)	.5	57
(Sunshine)	1 piece (.4 oz.)	.5	45
Fruit, iced (Nabisco)	1 piece (.6 oz.)	.9	71
Fudge:			
(Sunshine)	1 piece (.5 oz.)	.7	72
Chip (Pepperidge Farm)	1 piece (.4 oz.)	.6	51
Fudge Stripes (Keebler)	1 piece (.4 oz.)	.6	57
Gingersnap:			
(USDA)	1 oz.	1.6	119
(USDA) crumbs	1 cup; (4.1 oz.)	6.3	483
(Keebler)	1 piece (6 grams)	.3	24
(Nabisco) old fashion	1 piece (7 grams)	.4	29
(Sunshine)	1 piece (6 grams)	.3	24
Zu Zu (Nabisco)	1 piece (4 grams)	.2	16
Golden Bars (Stella D'oro)	1 piece (1 oz.)	1.7	123
Graham Cracker (see **CRACKERS,** Graham)			
Hermit bar, frosted (Tastykake)	1 pkg. (2 oz.)	3.2	321
Home Plate (Keebler)	1 piece (.5 oz.)	.9	58
Hostess With The Mostest (Stella D'oro)	1 piece (8 grams)	.5	39
Hydrox (Sunshine):			
Regular or mint	1 piece (.4 oz.)	.4	48
Vanilla	1 piece (.4 oz.)	.4	50
Jan Hagel (Keebler)	1 piece (1.0 grams)	.6	44
Jumble (Drake's)	1 piece (.4 oz.)	.6	52
Keebies (Keebler)	1 piece (.4 oz.)	.4	51
Ladyfingers (USDA)	.4-oz. ladyfinger (3¼″ × 1⅜″ × 1⅛″)	.9	40
Lemon:			
(Sunshine)	1 piece (.5 oz.)	.9	76
Jumble rings (Nabisco)	1 piece (.5 oz.)	.8	68
Lemon Coolers (Sunshine)	1 piece (6 grams)	.3	29
Nut Crunch (Pepperidge Farm)	1 piece (.4 oz.)	.7	57
Snaps (Nabisco)	1 piece (4 grams)	.2	17
Lido (Pepperidge Farm)	1 piece (.6 oz.)	1.0	91
Lisbon (Pepperidge Farm)	1 piece (5 grams)	.3	28
Macaroon:			
(USDA)	1 oz.	1.5	135
Almond (Tastykake)	2-oz. pkg. (2 pieces)	3.3	336
Coconut (Nabisco) *Bake Shop*	1 piece (.7 oz.)	.6	87

(USDA): United States Department of Agriculture

Food and Description	Measure or Quantity	Protein (grams)	Calories
Sandwich (Nabisco)	1 piece (.5 oz.)	.6	71
Margherite, chocolate (Stella D'oro)	1 piece (.6 oz.)	.9	73
Margherite, vanilla (Stella D'oro)	1 piece (.6 oz.)	1.0	73
Marquisette (Pepperidge Farm)	1 piece (8 grams)	.5	45
Marshmallow:			
(USDA)	1 oz.	1.1	116
Fancy Crests (Nabisco)	1 piece (.5 oz.)	.5	53
Mallowmars (Nabisco)	1 piece (.5 oz.)	.6	60
Mallo Puffs (Sunshine)	1 piece (.6 oz.)	.5	63
Minarets (Nabisco)	1 piece (10 grams)	.5	46
Puffs (Nabisco)	1 piece (.7 oz.)	.9	94
Sandwich (Nabisco)	1 piece (8 grams)	.4	32
Twirls (Nabisco)	1 piece (1.1 oz.)	1.2	133
Milano (Pepperidge Farm)	1 piece (.4 oz.)	.7	62
Milano, mint (Pepperidge Farm)	1 piece (.5 oz.)	.7	76
Mint sandwich (Nabisco) *Mystic*	1 piece (.6 oz.)	.9	88
Molasses (USDA)	1 oz.	1.8	120
Molasses & Spice (Sunshine)	1 piece (.6 oz.)	.9	67
Naples (Pepperidge Farm)	1 piece (.6 grams)	.4	33
Nassau (Pepperidge Farm)	1 piece (.6 oz.)	1.1	83
Oatmeal:			
(Drake's)	1 piece (.5 oz.)	1.0	69
(Keebler) old fashioned	1 piece (.6 oz.)	1.1	79
(Nabisco)	1 piece (.6 oz.)	1.0	82
(Nabisco) *Family Favorites*	1 piece (5 grams)	.3	24
(Sunshine)	1 piece (.4 oz.)	.7	58
Iced (Sunshine)	1 piece (.6 oz.)	.8	69
Irish (Pepperidge Farm)	1 piece (.4 oz.)	.6	50
Peanut butter (Sunshine)	1 piece (.6 oz.)	1.2	79
Raisin (USDA)	1 oz.	1.8	128
Raisin (Nabisco) *Bake Shop*	1 piece (.6 oz.)	1.0	77
Raisin (Pepperidge Farm)	1 piece (.4 oz.)	.8	55
Raisin bar (Tastykake)	1 pkg. (2¼ oz.)	4.5	298
Old Country Treats (Stella D'oro)	1 piece (.5 oz.)	2.4	64
Orleans (Pepperidge Farm)	1 piece (6 grams)	.3	30
Peach-apricot pastry (Stella D'oro)	1 piece (.8 oz.)	1.3	99
Peanut & peanut butter:			
(USDA)	1 oz.	2.8	134

(USDA): United States Department of Agriculture

Food and Description	Measure or Quantity	Protein (grams)	Calories
Bars, cocoa-covered (Nabisco)			
Crowns	1 piece (.6 oz.)	1.5	92
Caramel logs (Nabisco)			
Heydays	1 piece (.8 oz.)	2.4	122
Creme patties (Nab)	3 pieces (½-oz. pkg.)	1.5	74
Creme patties (Nab)	6 pieces (1-oz. pkg.)	30.4	148
Creme patties (Nabisco)	1 piece (7 grams)	.7	34
Creme patties, cocoa-covered			
(Nabisco) *Fancy*	1 piece (.4 oz.)	1.1	60
Patties (Sunshine)	1 piece (7 grams)	.8	33
Sandwich (Nabisco) *Nutter*			
Butter	1 piece (.5 oz.)	1.2	69
Pecan Sandies (Keebler)	1 piece (.6 oz.)	.8	85
Penguins (Keebler)	1 piece (.8 oz.)	.9	111
Pirouette (Pepperidge Farm):			
Chocolate laced	1 piece (7 grams)	.3	38
Lemon or original	1 piece (7 grams)	.3	37
Pitter Patter (Keebler)	1 piece (.6 oz.)	1.7	84
Pizzelle, Carolines (Stella			
D'oro)	1 piece (.4 oz.)	1.1	49
Raisin:			
(USDA)	1 oz.	1.2	107
Fruit biscuit (Nabisco)	1 piece (.5 oz.)	.7	58
Rich 'n Chips (Keebler)	1 piece (.5 oz.)	.7	73
Rochelle (Pepperidge Farm)	1 piece (.6 oz.)	.9	81
Sandwich, creme:			
(USDA)	1 oz.	1.4	140
(Tom Houston)	1 piece (.5 oz.)	2.3	74
Cameo (Nabisco)	1 piece (.5 oz.)	.7	68
Chocolate chip (Nabisco)	1 piece (.5 oz.)	.6	73
Chocolate fudge:			
(Keebler)	1 piece (.7 oz.)	1.1	99
Assorted (Nabisco) *Cookie*			
Break	1 piece (.4 oz.)	.5	52
Chocolate (Nabisco) *Cookie*			
Break	1 piece (.4 oz.)	.5	53
Orbit (Sunshine)	1 piece (.4 oz.)	.4	51
Oreo (Nab)	4 pieces (1-oz. pkg.)	1.4	140
Oreo (Nab)	6 pieces (1⅝-oz. pkg.)	2.3	228

(USDA): United States Department of Agriculture

Food and Description	Measure or Quantity	Protein (grams)	Calories
Oreo (Nab)	6 pieces (2⅛-oz. pkg.)	3.0	298
Oreo (Nabisco)	1 piece (.4 oz.)	.5	51
Oreo & Swiss (Nab)	6 pieces (1⅝-oz. pkg.)	2.4	230
Oreo & Swiss (Nab)	6 pieces (2¼-oz. pkg.)	3.3	319
Oreo & Swiss, assortment	1 piece (.4 oz.)	.5	51
Pride (Nabisco)	1 piece (.4 oz.)	.6	55
Social Tea (Nabisco)	1 piece (.4 oz.)	.5	51
Swiss (Nab)	4 pieces (1-oz. pkg.)	1.6	144
Swiss (Nab)	6 pieces (1¾-oz. pkg.)	2.9	252
Swiss (Nabisco)	1 piece (.4 oz.)	.6	52
Vanilla (Keebler)	1 piece (.6 oz.)	.8	82
Vanilla (Nabisco)	1 piece (.4 oz.)	.5	52
Vienna Finger (Sunshine)	1 piece (.5 oz.)	.7	71
Sesame, Regina (Stella D'oro)	1 piece (.4 oz.)	.7	51
Shortbread or shortcake:			
(USDA)	1 oz.	2.0	141
(USDA)	1¾" square (8 grams)	.6	40
(Nabisco) Dandy	1 piece (.4 oz.)	.5	46
(Pepperidge Farm)	1 piece (.5 oz.)	.8	72
Lorna Doone (Nab)	4 pieces (1-oz. pkg.)	1.8	138
Lorna Doone (Nab)	6 pieces (1½-oz. pkg.)	2.6	207
Lorna Doone (Nabisco)	1 piece (8 grams)	.5	37
Pecan (Nabisco)	1 piece (.5 oz.)	.6	80
Scotties (Sunshine)	1 piece (8 grams)	.6	39
Striped (Nabisco)	1 piece (10 grams)	.5	50
Vanilla (Tastykake)	6 pieces (2¼-oz. pkg.)	2.4	352
Social Tea Biscuit (Nabisco)	1 piece (5 grams)	.3	21
Spiced wafers (Nabisco)	1 piece (10 grams)	.5	41
Sprinkles (Sunshine)	1 piece (.6 oz.)	.6	57
Sugar cookie:			
(Keebler) old fashioned	1 piece (.6 oz.)	.9	79
(Pepperidge Farm)	1 piece (.4 oz.)	.6	51
(Sunshine)	1 piece (.6 oz.)	1.0	86

(USDA): United States Department of Agriculture

Food and Description	Measure or Quantity	Protein (grams)	Calories
Brown (Nabisco) *Family Favorite*	1 piece (5 grams)	.2	25
Brown (Pepperidge Farm)	1 piece (.4 oz.)	.8	48
Rings (Nabisco)	1 piece (.5 oz.)	.8	69
Sugar wafer:			
(USDA)	1 oz.	1.4	137
(Nab) *Biscos*	3 pieces (⅞-oz. pkg.)	.8	128
(Nabisco) *Biscos*	1 piece (4 grams)	.1	19
(Sunshine)	1 piece (9 grams)	.3	43
Krisp Kreem (Keebler)	1 piece (6 grams)	.1	31
Lemon (Sunshine)	1 piece (9 grams)	.3	44
Swedish Kreme (Keebler)	1 piece (5.7 oz.)	.7	98
Tahiti (Pepperidge Farm)	1 piece (.5 oz.)	.9	84
Toy (Sunshine)	1 piece (3 grams)	.2	13
Vanilla creme (Wise)	1 piece (7 grams)	.5	33
Vanilla snap (Nabisco)	1 piece (3 grams)	.2	13
Vanilla wafer:			
(USDA)	1 oz.	1.5	131
(Keebler)	1 piece (4 grams)	.2	19
(Nabisco) *Nilla*	1 piece (4 grams)	.2	18
(Sunshine) small	1 piece (3 grams)	.2	15
Venice (Pepperidge Farm)	1 piece (.4 oz.)	.6	57
Waffle creme (Nabisco) *Biscos*	1 piece (8 grams)	.2	42
Yum Yums (Sunshine)	1 piece (.5 oz.)	.5	83
COOKIE, DIETETIC:			
Angel puffs (Stella D'oro)	1 piece (3 grams)	.5	17
Apple pastry (Stella D'oro)	1 piece (.8 oz.)	1.1	94
Assorted wafers (Estee)	1 piece (5 grams)	.4	26
Banana wafers (Estee)	1 piece	1.9	123
Bittersweet chocolate wafer (Estee)	1 piece (7 oz.)	1.6	115
Choco Chip (Dia-Mel)	1 piece (9 grams)	8.0	40
Chocolate chip (Estee)	1 piece (7 grams)	.7	32
Chocolate wafer (Estee)	1 piece (5 grams)	.4	27
Fig pastry (Stella D'oro)	1 piece (.9 oz.)	1.4	100
Fruit wafer (Estee)	1 piece (5 grams)	.4	27
Have-A-Heart (Stella D'oro)	1 piece (.7 oz.)	1.4	97
Kichel (Stella D'oro)	1 piece (1 gram)	.2	8
Milk chocolate wafer (Estee)	1 piece (7 oz.)	1.8	110
Oatmeal raisin (Estee)	1 piece (7 grams)	.8	34

(USDA): United States Department of Agriculture

Food and Description	Measure or Quantity	Protein (grams)	Calories
Peach-apricot pastry (Stella D'oro)	1 piece (.8 oz.)	1.3	104
Prune pastry (Stella D'oro)	1 piece (.8 oz.)	1.3	92
Royal Nuggets (Stella D'oro)	1 piece (< 1 gram)	<.1	2
Sandwich (Estee)	1 piece (.4 oz.)	.6	55
Sandwich, lemon (Estee)	1 piece (.5 oz.)	.7	57
Vanilla Holland-filled wafer (Estee)	1 piece (5 grams)	.3	22
COOKIE DOUGH. refrigerated:			
Unbaked, plain (USDA)	1 oz.	1.0	127
Baked, plain (USDA)	1 oz.	1.1	141
Almond, chewy (Pillsbury)	1 cookie	.3	47
Apple-cinnamon (Pillsbury)	1 cookie	.3	53
Brownie, fudge (Pillsbury)	2″ square	2.0	140
Butterscotch nut (Pillsbury)	1 cookie	.3	53
Chocolate, Swiss style (Pillsbury)	1 cookie	.3	53
Chocolate chip (Pillsbury)	1 cookie	.3	57
Cinnamon sugar (Pillsbury)	1 cookie	.3	50
Oatmeal and chocolate chip (Pillsbury)	1 cookie	.7	57
Oatmeal raisin (Pillsbury)	1 cookie	.7	60
Peanut butter (Pillsbury)	1 cookie	.7	53
Peanut butter & chocolate chip (Pillsbury)	1 cookie	.7	50
Sugar (Pillsbury)	1 cookie	.3	53
COOKIE, HOME RECIPE:			
Brownie with nuts:			
(USDA)	1 oz.	1.8	137
(USDA)	.7-oz. piece (1¾″ × 1¾″ × ⅞″)	1.3	97
Chocolate chip (USDA)	1 oz.	1.5	146
Sugar, soft, thick (USDA)	1 oz.	1.7	126
COOKIE MIX:			
Plain, dry (USDA)	1 oz.	1.0	140
*Plain, prepared with egg & water (USDA)	1 oz.	1.4	140
*Plain, prepared with milk (USDA)	1 oz.	1.0	139

(USDA): United States Department of Agriculture
*Prepared as Package Directs

Food and Description	Measure or Quantity	Protein (grams)	Calories
*Apple Crunch Bar (Pillsbury)	2″ sq.	2.0	140
*Blueberry Crunch Bar (Pillsbury)	2″ sq.	2.0	140
Brownie:			
Dry, with egg (USDA)	1 oz.	1.4	119
Dry, without egg (USDA)	1 oz.	1.1	125
*Dry, with egg, prepared with water & nuts (USDA)	1 oz.	1.4	114
*Dry, without egg, prepared with egg, water & nuts (USDA)	1 oz.	1.4	121
*Butterscotch (Betty Crocker)	1½″ sq.	.7	59
*"Cake like," family size (Duncan Hines)	¹⁄₂₄ of pan (1.2 oz.)	2.3	148
*"Cake like," regular size (Duncan Hines)	¹⁄₁₆ of pan (1.2 oz.)	2.4	152
*Fudge (Betty Crocker)	1½″ sq.	.6	58
*Fudge, supreme (Betty Crocker)	1½″ sq.	.6	59
*Fudge, chewy, family size (Duncan Hines)	¹⁄₂₄ of pan (1.1 oz.)	1.9	142
*Fudge, chewy, regular size (Duncan Hines)	¹⁄₁₆ of pan (1.2 oz.)	2.0	147
*Fudge (Pillsbury)	1½″ sq.	.5	60
*German chocolate (Betty Crocker)	1½″ sq.	.7	70
*Walnut (Betty Crocker)	1½″ sq.	.8	63
*Walnut (Pillsbury)	1½″ sq.	.5	65
Chocolate mint (Nestlé's)	1 oz.	1.2	133
*Cherry Fruit 'N Crunch (Pillsbury)	2″ sq.	2.0	150
*Date bar (Betty Crocker)	2″ × 1″ bar	.6	58
*Macaroon, coconut (Betty Crocker)	1 macaroon (1¾″)	1.1	73
Lemon (Nestlé's)	1 oz.	1.2	132
Peach Fruit 'N Crunch (Pillsbury)	2″ sq.	2.0	140
Pecan bar (Pillsbury)	2″ × 1¼″ piece	.5	95

(USDA): United States Department of Agriculture
*Prepared as Package Directs

Food and Description	Measure or Quantity	Protein (grams)	Calories
*Strawberry Fruit 'N Crunch (Pillsbury)	2" sq.	2.0	140
Sugar (Nestlé's)	1 oz.	1.2	132
Toll House (Nestlé's)	1 oz.	1.2	132
*Vienna Dream bar (Betty Crocker)	1 bar (2" × 1⅓")	1.0	87

COOKING FATS (see **FATS**)

COOL'N CREAMY (Birds Eye)	½ cup (4.4 oz.)	2.6	172

CORN:
Fresh, white or yellow (USDA):

Food and Description	Measure or Quantity	Protein (grams)	Calories
Raw, untrimmed, on cob	1 lb. (weighed in husk)	5.7	157
Raw, trimmed, on cob	1 lb. (husk removed)	8.7	240
Raw, kernels	4 oz.	4.0	109
Boiled, kernels, cut from cob, drained	1 cup (5.9 oz.)	5.3	138
Boiled, whole	1 ear (5" × 1¾", 4.9 oz.)	2.5	70

Canned, regular pack:
Golden or yellow, whole kernel:

Food and Description	Measure or Quantity	Protein (grams)	Calories
Solids & liq., vacuum pack (USDA)	½ cup (3.7 oz.)	2.6	88
Solids & liq., wet pack (USDA)	½ cup (4.5 oz.)	2.4	84
Drained solids, wet pack (USDA)	½ cup (3 oz.)	2.2	72
Drained liq., wet pack (USDA)	4 oz.	.6	29
Drained solids (Butter Kernel)	½ cup (4.1 oz.)	2.2	75
Solids & liq., vacuum pack (Del Monte)	½ cup (3.7 oz.)	2.6	91
Drained solids, wet pack (Del Monte) Family Style	½ cup (3 oz.)	2.2	78
Shoe Peg, *Le Sueur*	¼ of 17-oz. can	2.2	79
Solids & lib. (Green Giant)	½ of 8.5-oz. can	2.2	80
Solids & lig., wet pack (Stokely-Van Camp)	½ cup (4.5 oz.)	2.1	74

(USDA): United States Department of Agriculture
*Prepared as Package Directs

Food and Description	Measure or Quantity	Protein (grams)	Calories
Vacuum pack, *Niblets*	⅓ of 12-oz. can	2.7	95
With peppers, solids & liq. (Del Monte)	½ cup (3.7 oz.)	2.6	88
With peppers, vacuum pack *Mexicorn*	⅓ of 12-oz. can	2.7	95
White, whole kernel:			
Solids & liq., wet pack (USDA)	½ cup (4.5 oz.)	2.4	84
Drained solids, wet pack (USDA)	½ cup (2.8 oz.)	2.1	67
Drained liq., wet pack (USDA)	4 oz.	.6	29
Vacuum pack (Green Giant)	⅓ of 12-oz. can	2.6	100
Canned, white or yellow, dietetic pack:			
Solids & liq., wet pack (USDA)	4 oz.	2.2	65
Drained solids (USDA)	4 oz.	2.8	86
Drained liq. (USDA)	4 oz. (by wt.)	.6	19
Solids & liq. (Blue Boy)	4 oz.	2.3	78
Solids & liq. (Diet Delight)	½ cup (4.4 oz.)	2.5	70
Solids & liq. (S and W) *Nutradiet,* unseasoned	4 oz.	1.8	59
Solids & liq. (Tillie Lewis)	½ cup	2.4	82
Canned, creamed style, white or yellow, regular pack:			
Solids & liq. (USDA)	½ cup (4.4 oz.)	2.6	102
(Butter Kernel)	½ cup (4.1 oz.)	2.4	92
Golden, solids & liq. (Del Monte)	½ cup (4.4 oz.)	2.8	112
Golden, solids & liq. (Green Giant)	½ of 8.5-oz. can	2.0	107
Solids & liq. (Stokely-Van Camp)	½ cup (4.1 oz.)	2.4	94
Canned, cream style, dietetic pack:			
Solids & liq. (USDA)	4 oz.	2.9	93
Solids & liq. (Blue Boy)	4 oz.	2.9	105
Solids & liq. (S and W) *Nutradiet*	4 oz.	2.9	95

(USDA): United States Department of Agriculture

Food and Description	Measure or Quantity	Protein (grams)	Calories
Frozen:			
On the cob:			
Not thawed (USDA)	4 oz.	4.1	111
Boiled, drained (USDA)	4 oz.	4.0	107
(Birds Eye)	1 ear (3.5 oz.)	3.6	98
Niblet Ears	1 ear (4.9 oz.)	4.2	167
Kernel, cut off cob:			
Not thawed (USDA)	4 oz.	3.5	93
Boiled, drained (USDA)	½ cup (3.2 oz.)	2.7	72
(Birds Eye)	½ cup (3.3 oz.)	2.8	77
Sweet white (Birds Eye)	½ cup (3.3 oz.)	2.9	77
Cream style (Green Giant)	⅓ of 10-oz. pkg.	1.9	74
In butter sauce:			
& peppers, *Mexicorn*	⅓ of 10-oz. pkg.	2.0	97
Yellow, *Niblets*	⅓ of 10-oz. pkg.	2.1	90
White (Green Giant)	⅓ of 12-oz. pkg.	2.6	112
Scalloped (Green Giant)	⅓ of 10-oz. pkg.	4.3	132
With peas & tomatoes (Birds Eye)	⅓ of 10-oz. pkg.	3.0	67
With Swiss cheese, casserole (Green Giant)	⅓ of 10-oz. pkg.	3.8	93
CORNBREAD:			
Cornpone, home recipe, prepared with white, whole-ground cornmeal (USDA)	4 oz.	5.1	231
Corn sticks, frozen (Aunt Jemima)	3 pieces (1¾ oz.)	2.7	145
Johnnycake, home recipe, prepared with yellow, degermed cornmeal (USDA)	4 oz.	9.9	303
Southern-style, home recipe, prepared with degermed cornmeal (USDA)	2½″ × 2½″ × 1⅝″ piece (2.9 oz.)	5.9	186
Southern-style, home recipe, prepared with whole-ground cornmeal (USDA)	4 oz.	8.4	235

(USDA): United States Department of Agriculture

Food and Description	Measure or Quantity	Protein (grams)	Calories
Spoonbread, home recipe, prepared with white, whole-ground cornmeal (USDA)	4 oz.	7.6	221
Sweet (Pillsbury) *Hungry Jack*	1 piece	1.5	95
CORNBREAD MIX:			
Dry (USDA)	1 oz.	2.1	122
*Prepared with egg & milk:			
(USDA)	4 oz.	6.9	264
(USDA)	2⅜" muffin (1.4 oz.)	2.4	93
(USDA)	2½" × 2½" × 1⅜" piece (1.9 oz.)	3.4	128
Dry (Albers)	1 oz.	1.9	109
*(Aunt Jemima)	⅙ of cornbread (2.4 oz.)	5.1	228
*(Dromedary)	2" × 2" piece (1.4 oz.)	2.5	125
*(Pillsbury) *Ballard*	⅛ of recipe	4.0	160
CORN CHEX, cereal, dry	1¼ cups (1 oz.)	1.8	111
CORN CHIPS (see **CRACKERS**)			
CORN CHOWDER, New England (Snow)	8 oz.	6.0	159
CORNED BEEF:			
Uncooked, boneless, medium fat (USDA)	1 lb.	71.7	1329
Cooked, boneless, medium fat (USDA)	4 oz.	26.0	422
Canned:			
Lean (USDA)	4 oz.	30.0	210
Medium fat (USDA)	4 oz.	28.7	245
Fat (USDA)	4 oz.	26.6	298
(Armour Star)	12-oz. can	92.6	967
(Hormel) *Dinty Moore*	4 oz.	30.8	253
Brisket (Wilson) *Tender Made*	4 oz.	19.7	180
Packaged (Oscar Mayer)	5-gram slice (16 to 3 oz.)	1.1	7

(USDA): United States Department of Agriculture
*Prepared as Package Directs

Food and Description	Measure or Quantity	Protein (grams)	Calories
CORNED BEEF HASH, canned:			
With potato (USDA)	4 oz.	10.0	205
(Armour Star)	15½-oz. can	36.4	831
(Hormel) *Mary Kitchen*	7½ oz.	19.6	400
(Nalley's)	4 oz.	10.0	179
(Van Camp)	½ cup (4.1 oz.)	10.1	208
(Wilson)	15½-oz. can	35.6	792
CORNED BEEF HASH DINNER, frozen (Banquet)	10-oz. dinner	19.9	372
CORNED BEEF SPREAD:			
(Underwood)	4½-oz. can	17.2	248
(Underwood)	1 T. (.5 oz.)	1.9	27
CORN FLAKES, cereal:			
Whole (USDA)	1 cup (1 oz.)	2.3	112
Crushed (USDA)	1 cup (2.5 oz.)	5.5	270
Frosted (USDA)	1 cup (1.4 oz.)	1.8	154
Country (General Mills)	1¼ cups (1 oz.)	2.4	111
(Kellogg's)	1⅓ cups (1 oz.)	2.2	106
(Ralston Purina)	1 cup (1 oz.)	1.8	109
(Van Brode)	1 oz.	2.2	106
CORN FRITTER:			
Home recipe (USDA)	4 oz.	8.8	428
Frozen (Mrs. Paul's)	8-oz. pkg.	10.8	562
CORN GRITS (see **HOMINY**)			
CORNMEAL MIX:			
Bolted (Aunt Jemina/Quaker)	¼ cup (1 oz.)	2.4	96
Degermed (Aunt Jemina/Quaker)	¼ cup (1 oz.)	2.2	94
CORNMEAL, WHITE or YELLOW:			
Dry:			
Bolted (USDA)	1 cup (4.3 oz.)	11.0	442
Degermed (USDA)	1 cup (4.9 oz.)	10.9	502
Degermed (Albers)	1 cup (6 oz.)	12.1	576
Self-rising, bolted (Aunt Jemima)	1 cup (6 oz.)	14.4	594

(USDA): United States Department of Agriculture

Food and Description	Measure or Quantity	Protein (grams)	Calories
Self-rising, degermed (USDA)	1 cup (5 oz.)	10.9	491
Self-rising, degermed (Aunt Jemima)	1 cup (6 oz.)	13.8	582
Self-rising, whole-ground (USDA)	1 cup (5 oz.)	12.1	489
Whole-ground, unbolted (USDA)	1 cup (4.3 oz.)	11.2	433
Cooked:			
*Degermed (USDA)	1 cup (8.5 oz.)	2.6	120
*Degermed (Albers)	1 cup	2.6	119
CORN PUDDING, home recipe (USDA)	1 cup (8.6 oz.)	9.8	255
CORN SALAD, raw (USDA):			
Untrimmed	1 lb. (weighed untrimmed)	8.7	91
Trimmed	4 oz.	2.3	24
CORN SOUFFLE, frozen (Stouffer's)	12-oz. pkg.	16.5	492
CORNSTARCH:			
(USDA)	1 cup (4.5 oz.)	.4	463
(USDA)	1 T. (8 grams)	Tr.	29
(Argo)	1 T. (8 grams)	Tr.	34
(Kingsford's)	1 T. (8 grams)	Tr.	34
(Duryea's)	1 T. (8 grams)	Tr.	34
CORNSTARCH PUDDING (see **VANILLA PUDDING**)			
CORN STICK (see **CORNBREAD**)			
CORN SYRUP (USDA)	1 T. (.7 oz.)	0.	61
CORN TOTAL, cereal (General Mills)	1¼ cups (1 oz.)	2.4	111
COTTAGE PUDDING, home recipe (USDA):			
Without sauce	2 oz.	3.6	195

(USDA): United States Department of Agriculture
*Prepared as Package Directs

[131]

Food and Description	Measure or Quantity	Protein (grams)	Calories
With chocolate sauce	2 oz.	3.0	180
With strawberry sauce	2 oz.	2.9	166
COUGH DROP:			
(Estee)	1 drop	Tr.	12
(Pine Bros.)	1 drop (3 grams)	0.	9
COUNT CHOCULA, cereal			
(General Mills)	1 cup (1 oz.)	1.4	106
COUNTRY-STYLE SAUSAGE,			
smoked links (USDA)	1 oz.	4.3	98
COWPEA, including black-eyed peas (USDA):			
Immature seeds:			
Raw, whole	1 lb. (weighed in pods)	22.5	317
Raw, shelled	½ cup (2.5 oz.)	6.5	91
Boiled, drained	½ cup (2.9 oz.)	6.6	88
Canned, solids & liq.	4 oz.	5.7	79
Frozen (see **BLACK-EYED PEA,** frozen)			
Young pods with seeds:			
Raw, whole	1 lb. (weighed untrimmed)	13.6	182
Boiled, drained	4 oz.	2.9	39
Mature seeds, dry:			
Raw	½ cup (3 oz.)	19.2	288
Boiled, drained	½ cup (4.4 oz.)	6.3	94
CRAB, all species:			
Fresh (USDA):			
Steamed, whole	1 lb. (weighed in shell)	37.7	202
Steamed, meat only	1 cup (4.4 oz.)	21.6	116
Canned:			
Drained solids (USDA)	1 packed cup (5.6 oz.)	27.8	162
(Del Monte) Alaska King	7½-oz. can	42.4	202
Frozen (Wakefield's) Alaska King, thawed & drained	4 oz.	19.3	96

(USDA): United States Department of Agriculture

Food and Description	Measure or Quantity	Protein (grams)	Calories
CRAB APPLE, fresh (USDA):			
Whole	1 lb. (weighed whole)	1.7	284
Flesh only	4 oz.	.5	77
CRAB CAKE, frozen, thins (Mrs. Paul's)	2½-oz. cake	6.6	158
CRAB, DEVILED:			
Home recipe, made with bread cubes, butter, parsley, eggs, lemon juice & catsup (USDA)	1 cup (8.5 oz.)	27.4	451
Frozen, breaded & fried (Mrs. Paul's)	3-oz. crab	7.9	173
CRAB IMPERIAL, home recipe, made with butter, flour, milk, onion, green pepper, eggs & lemon juice (USDA)	1 cup (7.8 oz.)	32.1	323
CRAB NEWBURG frozen (Stouffer's)	12-oz. pkg.	24.4	562
CRACKER, PUFFS and CHIPS:			
American Harvest (Nabisco)	1 piece (3 grams)	< .1	16
Arrowroot biscuit (Nabisco)	1 piece (5 grams)	.3	22
Bacon flavored thins (Nabisco)	1 piece (2 grams)	.2	11
Bacon Nips	1 oz.	2.7	147
Bacon rinds (Wonder)	1 oz.	19.4	146
Bacon toast (Keebler)	1 piece (3 grams)	.3	15
Bakon Tasters (Old London)	½-oz. bag	2.4	62
Bugles (General Mills)	15 pieces (½ oz.)	.8	81
Butter (USDA)	1 oz.	2.0	130
Butter thins (Nabisco)	1 piece (3 grams)	.2	15
Cheese flavored (see also individual brand names in this grouping):			
(USDA)	1 oz.	3.2	136
Cheese 'n Bacon, sandwich (Nab)	6 pieces (1¼-oz. pkg.)	4.2	179
Cheese 'n cracker (Kraft)	4 crackers & ¾ oz. cheese	1.0	138

(USDA): United States Department of Agriculture

[133]

Food and Description	Measure or Quantity	Protein (grams)	Calories
Cheese *Nips* (Nab)	24 pieces (⅞-oz. pkg.)	2.3	114
Cheese *Nips* (Nabisco)	1 piece (1 gram)	.1	5
Cheese on Rye, sandwich (Nab)	6 pieces (1¼-oz. pkg.)	3.6	191
Cheese Pixies (Wise)	1-oz. bag	2.0	155
Chee-Tos, cheese-flavored puffs	1 oz.	2.3	156
Chee-Tos, fried	1 oz.	2.2	156
Cheez Doodles (Old London)	1-oz. bag	2.0	155
Cheez-Its (Sunshine)	1 piece (1 gram)	.1	6
Cheez Waffles (Old London)	1 piece (2 grams)	.3	11
Che-zo (Keebler)	1 piece (<1 gram)	.1	5
Ritz (Nabisco)	1 piece (3 grams)	.3	17
Sandwich (Nab)	6 pieces (1¼-oz. pkg.)	3.9	185
Shapies, dip delights (Nabisco)	1 piece (2 grams)	.2	9
Shapies, shells (Nabisco)	1 piece (2 grams)	<.1	10
Thins (Pepperidge Farm)	1 piece (3 grams)	.5	12
Thins, dietetic (Estee)	1 piece	.2	6
Tid-Bit (Nab)	32 pieces (1⅛-oz. pkg.)	2.7	150
Tid-Bit (Nabisco)	1 piece (<1 gram)	<.1	4
Toast (Keebler)	1 piece (3 grams)	.3	16
Twists (Nalley)	1 oz.	1.7	155
Twists (Wonder)	1 oz.	2.4	154
Cheese & peanut butter sandwich:			
(USDA)	1 oz.	4.3	139
(Austin's)	1⅜-oz. pkg.	7.0	194
(Nab) *O-So-Gud*	4 pieces (1-oz. pkg.)	3.6	141
(Nab) squares	4 pieces (1-oz. pkg.)	4.0	139
(Nab) squares	6 pieces (1½-oz. pkg.)	6.0	208
(Nab) squares	6 pieces (1¾-oz. pkg.)	7.1	243
(Nab) variety pack	6 pieces (1½-oz. pkg.)	5.9	209

(USDA): United States Department of Agriculture

Food and Description	Measure or Quantity	Protein (grams)	Calories
(Nab) variety pack	6 pieces (1¾-oz. pkg.)	6.9	244
Chicken in a Biskit (Nabisco)	1 piece (2 grams)	.1	10
Chippers (Nabisco)	1 piece (3 grams)	.2	14
Chipsters (Nabisco)	1 piece (<1 gram)	<.1	2
Clam flavored crisps (Snow)	1 oz.	2.6	147
Club (Keebler)	1 piece (3 grams)	.2	15
Corn Capers (Wonder)	1 oz.	1.6	158
Corn cheese (Tom Houston)	10 pieces (5 grams)	.5	29
Corn chips:			
Cornettes	1 oz.	2.1	153
Fritos	1 oz.	1.9	159
Fritos, barbecue	1 oz.	2.1	156
Korkers (Nabisco)	1 piece (2 grams)	<.1	8
(Old London)	1-oz. bag	1.8	162
(Wonder)	1 oz.	1.5	162
Barbecue (Wise)	1¾-oz. bag	3.1	274
Rippled (Wise)	1-oz. bag	1.8	162
Corn Diggers (Nabisco)	1 piece (<1 gram)	<.1	4
Crown Pilot (Nabisco)	1 piece (.6 oz.)	1.5	73
Dipsy Doodles (Old London)	1-oz. bag	1.8	162
Doo Dads (Nabisco)	1 piece (<1 gram)	<.1	2
Escort (Nabisco)	1 piece (4 grams)	.3	20
Flings, cheese-flavored curls (Nabisco)	1 piece (2 grams)	.2	11
Flings, Swiss 'n ham (Nabisco)	1 piece (2 grams)	.2	10
Goldfish (Pepperidge Farm):			
Cheddar cheese	10 pieces (6 grams)	.7	28
Lightly salted	10 pieces (6 grams)	.5	28
Parmesan cheese	10 pieces (6 grams)	.8	28
Pizza	10 pieces (6 grams)	.5	29
Pretzel	10 pieces (7 grams)	.7	29
Onion	10 pieces (6 grams)	.4	28

Food and Description	Measure or Quantity	Protein (grams)	Calories
Sesame garlic	10 pieces (6 grams)	.5	29
Graham:			
(USDA)	2½″ sq. (7 grams)	.6	27
(Nabisco)	1 piece (7 grams)	.5	30
Chocolate or cocoa-covered:			
(USDA)	1 oz.	1.4	135
(Keebler) Deluxe	1 piece (9 grams)	.4	42
(Nabisco)	1 piece (.4 oz.)	.6	55
(Nabisco) *Fancy*	1 piece (.5 oz.)	.7	68
(Nabisco) *Pantry*	1 piece (.4 oz.)	.6	62
Sweet-Tooth (Sunshine)	1 piece (.4 oz.)	.4	45
Sugar-honey coated (USDA)	1 oz.	1.9	117
Sugar-honey coated (Nabisco)			
Honey Maid	1 piece (7 grams)	.5	30
Hi-Ho (Sunshine)	1 piece (4 grams)	.2	18
Hot Potatas (Old London)	⅝-oz. bag	1.1	82
Matzo (see **MATZO**)			
Melba toast (see **MELBA**)			
Milk lunch (Nabisco) *Royal Lunch*	1 piece (.4 oz.)	.8	55
Munchos	1 oz.	1.6	159
Onion flavored:			
Crisps (Snow)	1 oz.	2.7	157
French (Nabisco)	1 piece (2 grams)	.2	12
Funyuns (Frito-Lay)	1 oz.	2.2	137
Meal Mates (Nabisco)	1 piece (4 grams)	.4	19
Onyums (General Mills)	30 pieces (.5 oz.)	.8	79
Rings (Old London)	½-oz. bag	.2	68
Rings (Wise)	½-oz. bag	.2	65
Rings (Wonder)	1 oz.	<.1	133
Thins (Pepperidge Farm)	1 piece (3 grams)	.3	12
Toast (Keebler)	1 piece (3 grams)	.2	18
Oyster:			
(USDA)	10 pieces (.4 oz.)	.9	44
(USDA)	1 cup (1 oz.)	2.6	124
(Keebler)	1 piece (<1 gram)	<.1	2
Dandy (Nabisco)	1 piece (<1 gram)	<.1	3
Mini (Sunshine)	1 piece (<1 gram)	.1	3

(USDA): United States Department of Agriculture

Food and Description	Measure or Quantity	Protein (grams)	Calories
Oysterettes (Nabisco)	1 piece (< 1 gram)	< .1	3
Peanut butter 'n cheez crackers (Kraft)	4 crackers & ¾-oz. peanut butter	6.4	191
Peanut butter sandwich:			
Adora (Nab)	6 pieces (1½-oz. pkg.)	4.1	201
Cheese crackers (Wise)	1 piece (6 grams)	.7	30
Malted milk (Nab)	4 pieces (1-oz. pkg.)	3.3	137
Malted milk (Nab)	6 pieces (1⅜-oz. pkg.)	4.5	189
Toasted crackers (Wise)	1 piece (6 grams)	.6	30
Pizza Spins (General Mills)	32 pieces (½ oz.)	1.5	72
Pizza Wheels (Wise)	¾-oz. bag	1.4	90
Potato crisps (General Mills)	16 pieces (½ oz.)	.8	78
Ritz, plain (Nabisco)	1 piece (3 grams)	.2	16
Roman Meal Wafers	1 piece (4 grams)	.4	15
Rye thins (Pepperidge Farm)	1 piece (3 grams)	.4	10
Rye toast (Keebler)	1 piece (4 grams)	.3	17
Rye wafers, whole grain (USDA)	1⅞" × 3½" piece (6 grams)	.8	22
Rye wafers (Nabisco) *Meal Mates*	1 piece (4 grams)	.4	19
Ry-Krisp:			
Seasoned	1 triple cracker (6 grams)	.7	26
Traditional	1 triple cracker (6 grams)	.7	24
Saltine:			
(USDA)	4 crackers (.4 oz.)	1.0	48
Krispy (Sunshine) salted tops	1 piece (3 grams)	.2	11
Krispy (Sunshine) unsalted tops	1 piece (3 grams)	.3	12
Premium (Nab)	8 pieces (¾-oz. pkg.)	1.9	91
Premium (Nabisco)	1 piece (3 grams)	.2	12
Zesta (Keebler)	1 section (3 grams)	.3	12
Sea toast (Keebler)	1 piece (.5 oz.)	1.3	62

(USDA): United States Department of Agriculture

Food and Description	Measure or Quantity	Protein (grams)	Calories
Sesame:			
(Sunshine) *La Lanne*	1 piece (3 grams)	.3	15
Buttery flavored (Nabisco)	1 piece (3 grams)	.3	16
Wafer (Keebler)	1 piece (3 grams)	.3	16
Wafer (Nabisco) *Meal Mates*	1 piece (5 grams)	.5	21
Sip 'N Chips (Nabisco)	1 piece (2 grams)	.1	9
Sociables (Nabisco)	1 piece (2 grams)	.2	10
Soda:			
(USDA)	1 oz.	2.6	124
(USDA)	2½" sq. (6 grams)	.5	24
(Nabisco) *Premium*, unsalted tops	1 piece (3 grams)	.3	12
(Sunshine)	1 piece (4 grams)	.4	20
Soya (Sunshine) *La Lanne*	1 piece (3 grams)	.2	16
Star Lites (Wise)	1 cup (.5 oz.)	.9	63
Swedish rye wafer (Keebler)	1 piece (5 grams)	.6	5
Taco corn chips (Old London)	1¼-oz. bag	3.0	167
Taco tortilla chips (Frito-Lay) *Doritos*	1 oz.	2.8	150
Taco tortilla chips (Wonder)	1 oz.	1.7	144
Tortilla Chips (Frito-Lay) *Doritos*	1 oz.	2.4	137
Tortilla chips (Old London)	1½-oz. bag	3.1	207
Tortilla chips (Wonder)	1 oz.	1.8	148
Town House	1 piece (3 grams)	.2	18
Triangle Thins (Nabisco)	1 piece (2 grams)	.2	8
Triscuit (Nabisco)	1 piece (4 grams)	.4	21
Twigs, sesame & cheese (Nabisco)	1 piece (3 grams)	.3	14
Uneeda Biscuit (Nabisco) unsalted tops	1 piece (5 grams)	.5	22
Wafer-ets (Hol-Grain):			
Rice	1 piece (3 grams)	.2	12
Wheat	1 piece (2 grams)	.1	7
Waldorf, low salt (Keebler)	1 piece (3 grams)	.3	14
Waverly wafer (Nabisco)	1 piece (4 grams)	.2	18
Wheat chips (General Mills)	12 pieces (.5 oz.)	1.0	73
Wheat thins (Nabisco)	1 piece (2 grams)	.1	9
Wheat toast (Keebler)	1 piece (3 grams)	.3	16
Whistles (General Mills)	17 pieces (.5 oz.)	1.3	71
White thins (Pepperidge Farm)	1 piece (3 grams)	.4	12
Whole-wheat (USDA)	1 oz.	2.4	114

(USDA): United States Department of Agriculture

Food and Description	Measure or Quantity	Protein (grams)	Calories
Whole-wheat, natural (Froumine)	1 piece (.4 oz.)	1.4	46
CRACKER CRUMBS:			
Graham (USDA)	1 cup (3 oz.)	6.9	330
Graham (Keebler)	3 oz.	5.6	368
Graham (Nabisco)	1½ cups (4.6 oz.) or 9″ pie shell)	9.3	563
Graham (Sunshine)	3 oz.	6.1	357
CRACKER JACK (see POPCORN)			
CRACKER MEAL:			
(USDA)	3 oz.	7.8	373
(USDA)	1 T. (.4 oz.)	.9	44
(Keebler):			
Fine, medium or coarse	3 oz.	8.4	316
Zesty	3 oz.	7.7	363
(Sunshine)	3 oz.	9.0	340
Salted (Nabisco)	1 cup (3 oz.)	8.4	309
Unsalted (Nabisco)	1 cup (3 oz.)	8.7	319
CRACKER PIE CRUST MIX (see **PIECRUST MIX**)			
CRANAPPLE (Ocean Spray):			
Regular	½ cup (4.5 oz.)	<.1	94
*Frozen	½ cup (4.4 oz.)	Tr.	91
Low calorie	½ cup (4.2 oz.)	<.1	19
CRANBERRY:			
Fresh:			
Untrimmed (USDA)	1 lb. (weighed with stems)	1.7	200
Stems removed (USDA)	1 cup (4 oz.)	.5	52
(Ocean Spray)	1 oz.	.2	15
Dehydrated (USDA)	1 oz.	.8	104
CRANBERRY JUICE COCKTAIL:			
(USDA)	½ cup (4.4 oz.)	.1	81
Regular (Ocean Spray)	½ cup (4.4 oz.)	<.1	83

(USDA): United States Department of Agriculture
*Prepared as Package Directs

Food and Description	Measure or Quantity	Protein (grams)	Calories
*Frozen (Ocean Spray)	½ cup (4.4 oz.)	Tr.	75
Low calorie (Ocean Spray)	½ cup (4.4 oz.)	<.1	24
*CRANBERRY-ORANGE DRINK (Ocean Spray)	½ cup (4.4 oz.)	.2	67
CRANBERRY-ORANGE RELISH:			
Uncooked (USDA)	4 oz.	.5	202
(Ocean Spray)	4 oz.	.4	209
CRANBERRY PIE (Tastykake)	4-oz. pie	3.4	376
CRANBERRY SAUCE:			
Home recipe, sweetened, unstrained (USDA)	4 oz.	.2	202
Canned:			
Sweetened, strained (USDA)	½ cup (4.8 oz.)	.1	199
Jellied (Ocean Spray)	4 oz.	.1	184
Whole berry (Ocean Spray)	4 oz.	<.1	191
CRANPRUNE (Ocean Spray)	½ cup (4.4 oz.)	.1	82
CRAPPIE, white, raw, meat only (USDA)	4 oz.	19.1	90
CRAYFISH, freshwater (USDA):			
Raw, in shell	1 lb. (weighed in shell)	7.9	39
Raw, meat only	4 oz.	16.6	82
CREAM:			
Half & half:			
(USDA)	1 cup (8.5 oz.)	7.7	324
(USDA)	1 T. (.5 oz.)	.5	20
(Dean)	1 cup (8.6 oz.)	7.8	344
10.5% fat (Sealtest)	1 cup (8.5 oz.)	7.2	296
12% fat (Sealtest)	1 cup (8.5 oz.)	7.0	322
Light, table or coffee:			
(USDA)	1 cup (8.5 oz.)	7.2	506
(USDA)	1 T. (.5 oz.)	.4	32
16% fat (Sealtest)	1 T. (.5 oz.)	.4	26
18% fat (Sealtest)	1 T. (.5 oz.)	.4	28

(USDA): United States Department of Agriculture
*Prepared as Package Directs

Food and Description	Measure or Quantity	Protein (grams)	Calories
25% fat (Sealtest)	1 T. (.5 oz.)	.4	37
Light whipping:			
(USDA)	1 cup (8.4 oz.)	6.0	717
(USDA)	1 T. (.5 oz.)	.4	45
30% fat (Sealtest)	1 T. (.5 oz.)	.4	44
Whipped topping, pressurized:			
(USDA)	1 cup (2.1 oz.)	2.0	155
(USDA)	1 T. (3 grams)	.1	10
Heavy whipping:			
Unwhipped (USDA)	1 cup (8.4 oz., or 2 cups whipped)	5.2	838
Unwhipped (USDA)	1 T. (.5 oz.)	.3	53
(Dean)	1 T. (.5 oz.)	.3	51
36% fat (Sealtest)	1 T. (.5 oz.)	.3	52
Sour:			
(USDA)	1 cup (8.1 oz.)	6.9	485
(USDA)	1 T. (.4 oz.)	.4	25
(Axelrod)	8-oz. container	7.4	433
(Borden)	1 cup (8.6 oz.)	6.7	454
(Borden)	1 T. (.5 oz.)	.4	28
(Breakstone)	8-oz. container	7.3	464
(Breakstone)	1 T. (.5 oz.)	.5	29
(Dean)	1 T. (.5 oz.)	.5	28
(Sealtest)	1 T. (.5 oz.)	.4	28
Half & half (Sealtest)	1 T. (.5 oz.)	.4	20
Imitation:			
(Dean) lowfat, *Sour Slim*	1 T. (1.1 oz.)	.9	30
(Sealtest) non dairy	1 T. (.5 oz.)	.3	30
Sour Treat (Delite)	1 T. (.5 oz.)	.5	25
Zest (Borden) 13.5% vegetable fat	1 T.	.6	24
Sour cream, dried (Data from General Mills)	1 oz.	2.6	188
Sour dairy dressing (Sealtest)	1 T. (.5 oz.)	.5	22
Sour dressing or sour cream, made with nonfat dry milk:			
(USDA)	1 cup (8.3 oz.)	9.0	440
(USDA)	1 T. (.4 oz.)	.6	20
Sour dressing, cultured (Breakstone)	1 T. (.5 oz.)	.5	27

(USDA): United States Department of Agriculture

Food and Description	Measure or Quantity	Protein (grams)	Calories
CREAMIES (Tastykake):			
Banana cake	1⅞-oz. pkg.	2.4	238
Chocolate	1⅞-oz. pkg.	3.4	290
Koffee Kake	1⅞-oz. pkg.	5.3	303
Vanilla	1⅞-oz. pkg.	3.6	292
CREAM PUFF, home recipe, with custard filling (USDA)	3½" × 2" (4.6 oz.)	8.4	303
***CREAM OF RICE,** cereal	4 oz.	1.7	82
CREAMSICLE	2-fl.-oz. pop. (1.9 oz.)	1.0	78
CREAM SUBSTITUTE:			
Liquid, frozen (USDA)	1 cup (8.6 oz.)	3.0	345
Liquid, frozen (USDA)	1 tsp. (5 grams)	Tr.	7
Powdered (USDA)	1 cup (3.3 oz.)	4.0	505
Powdered (USDA)	1 tsp. (2 grams)	Tr.	10
Coffee-mate (Carnation)	1 tsp. (2 grams)	<.1	11
Coffee-mate (Carnation)	1 packet (3 grams)	.1	17
Coffee Rich	1 tsp. (5 grams)	Tr.	8
Coffee Twin (Sealtest)	½ fl. oz. (.5 oz.)	.1	15
Cremora (Borden)	1 tsp. (2 grams)	.1	11
Perx	1 tsp. (5 grams)	<.1	8
Poly Perx	1 oz.	.3	40
CREAM OF WHEAT, cereal:			
Instant, dry	1 oz. (¾ cup cooked)	3.0	99
Mix 'n Eat:			
Regular, dry	3½ T. (1 oz.)	2.9	99
Baked apple & cinnamon, dry	3¾ T. (1¼ oz.)	2.8	129
Maple & brown sugar, dry	3¾ T. (1¼ oz.)	2.9	128
Quick, dry	1 oz. (¾ cup cooked)	3.0	99
Regular, dry	1 oz. (¾ cup cooked)	3.0	102

(USDA): United States Department of Agriculture
*Prepared as Package Directs

Food and Description	Measure or Quantity	Protein (grams)	Calories
CRESS, GARDEN (USDA):			
Raw, whole	1 lb. (weighed untrimmed)	8.4	103
Boiled in small amount of water, short time, drained	1 cup (6.3 oz.)	3.4	41
Boiled in large amount of water, long time, drained	1 cup (6.3 oz.)	3.2	40
CRISP RICE, cereal (Van Brode)	1 oz.	1.6	106
CRISPY CRITTERS, cereal (Post)	1 cup (1 oz.)	2.2	113
CRISPY RICE, cereal (Ralston Purina)	1 cup (1 oz.)	1.8	109
CROAKER (USDA):			
Atlantic:			
Raw, whole	1 lb. (weighed whole)	27.4	148
Raw, meat only	4 oz.	20.2	109
Baked	4 oz.	27.6	151
White, raw, meat only	4 oz.	20.4	95
Yellowfin, raw, meat only	4 oz.	21.8	101
CRULLER (see **DOUGHNUT**)			
CUCUMBER, fresh (USDA):			
Eaten with skin	½ lb. (weighed with skin)	2.0	32
Eaten without skin	½ lb. (weighed with skin)	1.0	23
Not pared, 10-oz. cucumber	7½″ × 2″ pared cucumber (7.3 oz.)	1.2	29
Pared	6 slices (2″ × ⅛″; 1.8 oz.)	.3	7
Pared & diced	½ cup (2.5 oz.)	.4	10
CUPCAKE:			
Home recipe (USDA):			
Without icing	2¾″ cupcake (1.4 oz.)	1.8	146

(USDA): United States Department of Agriculture

Food and Description	Measure or Quantity	Protein (grams)	Calories
With chocolate icing	2¾" cupcake (1.8 oz.)	2.1	184
With boiled white icing	2¾" cupcake (1.8 oz.)	1.9	176
With uncooked white icing	2¾" cupcake	1.7	184
Commercial:			
Chocolate (Tastykake)	1 cupcake (1 oz.)	2.2	192
Chocolate, chocolate creme filled (Tastykake)	1 cupcake (1¼ oz.)	2.1	128
Coconut (Tastykake)	1 cupcake (¾ oz.)	1.0	92
Creme filled, chocolate butter cream (Tastykake)	1 cupcake (1⅛ oz.)	2.1	161
Devil's food cake (Hostess)	1 cupcake (1½ oz.)	1.4	162
Lemon creme filled (Tastykake)	1 cupcake (⅞ oz.)	1.1	124
Orange (Hostess)	1 cupcake (1½ oz.)	1.5	166
Orange creme filled (Tastykake)	1 cupcake (⅞ oz.)	1.1	133
Vanilla creme filled (Tastykake)	1 cupcake (⅞ oz.)	1.1	123
Vanilla *Triplets* (Tastykake)	1 cupcake (.8 oz.)	1.0	101
CUPCAKE MIX:			
(USDA)	4 oz.	4.2	497
*Prepared with eggs, milk, without icing (USDA)	2½" cupcake (.9 oz.)	1.2	88
*Prepared with eggs, milk, with chocolate icing (USDA)	2½" cupcake (1.2 oz.)	1.6	129
*(Flako)	1 large cupcake (1.3 oz., 1/12 of pkg.)	1.9	140
*Devil's food (Pillsbury)	1 cupcake	2.0	170
*Fudge (Pillsbury)	1 cupcake	2.0	180
*Yellow (Pillsbury)	1 cupcake	2.0	170

(USDA): United States Department of Agriculture
*Prepared as Package Directs

Food and Description	Measure or Quantity	Protein (grams)	Calories
CURRANT:			
Fresh (USDA):			
Black European:			
Whole	1 lb. (weighed with stems)	7.6	240
Stems removed	4 oz.	1.9	61
Red & white:			
Whole	1 lb. (weighed with stems)	6.2	220
Stems removed	1 cup (3.9 oz.)	1.5	55
Dried, Zante (Del Monte)	½ cup (2.5 oz.)	2.4	205
CUSK (USDA):			
Raw, drawn	1 lb. (weighed drawn, head & tail on)	45.3	197
Raw, meat only	4 oz.	19.5	85
Steamed	4 oz.	26.5	120
CUSTARD:			
Home recipe, baked (USDA)	½ cup (4.7 oz.)	7.1	152
Chilled (Sealtest)	4 oz.	4.3	149
CUSTARD APPLE, bullock's heart, fresh (USDA):			
Whole	1 lb. (weighed with skin & seeds)	4.5	266
Flesh only	4 oz.	1.9	115
CUSTARD, FROZEN (see **ICE CREAM**)			
CUSTARD PIE:			
Home recipe (USDA)	⅙ of 9" pie (5.4 oz.)	9.3	331
Frozen (Banquet)	5 oz.	6.8	274
Frozen (Mrs. Smith's)	⅙ of 8" pie (4 oz.)	7.5	296
CUSTARD PUDDING MIX:			
Dry, with vegetable gum base (USDA)	1 oz.	0.	109

(USDA): United States Department of Agriculture

Food and Description	Measure or Quantity	Protein (grams)	Calories
*Prepared with whole milk (USDA)	4 oz.	3.5	149
*No egg yolk (Jell-O)	½ cup (5 oz.)	5.8	165
Real egg (Lynden Farms)	4-oz. pkg.	28.4	441
*Regular (Royal)	½ cup (5.1 oz.)	4.3	132

D

DAMSON PLUM (see **PLUM**)

DANDELION GREENS (USDA):
Raw, trimmed	1 lb.	12.2	204
Boiled, drained	½ cup (3.2 oz.)	1.8	30

DANISH PASTRY (see **COFFEE CAKE**)

DANISH-STYLE VEGETABLES,
Frozen (Birds Eye)	⅓ of 10-oz. pkg.	1.7	92

DANNY (see **YOGURT**)

DATE, dry:
Domestic:
With pits (USDA)	1 lb. (weighed with pits)	8.7	1081
Without pits (USDA)	4 oz.	2.5	311
Without pits, chopped (USDA)	1 cup (6.1 oz.)	3.8	477
Whole (Cal-Date)	1 date (.8 oz.)	.5	62
Diced (Cal-Date)	4 oz.	2.5	322
Chopped (Dromedary)	1 cup (5 oz.)	2.7	493
Pitted (Dromedary)	1 cup (5 oz.)	2.4	470
Imported, Iraq (Bordo):			
Whole	4 average dates (.9 oz.)	.5	73
Diced	¼ cup (2 oz.)	1.1	159

DESSERT CUP (Del Monte):
Pudding 'n apricot	5-oz. container	1.1	173
Pudding 'n peach	5-oz. container	1.0	173
Pudding 'n pineapple	5-oz. container	1.0	173

(USDA): United States Department of Agriculture
*Prepared as Package Directs

Food and Description	Measure or Quantity	Protein (grams)	Calories
DEVIL DOG (Drake's):			
Regular	1 piece (1.6 oz.)	2.3	178
Senior	1 piece (2.3 oz.)	2.9	244
DEVIL'S FOOD CAKE:			
Home recipe (USDA):			
Without icing	3″ × 3″ × 1½″ (1.9 oz.)	2.6	201
With chocolate icing, 2 layer	1/16 of 9″ cake (2.6 oz.)	3.4	277
With chocolate icing, 2 layer	1/16 of 10″ cake (4.2 oz.)	5.4	443
With uncooked white icing	1/16 of 10″ cake (4.2 oz.)	4.6	443
Commercial, frozen:			
With chocolate icing (USDA)	2 oz.	2.4	215
With whipped-cream filling & chocolate icing (USDA)	2 oz.	2.0	210
(Pepperidge Farm)	1/6 of cake (3.1 oz.)	3.1	326
Layers only (Mrs. Smith's)	1/6 of 16-oz. pkg.	6.0	248
DEVIL'S FOOD CAKE MIX:			
Dry (USDA)	1 oz.	1.4	115
*With chocolate icing (USDA)	1/16 of 9″ cake (2.4 oz.)	3.0	234
*(Duncan Hines)	1/12 of cake (2.7 oz.)	3.0	205
*(Swans Down)	1/12 of cake (2.4 oz.)	3.3	184
*Butter (Betty Crocker)	1/12 of cake	3.4	269
*Layer (Betty Crocker)	1/12 of cake	2.9	199
*Red Devil (Pillsbury)	1/12 of cake	3.0	210
DEWBERRY, fresh (see **BLACKBERRY,** fresh)			
DEWBERRY PRESERVE (Bama)	1 T. (.7 oz.)	.1	54
DING DONG (Hostess)	1 cake (1.3 oz.)	1.3	185

(USDA): United States Department of Agriculture
*Prepared as Package Directs

Food and Description	Measure or Quantity	Protein (grams)	Calories
DINNER, frozen (see individual listings such as **BEEF DINNER, CHICKEN DINNER, CHINESE DINNER, ENCHILADA DINNER,** etc.)			
DIP:			
Bacon & horseradish:			
(Borden)	1 oz.	1.0	58
(Breakstone)	2 Tbs. (1.1 oz.)	1.8	62
(Kraft) *Teez*	1 oz.	1.5	57
(Kraft) *Ready Dip,* Neufchâtel cheese	1 oz.	1.8	71
Bacon & smoke (Sealtest) *Dip'n Dressing*	1 oz.	1.3	47
Barbecue (Borden)	1 oz.	1.0	58
Blue cheese:			
(Breakstone)	1 T. (.6 oz.)	.8	32
(Dean) Tang	1 oz.	1.5	61
(Kraft) *Ready Dip,* Neufchâtel cheese	1 oz.	2.2	69
(Kraft) *Teez*	1 oz.	1.2	51
(Sealtest) *Dip 'n Dressing*	1 oz.	1.4	49
Casino (Sealtest) *Dip 'n Dressing*	1 oz.	.9	44
Chipped beef (Sealtest) *Dip 'n Dressing*	1 oz.	1.4	46
Clam:			
(Kraft) *Ready Dip,* Neufchâtel cheese	1 oz.	1.7	67
(Kraft) *Teez*	1 oz.	1.0	45
& lobster (Borden)	1 oz.	1.0	58
Cucumber & onion (Breakstone)	1 T. (.5 oz.)	.6	27
Dill pickle (Kraft) *Ready Dip,* Neufchâtel cheese	1 oz.	1.2	67
Garlic (Dean)	1 oz.	1.3	58
Garlic (Kraft) *Teez*	1 oz.	.7	47
Green goddess (Kraft) *Teez*	1 oz.	.7	46
Jalapeño bean (Fritos)	1 oz.	1.6	34
Onion:			
(Borden) French	1 oz.	1.0	58

Food and Description	Measure or Quantity	Protein (grams)	Calories
(Breakstone)	1 T.	.6	29
(Dean) French onion	1 oz.	1.3	58
(Kraft) French onion, *Teez*	1 oz.	.6	43
(Kraft) *Ready Dip*, Neufchâtel cheese	1 oz.	1.2	68
(Sealtest) French onion, *Dip'n Dressing*	1 oz.	.9	46
& garlic (Sealtest) *Dip'n Dressing*	1 oz.	.9	46
Pizza (Borden)	1 oz.	1.0	58
Skinny Dip (Dean)	1 oz.	1.1	27
Tasty Tartar (Borden)	1 oz.	1.2	48
Western Bar B-Q (Borden)	1 oz.	1.2	48
DIP MIX:			
Green onion (Lawry's)	1 pkg. (.6 oz.)	1.8	50
Guacamole (Lawry's)	1 pkg. (.6 oz.)	1.6	60
Toasted onion (Lawry's)	1 pkg. (.6 oz.)	1.8	48
DISTILLED LIQUOR (USDA) 86 proof	1 fl. oz. (1 oz.)	0.	70
DOCK, including SHEEP SORREL (USDA):			
Raw, whole	1 lb. (weighed untrimmed)	6.7	89
Raw, trimmed	4 oz.	2.4	32
Boiled, drained	4 oz.	1.8	22
DOGFISH, Spiny, raw, meat only (USDA)	4 oz.	20.0	177
DOLLY VARDEN, raw, meat & skin (USDA)	4 oz.	22.6	163
DOUGHNUT:			
Cake type:			
(USDA)	1 piece (1.1 oz.)	1.5	125
(Hostess) 10 to pkg.	1 piece (1¼ oz.)	1.6	139
Powdered, frozen (Morton)	1 piece (.6 oz.)	.8	82
Sugar & spice, frozen (Morton)	1 piece (.6 oz.)	.8	82
Yeast-leavened (USDA)	2 oz.	3.6	235

(USDA): United States Department of Agriculture

Food and Description	Measure or Quantity	Protein (grams)	Calories
DRUM, raw (USDA):			
Freshwater:			
Whole	1 lb. (weighed whole)	20.4	143
Meat only	4 oz.	19.6	137
Red:			
Whole	1 lb. (weighed whole)	33.5	149
Meat only	4 oz.	20.4	91
DUCK, raw (USDA):			
Domesticated:			
Ready-to-cook	1 lb. (weighed with bones)	59.5	1213
Meat & skin	4 oz.	18.1	370
Meat only	4 oz.	24.3	187
Wild:			
Dressed	1 lb. (weighed dressed)	55.5	613
Meat, skin and giblets	4 oz.	23.9	264
Meat only	4 oz.	24.2	156

E

ECLAIR, home recipe, with custard filling & chocolate icing (USDA)	4 oz.	7.0	271
EEL (USDA):			
Raw, meat only	4 oz.	18.0	264
Smoked, meat only	4 oz.	21.1	374
EGG BEATERS (Fleischmann's)	¼ cup (2.1 oz.)	6.6	100
EGG, CHICKEN (USDA)			
Raw:			
White only	1 large egg (1.2 oz.)	3.6	17
White only	1 cup (9 oz.)	27.8	130
Yolk only	1 large egg (.6 oz.)	2.7	59
Yolk only	1 cup (8.5 oz.)	38.4	835
Whole, small	1 egg (1.3 oz.)	4.8	60

(USDA): United States Department of Agriculture

Food and Description	Measure or Quantity	Protein (grams)	Calories
Whole, medium	1 egg (1.5 oz.)	5.6	71
Whole	1 cup (8.8 oz.)	32.4	409
Whole, large	1 egg (1.8 oz.)	6.4	81
Whole, extra large	1 egg (2 oz.)	7.4	94
Whole, jumbo	1 egg (2.3 oz.)	8.3	105
Cooked:			
Boiled	1 large egg (1.8 oz.)	6.4	81
Fried in butter	1 large egg (1.6 oz.)	6.3	99
Omelet, mixed with milk & cooked in fat	1 large egg (2.2 oz.)	6.9	107
Poached	1 large egg (1.7 oz.)	6.1	78
Scrambled, mixed with milk & cooked in fat	1 cup (7.8 oz.)	24.6	381
Scrambled, mixed with milk & cooked in fat	1 large egg (2.3 oz.)	7.2	111
Dried:			
White, flakes	1 oz.	21.3	99
White, powder	1 oz.	22.7	105
Yolk	1 cup (3.4 oz.)	31.9	637
Whole	1 cup (3.8 oz.)	50.8	639
Whole, glucose reduced	1 oz.	13.9	173
Frozen, whole, raw	1 oz.	3.7	46
Preserved, limed	1 egg (1.8 oz.)	6.6	80
EGG, DUCK, raw (USDA)	1 egg (2.8 oz.)	10.7	153
EGG, GOOSE, raw (USDA)	1 egg (5.8 oz.)	22.8	303
EGG McMUFFIN (McDonald's)	1 piece (4.5 oz.)	17.6	313
EGG MIX:			
Plain (Durkee)	1 pkg. (.8 oz.)	7.9	124
Western (Durkee)	1 pkg. (1¼ oz.)	10.5	171
With bacon (Durkee)	1 pkg. (1¼ oz.)	11.8	181
EGG, TURKEY, raw (USDA)	1 egg (3.1 oz.)	11.5	150
EGG NOG, dairy:			
(Borden) 4.69% fat	½ cup (4.2 oz.)	3.9	132

(USDA): United States Department of Agriculture

Food and Description	Measure or Quantity	Protein (grams)	Calories
(Borden) 6% fat	½ cup (4.2 oz.)	3.1	154
(Borden) 8% fat	½ cup (4.2 oz.)	3.4	175
(Sealtest) 6% butterfat	½ cup (4.6 oz.)	5.8	174
(Sealtest) 8% butterfat	½ cup (4.6 oz.)	5.3	192
EGGPLANT:			
Raw, whole (USDA)	1 lb. (weighed untrimmed)	4.4	92
Boiled, drained (USDA)	4 oz.	1.1	22
Boiled, drained, diced (USDA)	1 cup (7.1 oz.)	2.0	38
Frozen, breaded slices (Mrs. Paul's)	7-oz. pkg.	7.0	564
Frozen, breaded sticks (Mrs. Paul's)	7-oz. pkg.	9.6	517
Frozen, parmesan (Mrs. Paul's)	11-oz. pkg.	17.3	577
Frozen, parmigiana (Buitoni)	4 oz.	6.9	189
Frozen, slices, breaded & fried (Mrs. Paul's)	7-oz. pkg.	7.0	564
Frozen, sticks, breaded & fried (Mrs. Paul's)	7-oz. pkg.	9.6	517
EGG ROLL, frozen, shrimp (Hung's)	1 roll	5.6	131
EGG, SCRAMBLED, BREAKFAST, frozen (Swanson):			
With coffee cake	6½-oz. breakfast	19.3	507
With link sausage & coffee cake	5½-oz. breakfast	13.7	409
EGGSTRA (Tillie Lewis)	1 large egg (dry, 10 grams)	5.3	43
ELDERBERRY, fresh (USDA):			
Whole	1 lb. (weighed with stems)	11.1	307
Stems removed	4 oz.	2.9	82

(USDA): United States Department of Agriculture
*Prepared as Package Directs

Food and Description	Measure or Quantity	Protein (grams)	Calories
ENCHILADA, beef, frozen:			
With cheese & chili gravy (Banquet)	8 enchiladas (2 lbs.)	52.2	1297
With rice (Swanson)	9⅝-oz. pkg.	12.3	378
With sauce (Banquet)	6-oz. bag	8.2	259
ENCHILADA DINNER, frozen:			
Beef:			
(Banquet)	12-oz. dinner	17.0	479
(Morton)	12-oz. dinner	20.4	524
(Swanson)	15-oz. dinner	19.6	561
Cheese (Banquet)	12-oz. dinner	12.2	282
ENCHILADA MIX:			
(Durkee)	1 pkg. (1⅛ oz.)	2.6	89
(Lawry's)	1 pkg. (1.6 oz.)	4.4	144
ENDIVE, BELGIAN or FRENCH (see **CHICORY, WITLOOF**)			
ENDIVE, CURLY or ESCAROLE, raw (USDA):			
Untrimmed	1 lb. (weighed untrimmed)	6.8	80
Trimmed	½ lb.	3.9	45
Cut up or shredded	1 cup (2.5 oz.)	1.2	14
ESCAROLE (see **ENDIVE**)			
EULACHON or SMELT, raw, meat only (USDA)	4 oz.	16.6	134
EXTRACT (see individual listings)			

F

FARINA (see also **CREAM OF WHEAT**):

(USDA): United States Department of Agriculture

Food and Description	Measure or Quantity	Protein (grams)	Calories
Regular:			
Dry:			
(USDA)	1 cup (6 oz.)	19.3	627
Cream, enriched (H-O)	1 cup (6.1 oz.)	19.8	646
Pearls of Wheat (Albers)	1 cup	18.6	608
Cooked:			
(USDA)	1 cup (8.4 oz.)	3.1	100
(Quaker)	1 cup (1 oz. dry)	2.5	100
*Prepared with milk			
(Pillsbury)	¾ cup	9.0	230
*Prepared with water			
(Pillsbury)	¾ cup	3.0	100
Quick-cooking (USDA):			
Dry	1 oz.	3.2	103
Cooked	1 cup (8.6 oz.)	3.2	105
Instant-cooking (USDA):			
Dry	1 oz.	3.2	103
Cooked	8 oz.	3.9	125
FAT, COOKING (USDA)	1 T. (.5 oz.)	0.	115
FENNEL LEAVES, raw, (USDA):			
Untrimmed	1 lb. (weighed untrimmed)	11.8	118
Trimmed	4 oz.	3.2	32
FIG:			
Fresh:			
(USDA)	1 lb.	5.4	363
Small (USDA)	1.3-oz. fig (1½″)	.5	30
Candied (USDA)	1 oz.	1.0	85
Candied (Bama)	1 T. (.7 oz.)	.1	37
Canned, regular pack:			
Light syrup, solids & liq.			
(USDA)	4 oz.	.6	74
Heavy syrup, solids & liq.:			
(USDA)	½ cup (4.4 oz.)	.6	106
(USDA)	3 figs & 2 T. syrup	.6	96
(Del Monte)	½ cup (4.4 oz.)	.6	104
(Stokely-Van Camp)	½ cup (4.2 oz.)	.6	100

(USDA): United States Department of Agriculture
*Prepared as Package Directs

Food and Description	Measure or Quantity	Protein (grams)	Calories
Extra heavy syrup, solids & liq. (USDA)	4 oz.	.6	117
Canned, unsweetened or dietetic pack:			
Water pack, solids & liq. (USDA)	4 oz.	.6	54
Kadota, solids & liq. (Diet Delight)	½ cup (4.4 oz.)	.6	76
Solids & liq. (Tillie Lewis)	½ cup (4.5 oz.)	.6	64
Whole (S and W) *Nutradiet*, low calorie	6 whole figs (3.5 oz.)	.5	49
Dried:			
Chopped (USDA)	1 cup (6 oz.)	7.4	469
(USDA)	.7-oz. fig (2″ × 1″)	.9	58
Calimyrna (Del Monte)	1 cup (5.4 oz.)	4.0	386
Mission (Del Monte)	1 cup (5.4 oz.)	4.3	380
FIG PRESERVE, sweetened (Bama)	1 T.	.1	51
FILBERT or HAZELNUT (USDA):			
Whole	1 lb. (weighed in shell)	26.3	1323
Shelled	1 oz.	3.6	180
FINNAN HADDIE (see **HADDOCK, SMOKED**)			
FISH (see individual listings)			
FISH CAKE:			
Home recipe, made with canned flaked fish, potato & egg, fried (USDA)	2 oz.	8.3	98
Frozen:			
Fried, reheated (USDA)	2 oz.	5.2	153
Fried, breaded (Mrs. Paul's)	2-oz. cake	5.7	105
Fried, breaded, thins (Mrs. Paul's)	2½-oz. cake	6.9	138

(USDA): United States Department of Agriculture

Food and Description	Measure or Quantity	Protein (grams)	Calories
FISH & CHIPS, frozen:			
(Gorton)	½ of 1-lb. pkg.	21.0	395
(Mrs. Paul's)	14-oz. pkg.	35.1	706
(Swanson)	5-oz. pkg.	12.8	268
FISH CHOWDER, New England			
(Snow)	8 oz.	10.6	144
FISH DINNER, frozen:			
(Morton)	8¾-oz. dinner	22.3	374
Filet of ocean fish (Swanson)	11½-oz. dinner	29.2	397
With French fries (Swanson)	9¾-oz. dinner	26.0	429
With green beans & peach (Weight Watchers)	18-oz. dinner	40.3	266
With pineapple chunks (Weight Watchers)	9½-oz. luncheon	21.8	175
FISH FILLET:			
Sandwich (McDonald's)	1 piece (4.8 oz.)	15.3	407
Frozen:			
Breaded, fried (Mrs. Paul's)	2-oz. piece	6.9	104
Buttered (Mrs. Paul's)	2½-oz. piece	12.1	94
Crisps (Gorton)	½ of 8-oz. pkg.	19.0	240
FISH FLAKES, canned (USDA)	4 oz.	28.0	126
FISH LOAF, home recipe, made with canned flaked fish, bread crumbs, eggs, tomatoes, onion & fat (USDA)	4 oz.	16.0	141
FISH PUFFS, frozen (Gorton)	½ of 8-oz. pkg.	19.0	265
FISH STICK, frozen:			
Cooked, commercial, 3¾″ × 1″ × ½″ sticks (USDA)	10 sticks (8-oz. pkg.)	37.7	400
(Gorton)	8-oz. pkg.	34.0	400
Breaded, fried (Mrs. Paul's)	14-oz. pkg.	50.0	756
FLAN PUDDING:			
Chilled (Breakstone)	5-oz. container	3.4	185
*Mix, regular (Royal)	½ cup (4.9 oz.)	4.3	141

(USDA): United States Department of Agriculture
*Prepared as Package Directs

Food and Description	Measure or Quantity	Protein (grams)	Calories
FLOUNDER:			
Raw (USDA):			
Whole	1 lb. (weighed whole)	25.0	118
Meat only	4 oz.	18.9	90
Baked (USDA)	4 oz.	34.0	229
Frozen:			
(Gorton)	1-lb. pkg.	75.0	360
Dinner (Weight Watchers)	18-oz. dinner	41.5	269
& broccoli (Weight Watchers)	9½-oz. luncheon	18.0	170
FLOUR:			
Buckwheat, dark, sifted (USDA)	1 cup (3.5 oz.)	11.5	326
Buckwheat, light, sifted (USDA)	1 cup (3.5 oz.)	6.3	340
Carob or St. John's-bread (USDA)	1 oz.	1.3	51
Chestnut (USDA)	1 oz.	1.7	103
Corn, sifted (USDA)	1 cup (3.9 oz.)	8.6	405
Cottonseed (USDA)	1 oz.	13.6	101
Cottonseed (Data from General Mills)	1 oz.	14.7	93
Fish, from whole fish (USDA)	1 oz.	22.1	95
Fish, from fillets (USDA)	1 oz.	26.4	113
Fish, from fillet waste (USDA)	1 oz.	20.1	86
Lima bean (USDA)	1 oz.	6.1	97
Peanut, defatted (USDA)	1 oz.	13.6	105
Potato (USDA)	1 oz.	2.3	100
Rice, stirred, spooned (USDA)	1 cup (5.6 oz.)	11.7	574
Rye:			
Light (USDA):			
Unsifted, spooned	1 cup (3.6 oz.)	9.5	361
Sifted, spooned	1 cup (3.1 oz.)	8.3	314
Medium (USDA)	1 oz.	3.2	99
Dark (USDA):			
Unstirred	1 cup (4.5 oz.)	20.9	419
Stirred	1 cup (4.5 oz.)	20.7	415
Soybean (USDA):			
Defatted, stirred	1 cup (3.6 oz.)	47.5	329
Low fat, stirred	1 cup (3.1 oz.)	38.2	313
Full fat, stirred	1 cup (2.5 oz.)	26.4	303
High fat	1 oz.	11.7	108
Sunflower seed, partially defatted (USDA)	1 oz.	12.8	96

(USDA): United States Department of Agriculture

Food and Description	Measure or Quantity	Protein (grams)	Calories
Tapioca, unsifted, spooned			
(USDA)	1 cup (3.8 oz.)	.6	377
Wheat:			
All purpose:			
(USDA)	1 oz.	3.0	103
Unsifted, dipped (USDA)	1 cup (5 oz.)	15.0	521
Unsifted, spooned (USDA)	1 cup (4.4 oz.)	13.2	459
Sifted, spooned (USDA)	1 cup (4.1 oz.)	12.2	422
Bread:			
(USDA)	1 oz.	3.3	103
Unsifted, dipped (USDA)	1 cup (4.8 oz.)	16.0	496
Unsifted, spooned (USDA)	1 cup (4.3 oz.)	14.5	449
Sifted, spooned (USDA)	1 cup (4.1 oz.)	13.8	427
Cake or pastry:			
(USDA)	1 oz.	2.1	103
Unsifted, dipped (USDA)	1 cup (4.2 oz.)	8.9	433
Unsifted, spooned (USDA)	1 cup (3.9 oz.)	8.3	404
Sifted, spooned (USDA)	1 cup (3.5 oz.)	7.4	360
Gluten:			
(USDA)	1 oz.	11.7	107
Unsifted, dipped (USDA)	1 cup (5 oz.)	58.8	537
Unsifted, spooned (USDA)	1 cup (4.8 oz.)	55.9	510
Sifted, spooned (USDA)	1 cup (4.8 oz.)	56.3	514
Self-rising:			
(USDA)	1 oz.	2.6	100
Unsifted, dipped (USDA)	1 cup (4.6 oz.)	12.1	458
Unsifted, spooned (USDA)	1 cup (4.5 oz.)	11.8	447
Sifted, spooned (USDA)	1 cup (3.7 oz.)	9.9	373
Whole wheat:			
(USDA)	1 oz.	3.8	94
Stirred, spooned (USDA)	1 cup (4.8 oz.)	18.2	456
Aunt Jemima, self-rising			
(Quaker Oats)	1 cup (4 oz.)	9.6	384
Gold Medal (Betty Crocker):			
Regular	1 cup	14.2	483
Better-for-bread	1 cup	16.5	483
Self-rising	1 cup	13.6	476
Wondra	1 cup	13.4	483
Presto, self-rising	1 cup (3.9 oz.)	8.9	392
(Quaker)	1 cup (4 oz.)	10.4	400
Robin Hood, all purpose	1 cup (4 oz.)	12.0	401
Robin Hood, self-rising	1 cup (4 oz.)	11.0	381

(USDA): United States Department of Agriculture

Food and Description	Measure or Quantity	Protein (grams)	Calories
Softasilk for cakes (Betty Crocker)	1 cup	9.8	412
FRANKENBERRY, cereal (General Mills)	1 cup (1 oz.)	1.3	109
FRANKFURTER or WIENER:			
Raw:			
All kinds (USDA)	1.6-oz. frankfurter	5.7	140
All meat (USDA)	1.6-oz. frankfurter	5.9	134
With cereal (USDA)	1.6-oz. frankfurter	6.5	112
With nonfat dry milk (USDA)	1.6-oz. frankfurter	5.9	136
With nonfat dry milk & cereal (USDA)	1.6-oz. frankfurter	6.4	
(Armour Star) all meat	1.6-oz. frankfurter	4.7	155
(Eckrich):			
Fun Franks, with skin	1.2-oz. frankfurter	4.0	112
Fun Franks, skinless	1.6 oz. frankfurter	5.0	150
Fun Franks, beef	1-oz. frankfurter	5.5	152
Fun Franks, jumbo	2-oz. frankfurter	6.0	189
(Hormel) all beef:			
10 per 12-oz. pkg.	1.2-oz. frankfurter	4.6	105
10 per 1-lb. pkg.	1.6-oz. frankfurter	6.1	140
12 per 12-oz. pkg.	1-oz. frankfurter	3.8	85
(Hormel) all meat:			
10 per 12-oz. pkg.	1.2-oz. frankfurter	3.9	105
10 per 1-lb. pkg.	1.6-oz. frankfurter	5.2	140
12 per 12-oz. pkg.	1-oz. frankfurter	3.3	90
(Hygrade) *Ball Park*	2-oz. frankfurter	6.8	177
(Oscar Mayer) all meat, imperial size, 5 per lb.	3.2-oz. frankfurter	10.0	288

(USDA): United States Department of Agriculture

Food and Description	Measure or Quantity	Protein (grams)	Calories
(Oscar Mayer) all meat, Little Wiener, 16 per 5½ oz.	.7-oz. frankfurter	2.8	81
(Oscar Mayer) all meat, *1883*, 6 per lb.	1 frankfurter	9.1	217
(Oscar Mayer) all meat wieners:			
8 per lb.	2-oz. frankfurter	6.3	180
10 per lb.	1.6-oz. frankfurter	5.0	142
(Oscar Mayer) pure beef:			
8 per lb.	2-oz. frankfurter	6.3	181
10 per lb.	1.6-oz. frankfurter	5.0	143
(Oscar Mayer) pure beef, Machiaeh Brand	2-oz. frankfurter	6.2	175
(Wilson) all beef	1.6-oz. frankfurter	6.0	136
(Wilson) skinless, all meat	1.6-oz. frankfurter	5.6	140
Cooked, all kinds, 10 per lb.			
raw (USDA)	1 frankfurter	5.6	136
Canned (USDA)	2 oz.	7.6	125
Canned (Hormel)	12-oz. can	41.1	966
FRANKS & BEANS (see **BEANS & FRANKS**)			
FRANKS-N-BLANKETS, frozen (Durkee)	1 piece (.4 oz.)	1.8	45
FRENCH TOAST, frozen:			
(Aunt Jemima)	1 slice (1.5 oz.)	3.2	88
With link sausage (Swanson)	4½-oz. breakfast	17.8	290
FROG LEGS, raw (USDA):			
Bone in	1 lb. (weighed with bone)	48.3	215
Meat only	4 oz.	18.6	83
FROOT LOOPS, cereal (Kellogg's)	1 cup (1 oz.)	1.4	115

(USDA): United States Department of Agriculture

Food and Description	Measure or Quantity	Protein (grams)	Calories
FROSTED RICE KRINKLES, cereal (Post)	⅞ cup (1 oz.)	.9	111
FROSTED SHAKE, canned, any flavor (Borden)	9¼-fl.-oz. can	11.0	320
FROSTED TREAT (Weight Watchers)	1 serving (4¾ oz.)	8.5	140
FROSTING (see **CAKE ICING**)			
FROSTY O's (General Mills)	1 cup (1 oz.)	1.8	112
FROZEN CUSTARD (see **ICE CREAM**)			
FROZEN DESSERT:			
Charlotte Freeze (Borden):			
Chocolate	⅓ pt. (3 oz.)	3.1	150
Vanilla	⅓ pt. (3 oz.)	2.6	138
Cherry (SugarLo):			
4% fat, ice milk	⅓ pt. (3.5 oz.)	4.1	129
10% fat, ice cream	⅓ pt. (3.5 oz.)	4.3	187
Chocolate:			
(Borden)	⅓ pt. (3.4 oz.)	4.8	164
Chocolate chip, mint (Borden)	⅓ pt. (3.4 oz.)	4.7	177
(SugarLo) 4.6% fat, ice milk	⅓ pt. (3.5 oz.)	4.1	140
(SugarLo) 10.8% fat, ice cream	⅓ pt. (3.5 oz.)	4.3	197
Coffee (SugarLo):			
4% fat, ice milk	⅓ pt. (3.5 oz.)	4.1	129
10% fat, ice cream	⅓ pt. (3.5 oz.)	4.3	187
Lemon chiffon (SugarLo):			
4% fat, ice milk	⅓ pt. (3.5 oz.)	4.1	129
10% fat, ice cream	⅓ pt. (3.5 oz.)	4.3	187
Maple (SugarLo):			
4% fat, ice milk	⅓ pt. (3.5 oz.)	4.1	129
10% fat, ice cream	⅓ pt. (3.5 oz.)	4.3	187
Orange-Pineapple (SugarLo):			
3.6% fat, ice milk	⅓ pt. (3.5 oz.)	3.7	124
9% fat, ice cream	⅓ pt. (3.5 oz.)	3.9	174

Food and Description	Measure or Quantity	Protein (grams)	Calories
Raspberry, black (Borden)	⅓ pt. (3.4 oz.)	4.7	160
Shake (SugarLo) chocolate	⅓ pt. (3.2 oz.)	5.4	75
Shake (SugarLo) vanilla & strawberry	⅓ pt. (3.2 oz.)	5.4	75
Strawberry (SugarLo):			
3.6% fat, ice milk	⅓ pt. (3.5 oz.)	3.7	124
9% fat, ice cream	⅓ pt. (3.5 oz.)	3.9	174
Vanilla:			
(Borden)	⅓ pt. (3.4 oz.)	4.7	160
(SugarLo) 4% fat, ice milk	⅓ pt. (3.5 oz.)	4.1	129
(SugarLo) 10% fat, ice cream	⅓ pt. (3.5 oz.)	4.3	187
Chocolate spin (Borden)	⅓ pt. (3.4 oz.)	4.7	172
Fudge or raspberry swirl:			
(SugarLo) 3.9% fat, ice milk	⅓ pt. (3.5 oz.)	3.2	127
(SugarLo) 10% fat, ice cream	⅓ pt. (3.5 oz.)	3.7	184
Vanilla-coated:			
Ice cream bar (SugarLo)	2½-oz. bar	2.7	134
Ice milk bar (SugarLo)	2½-oz. bar	2.7	111
FRUIT CAKE (USDA):			
Dark, home recipe	1-lb. loaf	21.8	1719
Dark, home recipe	2″ × 2″ × ½″ slice (1.1 oz.)	1.4	114
Dark, home recipe	⅟₃₀ of 8″ loaf (.5 oz.)	.7	57
Light, home recipe	1-lb. loaf	27.2	1768
Light, home recipe	2″ × 2″ × ½″ slice (1.1 oz.)	1.8	117
Light, home recipe	⅟₃₀ of 8″ loaf (.5 oz.)	.9	58
FRUIT COCKTAIL:			
Canned, regular pack, solids & liq.:			
Light syrup (USDA)	4 oz.	.5	68
Heavy syrup (USDA)	½ cup (4.5 oz.)	.5	97
Heavy syrup (Del Monte)	½ cup (4.3 oz.)	.4	88
Heavy syrup (Dole)	½ cup (including 3 T. syrup)	.5	94
Heavy syrup (Hunt's)	½ cup (4.5 oz.)	.5	89

(USDA): United States Department of Agriculture

Food and Description	Measure or Quantity	Protein (grams)	Calories
Heavy syrup (Stokely-Van Camp)	½ cup (4 oz.)	.5	87
Extra heavy syrup (USDA)	4 oz.	.5	104
Canned, unsweetened or dietetic pack, solids & liq.:			
Water pack (USDA)	4 oz.	.5	42
(Diet Delight)	½ cup (4.4 oz.)	.6	67
(S and W) *Nutradiet,* low calorie	4 oz.	.6	41
(Tillie Lewis)	½ of 8-oz. can	.4	44
FRUIT CUP (Del Monte):			
Fruit cocktail, solids & liq.	5¼-oz. container	.6	106
Mixed fruits, solids & liq.	5-oz. container	.4	100
Peaches, diced, solids & liq.	5¼-oz. container	.4	107
Pineapple, in its own juice, solids & liq.	4¼-oz. container	.6	68
FRUIT, MIXED:			
Dried (Del Monte)	1 cup (6.2 oz.)	3.9	442
Frozen, quick thaw (Birds Eye)	½ cup (5 oz.)	.7	111
FRUIT PUNCH:			
(Alegre)	6 fl. oz. (6.6 oz.)	Tr.	98
(Del Monte) tropical	6 fl. oz.	.1	82
FRUIT SALAD:			
Bottled, chilled (Kraft)	4 oz.	.7	57
Canned, regular pack, solids & liq:			
Light syrup (USDA)	4 oz.	.3	67
Heavy syrup (USDA)	½ cup (4.3 oz.)	.3	85
Heavy syrup (Del Monte) fruits for salad	½ cup (4.3 oz.)	.4	78
Heavy syrup (Del Monte) tropical	½ cup (4.4 oz.)	.6	110
Heavy syrup (Stokely-Van Camp)	½ cup (4.2 oz.)	.4	90
Extra heavy syrup (USDA)	4 oz.	.3	102
Canned, unsweetened or dietetic pack, solids & liq.:			
Water pack (USDA)	4 oz.	.5	40
(Diet Delight)	½ cup (4.4 oz.)	.5	67

(USDA): United States Department of Agriculture

Food and Description	Measure or Quantity	Protein (grams)	Calories
(S and W) *Nutradiet,* low calorie	4 oz.	.5	40
(S and W) Nutradiet, unsweetened	4 oz.	.5	43
FRUIT TREATS, apple (Mott's):			
& apricots	½ cup (4.4 oz.)	.1	104
& cherries	½ cup (4.6 oz.)	.1	109
& pineapple	½ cup (5.4 oz.)	.2	127
& raspberries	½ cup (4.7 oz.)	.1	111
& strawberries	½ cup (4.4 oz.)	.1	104
FRUITY PEBBLES, cereal (Post)	⅞ cup (1 oz.)	.9	111
FUDGE CAKE MIX:			
*Butter recipe (Duncan Hines)	1/12 of cake (3.3 oz.)	3.7	283
*Cherry (Betty Crocker)	1/12 of cake	2.9	199
*Chocolate (Pillsbury)	1/12 of cake	3.0	210
*Dark chocolate (Betty Crocker)	1/12 of cake	3.2	198
*Macaroon (Pillsbury)	1/12 of cake	3.0	220
*Marble (Duncan Hines)	1/12 of cake (2.7 oz.)	3.0	202
*Marble (Pillsbury)	1/12 of cake	4.0	350
*Nut (Pillsbury) *Bundt*	1/12 of cake	4.0	300
*Sour cream	1/12 of cake	3.1	195
*Sour cream (Pillsbury)	1/12 of cake	4.0	210
FUDGE ICE BAR:			
Fudgsicle	2½-fl.-oz. bar (2.8 oz.)	3.0	102
(Sealtest)	2½-fl.-oz. bar (2.6 oz.)	3.8	91
FUDGE PUDDING, canned (Thank You)	½ cup (4.5 oz.)	3.7	202

*Prepared as Package Directs

Food and Description	Measure or Quantity	Protein (grams)	Calories
FUNNY BONES (Drake's)	1¼-oz. cake	2.1	153
FUNNY FACE, orange (Pillsbury)	6 fl. oz.	0.	68

G

Food and Description	Measure or Quantity	Protein (grams)	Calories
GARBANZO, dry (see **CHICKPEA,** dry)			
GARBANZO SOUP, canned (Hormel)	15-oz. can	27.6	459
GARLIC:			
Raw (USDA):			
Whole	2 oz. (weighed with skin)	3.1	68
Peeled	1 oz.	1.8	39
Dehydrated flakes (Gilroy)	1 tsp. (2 grams)	.3	6
Dehydrated, powder (Gilroy)	1 tsp. (2 grams)	.5	10
GARLIC SPREAD (Lawry's)	1 T. (.5 oz.)	.2	79
GEFILTE FISH (Manischewitz):			
2-lb. jar	1 piece (2.4 oz.)	7.3	64
24-oz. jar	1 piece (2.6 oz.)	8.0	70
1-lb. jar	1 piece (2.2 oz.)	6.8	60
4-piece can	1 piece (3.7 oz.)	11.4	100
2-piece can	1 piece (3.5 oz.)	10.9	95
Fish balls	1 piece (1.5 oz.)	4.6	40
Fishlets	1 piece (7 grams)	.8	7
Whitefish & pike:			
2-lb. jar	1 piece (1.7 oz.)	5.5	40
1-lb. jar	1 piece (1.5 oz.)	4.9	35
4-piece can	1 piece (3.8 oz.)	12.1	87
2-piece can	1 piece (3.5 oz.)	11.3	81
Fishlets	1 piece (7 grams)	.8	6
GELATIN, unflavored, dry:			
(USDA)	1 envelope (7 grams)	6.0	23
(Knox)	1 envelope (7 grams)	7.0	28

(USDA): United States Department of Agriculture

Food and Description	Measure or Quantity	Protein (grams)	Calories
GELATIN DESSERT POWDER:			
Regular:			
Dry (USDA)	3-oz. pkg.	8.0	315
Dry (USDA)	½ cup (3.3 oz.)	8.8	347
*Prepared with water (USDA)	½ cup (4.2 oz.)	1.8	71
*Prepared with fruit added (USDA)	½ cup (4.3 oz.)	1.6	81
*All flavors (Jells Best)	½ cup	1.4	80
*All flavors (Royal)	½ cup (4.2 oz.)	2.0	82
*Regular flavors (Jell-O)	½ cup (4.9 oz.)	1.7	81
*Wild flavors (Jell-O)	½ cup (4.9 oz.)	1.7	81
*Dietetic, all flavors:			
(Dia-Mel) *Gela-Thin*	1 envelope (7 grams)	4.8	19
(D-Zerta)	½ cup (4.3 oz.)	1.6	8
*(Louis Sherry) *Shimmer*	½ cup (4.4,oz.)	1.8	10
GELATIN DRINK (Knox):			
Flavored, dry	1 envelope (.7 oz.)	6.0	30
Plain	1 envelope (.7 oz.)	6.3	25
GEL CUP (Del Monte):			
Lemon-lime with pineapple	5-oz. container	.3	115
Orange with peaches	5-oz. container	.4	107
Strawberry with peaches	5-oz. container	.4	111
GERMAN DINNER, frozen (Swanson)	11-oz. dinner	25.9	405
GINGERBREAD, home recipe (USDA)	1.9-oz. piece (2″ × 2″ × 2″)	2.1	174
GINGERBREAD MIX:			
Dry (USDA)	1 oz.	1.5	120
*Prepared with water (USDA)	⅑ of 8″ sq. (2.2 oz.)	2.0	174
*(Betty Crocker)	⅑ of cake	2.4	171
*(Dromedary)	1.2-oz. piece (2″ × 2″)	1.2	100
*(Pillsbury)	3″ sq.	2.0	190
GINGER CANDIED (USDA)	1 oz.	<.1	96

(USDA): United States Department of Agriculture
*Prepared as Package Directs

Food and Description	Measure or Quantity	Protein (grams)	Calories
GINGER ROOT, fresh (USDA):			
With skin	1 oz.	.4	13
Without skin	1 oz.	.4	14
GOOD HUMOR:			
Chocolate eclair	1 bar (2.4 oz.)	2.9	224
Strawberry shortcake	1 bar (2.2 oz.)	1.8	179
Toasted almond	1 bar (2.4 oz.)	3.0	229
Vanilla	1 bar (2.1 oz.)	2.4	197
Whammy:			
Assorted sticks	1 bar (1.4 oz.)	1.6	136
Vanilla	1 bar (1.4 oz.)	1.7	137
GOOSE, domesticated (USDA):			
Raw, ready-to-cook	1 lb. (weighed with bones)	54.3	1172
Raw, total edible	4 oz.	18.6	401
Raw, meat & skin	4 oz.	18.0	421
Raw, meat only	4 oz.	25.3	180
Roasted, total edible	4 oz.	26.9	483
Roasted, meat & skin	4 oz.	26.0	500
Roasted, meat only	4 oz.	38.4	264
GOOSE, GIBLET, raw (USDA)	4 oz.	23.9	177
GOOSE GIZZARD, raw (USDA)	4 oz.	24.3	158
GOOSEBERRY (USDA):			
Fresh	1 lb.	3.6	177
Fresh	1 cup (5.3 oz.)	12.0	58
Canned, solids & liq.:			
Regular pack, heavy syrup	4 oz.	.6	102
Regular pack, extra heavy syrup	4 oz.	.6	133
Water pack	4 oz.	.6	29
GOULASH DINNER (Chef Boy-Ar-Dee)	7-⅓-oz. pkg.	10.2	262
GRANOLA:			
Basic (Pillsbury)	¼ cup	3.0	120
With coconut & cashew (Pillsbury)	¼ cup	4.0	140

(USDA): United States Department of Agriculture

[167]

Food and Description	Measure or Quantity	Protein (grams)	Calories
With raisins & almond (Pillsbury)	¼ cup	3.0	120
Sun Country:			
Honey almond	½ cup (2 oz.)	8.0	249
Raisin	½ cup (2 oz.)	7.0	252
Vita Crunch:			
Regular	½ cup (2.4 oz.)	8.2	295
Date	½ cup	8.2	284
Raisin	½ cup	8.4	289
Toasted almonds	½ cup	9.0	296
GRAPE:			
Fresh:			
American type (slip skin), Concord, Delaware, Niagara, Catawba & Scuppernong, pulp only:			
(USDA)	½ lb. (weighed with stem, skin & seeds)	1.8	98
(USDA)	½ cup (2.7 oz.)	1.0	52
(USDA)	3½″ × 3″ bunch (3.5 oz.)	.8	43
European type (adherent skin), Malaga, Muscat, Thompson seedless, Emperor & Flame Tokay:			
(USDA)	½ lb. (weighed with stems & seeds)	1.2	139
Whole (USDA)	20 grapes (¾″ dia.)	.5	54
Whole (USDA)	½ cup (3.1 oz.)	.5	58
Halves (USDA)	½ cup (3 oz.)	.5	58
Canned, solids & liq. (USDA):			
Thompson seedless, heavy syrup	4 oz.	.6	87
Thompson seedless, water pack	4 oz.	.6	58
GRAPEADE, chilled (Sealtest)	6 fl. oz. (6.5 oz.)	Tr.	96
GRAPE-BERRY (Ocean Spray)	½ cup (5 oz.)	.2	92

(USDA): United States Department of Agriculture

Food and Description	Measure or Quantity	Protein (grams)	Calories
GRAPE DRINK:			
(Del Monte)	6 fl. oz. (6.5 oz.)	.2	90
(Hi-C)	6 fl. oz. (4.2 oz.)	.1	98
GRAPE JAM (Bama)	1 T. (.7 oz.)	.1	54
GRAPE JELLY			
Sweetened (Smucker's)	1 T. (.7 oz.)	<.1	50
Low calorie:			
(Kraft)	1 oz.	<.1	34
(Diet Delight) Concord	1 T. (.6 oz.)	<.1	21
(S and W) *Nutradiet,*			
Concord	1 T. (.5 oz.)	<.1	10
(Slenderella)	1 T. (.6 oz.)	<.1	26
(Smucker's)	1 T. (.5 oz.)	<.1	6
(Tillie Lewis)	1 T. (.8 oz.)	<.1	11
GRAPE JUICE:			
Canned:			
(USDA)	½ cup (4.4 oz.)	.3	83
(Heinz)	5½-fl.-oz. can	.8	130
(S and W) *Nutradiet*	4 oz. (by wt.)	.4	68
Frozen, concentrate, sweetened:			
(USDA)	6 fl.-oz. can (7.6 oz.)	1.3	395
*Diluted with 3 parts water			
(USDA)	½ cup (4.4 oz.)	.2	66
*(Minute Maid)	½ cup (4.2 oz.)	.2	66
*(Snow Crop)	½ cup (4.2 oz.)	.2	66
GRAPE JUICE DRINK, canned (USDA) approximately 30% grape juice	1 cup (8.8 oz.)	.2	135
GRAPE-NUTS, cereal (Post)	¼ cup (1 oz.)	2.5	104
GRAPE NUTS FLAKES, cereal (Post)	⅔ cup (1 oz.)	2.5	101
GRAPE PIE (Tastykake)	4-oz. pie	3.9	369

(USDA): United States Department of Agriculture
*Prepared as Package Directs

Food and Description	Measure or Quantity	Protein (grams)	Calories
GRAPE SOFT DRINK (Yoo-Hoo)			
High Protein	6 fl. oz. (6.4 oz.)	6.0	100
GRAPEFRUIT:			
Fresh, pulp only:			
Pink & red:			
Seeded type (USDA)	1 lb. (weighed with seeds and skin)	1.1	87
Seeded type (USDA)	½ med. grapefruit (3¾", 8.5 oz.)	.6	46
Seedless type (USDA)	1 lb. (weighed with skin)	1.2	93
Seedless type (USDA)	½ med. grapefruit (8.5 oz.)	.6	49
White:			
Seeded type (USDA)	1 lb. (weighed with seeds & skin)	1.0	84
Seeded type (USDA)	½ med. grapefruit (3¾", 8.5 oz.)	.5	44
Seedless type (USDA)	1 lb. (weighed with skin)	1.1	87
Seedless type, sections (USDA)	1 cup (7 oz.)	1.0	78
Seedless type (USDA)	½ med. grapefruit (3¾", 8.5 oz.)	.6	46
(Sunkist)	½ grapefruit (8.5 oz.)	1.0	44
Bottled, chilled, sweetened sections (Kraft)	4 oz.	.6	53
Bottled, chilled, unsweetened sections (Kraft)	4 oz.	.5	40
Canned, sections, syrup pack, solids & liq.:			
(USDA)	½ cup (4.5 oz.)	.8	90
(Del Monte)	½ cup (4.5 oz.)	.8	69
Light syrup (Stokely-Van Camp)	½ cup (4 oz.)	.7	80

(USDA): United States Department of Agriculture

Food and Description	Measure or Quantity	Protein (grams)	Calories
Canned, sections, unsweetened or dietetic pack, solids & liq.:			
Water Pack (USDA)	½ cup (4.2 oz.)	.7	36
Juice pack (Del Monte)	½ cup (4.5 oz.)	.8	45
(Diet Delight) unsweetened	½ cup (4.3 oz.)	.7	41
(S and W) *Nutradiet,* low calorie	4 oz.	.7	40
(S and W) *Nutradiet,* unsweetened	4 oz.	.8	41
(Tillie Lewis)	½ cup (4.4 oz.)	.7	45
GRAPEFRUIT JUICE:			
Fresh, pink, red or white, all varieties (USDA)	½ cup (4.3 oz.)	.6	48
Bottled, chilled, sweetened (Kraft)	½ cup (4.3 oz.)	.6	60
Bottled, chilled, unsweetened (Kraft)	½ cup (4.3 oz.)	.6	48
Canned:			
Sweetened:			
(USDA)	½ cup (4.4 oz.)	.6	66
(Del Monte)	½ cup (4.3 oz.)	.8	48
(Heinz)	5½-fl.-oz. can	1.0	73
(Stokely-Van Camp)	½ cup (4.4 oz.)	.6	67
Unsweetened:			
(USDA)	½ cup (4.4 oz.)	.6	51
(Del Monte)	½ cup (4.3 oz.)	.8	46
(Diet Delight)	½ cup (4 oz.)	.7	39
(Heinz)	5½-fl.-oz. can	1.0	56
(Stokely-Van Camp)	½ cup (4.5 oz.)	.6	52
Frozen, concentrate:			
Sweetened:			
(USDA)	6-fl.-oz. can (7.4 oz.)	3.4	348
*Diluted with 3 parts water (USDA)	½ cup (4.4 oz.)	.5	58
Unsweetened:			
(USDA)	6-fl.-oz. can (7.3 oz.)	3.9	300
*Diluted with 3 parts water (USDA)	½ cup (4.4 oz.)	.6	51
*(Florida Diet)	½ cup (4.3 oz.)	.6	50
*(Minute Maid)	½ cup (4.2 oz.)	.7	50

(USDA): United States Department of Agriculture
*Prepared as Package Directs

Food and Description	Measure or Quantity	Protein (grams)	Calories
*(Snow Crop)	½ cup (4.2 oz.)	.7	50
Dehydrated, crystals:			
(USDA)	4-oz. can	5.4	429
*Reconstituted (USDA)	½ cup (4.4 oz.)	.6	50
GRAPEFRUIT-ORANGE JUICE (see **ORANGE-GRAPEFRUIT JUICE**)			
GRAPEFRUIT PEEL, CANDIED (USDA)	1 oz.	.1	90
GRAVY, canned:			
Beef (Franco-American)	¼ cup	3.0	44
Brown:			
Ready Gravy	¼ cup (2.2 oz.)	1.5	44
With onion			
(Franco-American)	¼ cup	1.0	24
Chicken (Franco-American)	¼ cup	1.4	51
Chicken giblet			
(Franco-American)	¼ cup	1.0	28
Mushroom (Franco-American)	¼ cup	1.0	27
Mushroom, brown, *Dawn Fresh*	5¾-oz. can	1.6	60
GRAVY MASTER	1 fl. oz. (1.3 oz.)	2.7	66
GRAVY with MEAT or TURKEY,			
Giblet & sliced turkey			
(Banquet):			
Cooking bag	5-oz. bag	12.2	129
Buffet	2-lb. pkg.	102.0	677
Sliced beef, frozen (Banquet):			
Cooking bag	5 oz.	19.8	158
Buffet	2 lbs.	105.5	956
Sliced beef (Morton House)	½ of 12½-oz. can	12.0	189
Sliced pork (Morton House)	½ of 12½-oz. can	12.1	193
Sliced turkey (Morton House)	½ of 12½-oz. can	13.2	140
GRAVY MIX:			
Au jus (Durkee)	1-oz. pkg.	2.5	67
*Au jus (Durkee)	2 cups (1-oz. pkg.)	3.2	80
Au jus (French's)	¾-oz. pkg.	1.0	44
*Au jus (French's)	¼ cup	.1	5

(USDA): United States Department of Agriculture
*Prepared as Package Directs

Food and Description	Measure or Quantity	Protein (grams)	Calories
Beef:			
(Swiss Products)	1¼-oz. pkg.	2.5	107
(Swiss Products)	⅞-oz. pkg.	1.8	75
*(Wyler's)	2-oz. serving	1.1	25
Brown:			
(Durkee)	.8-oz. pkg.	2.1	59
*(Durkee)	1 cup (.8-oz. pkg.)	2.4	56
(French's)	¾-oz. pkg.	2.5	72
*(French's)	¼ cup	.6	18
*(Kraft)	2-oz. serving	1.4	21
(Lawry's)	1¼-oz. pkg.	5.9	136
(McCormick)	⅞-oz. pkg.	1.1	100
*(McCormick)	2-oz. serving	.3	25
*(Pillsbury)	¼ cup	.0	10
Chicken:			
(Durkee)	1-oz. pkg.	3.4	96
*(Durkee)	1 cup (1-oz. pkg.)	3.2	96
(French's)	1½-oz. pkg.	7.1	130
*(French's)	¼ cup	1.8	37
*(Kraft)	2-oz. serving	.5	30
(Lawry's)	1-oz. pkg.	4.4	110
*(McCormick)	2-oz. serving	.7	20
*(Pillsbury)	¼ cup	1.0	30
(Swiss)	1¼-oz. pkg.	2.9	107
(Swiss)	⅞-oz. pkg.	2.0	75
*(Wyler's)	2-oz. serving	.7	25
*Herb (McCormick)	2-oz. serving	3.0	22
Home style (French's)	⅞-oz. pkg.	4.0	80
*Home style (French's)	¼ cup	1.0	20
*Home style (Pillsbury)	¼ cup	0.	10
Mushroom:			
(Durkee)	.8-oz. pkg.	2.0	60
*(Durkee)	1 cup (.8-oz. pkg.)	2.4	72
(French's)	¾-oz. pkg.	4.0	62
*(French's)	¼ cup	1.3	16
(Lawry's)	1.3-oz. pkg.	6.0	145
*(McCormick)	2-oz. serving	.9	17
*(Wyler's)	2-oz. serving	.8	15
Onion:			
(Durkee)	1-oz. pkg.	2.8	85
*(Durkee)	1 cup (1-oz. pkg.)	3.2	88
(French's)	1-oz. pkg.	2.8	72
*(French's)	¼ cup	.7	18

*Prepared as Package Directs

Food and Description	Measure or Quantity	Protein (grams)	Calories
*(Kraft)	2-oz. serving	.9	22
*(McCormick)	2-oz. serving	.2	29
*(Wyler's)	2-oz. serving	.8	17
Pork:			
(French's)	¾-oz. pkg.	3.3	74
*(French's)	¼ cup	1.1	19
Turkey:			
(French's)	⅞-oz. pkg.	3.9	90
*(French's)	¼ cup	1.0	23
GREAT HONEY CRUNCHERS, cereal (Nabisco):			
Rice	1 cup (1 oz.)	1.0	113
Wheat	1 cup (1 oz.)	1.9	115
GREEN PEA (see **PEA**)			
GRITS (see **HOMINY GRITS**)			
GROUND-CHERRY, Poha or Cape Gooseberry, fresh (USDA):			
Whole	1 lb. (weighed with husks & stems)	7.9	221
Flesh only	4 oz.	2.2	60
GROUPER, raw (USDA):			
Whole	1 lb. (weighed whole)	37.6	170
Meat only	4 oz.	21.9	99
GUAVA, COMMON, fresh (USDA):			
Whole	1 lb. (weighed untrimmed)	3.5	273
Whole	1 guava (2.8 oz.)	.6	48
Flesh only	4 oz.	.9	70
GUAVA JELLY (Smucker's)	1 T. (.7 oz.)	2.1	50

(USDA): United States Department of Agriculture
*Prepared as Package Directs

Food and Description	Measure or Quantity	Protein (grams)	Calories
GUAVA, STRAWBERRY, fresh (USDA):			
Whole	1 lb. (weighed untrimmed)	4.4	289
Flesh only	4 oz.	1.1	74
GUINEA HEN, raw (USDA):			
Ready-to-cook	1 lb.	88.0	594
Meat & skin	4 oz.	26.5	179
Giblets	2 oz.	11.8	89
GUM (see **CHEWING GUM**)			

H

Food and Description	Measure or Quantity	Protein (grams)	Calories
HADDOCK:			
Raw (USDA):			
Whole	1 lb. (weighed whole)	39.8	172
Meat only	4 oz.	20.8	90
Fried, dipped in egg, milk & bread crumbs (USDA)	4″ × 3″ × ½″ fillet (3.5 oz.)	19.6	165
Frozen (Gorton)	⅓ of 1-lb. pkg.	28.0	120
Smoked, canned or not (USDA)	4 oz.	26.3	117
HADDOCK MEALS, frozen:			
(Banquet)	8¾-oz. dinner	21.3	419
(Weight Watchers)	18-oz. dinner	43.4	256
& spinach (Weight Watchers)	9½-oz. luncheon	24.9	175
HAKE, raw (USDA):			
Whole	1 lb. (weighed whole)	32.2	144
Meat only	4 oz.	18.7	84
HALF & HALF, milk & cream (see **CREAM**)			

(USDA): United States Department of Agriculture

Food and Description	Measure or Quantity	Protein (grams)	Calories
HALIBUT:			
Atlantic & Pacific:			
Raw (USDA):			
Whole	1 lb. (weighed whole)	55.9	268
Meat only	4 oz.	23.7	113
Broiled with fat (USDA)	6½" × 2½" × ⅝" or 4" × 3" × ½" steak (4.4 oz.)	31.5	214
Smoked (USDA)	4 oz.	23.6	254
California, raw, meat only (USDA)	4 oz.	22.5	110
HAM (see also **PORK**):			
Boiled:			
Luncheon meat (USDA)	1 oz.	5.4	66
Luncheon meat, chopped (USDA)	1 cup (4.8 oz.)	25.8	318
Luncheon meat, diced (USDA)	1 cup (5 oz.)	26.8	330
(Hormel)	1 oz.	5.4	35
Chopped (Hormel)	1 oz.	4.6	70
Minced (Oscar Mayer)	1 slice (10 per ½ lb.)	3.9	56
Smoked (Oscar Mayer)	1 slice (8 per 6 oz.)	4.0	29
Smoked, thin sliced (Oscar Mayer)	1 slice (10 per 3 oz.)	1.5	11
Canned:			
(USDA)	1 oz.	5.2	55
(Armour Golden Star)	1 oz.	5.4	36
(Armour Star)	1 oz.	4.9	53
(Hormel)	1 oz. (8-lb. can)	4.6	57
(Hormel)	1 oz. (6-lb. can)	4.8	44
(Hormel)	1 oz. (4-lb. can)	4.8	48
(Hormel)	1 oz. (1-lb. 8-oz. can)	4.8	48
(Oscar Mayer) *Jubilee*, bone in	1 lb.	81.6	835
(Oscar Mayer) *Jubilee*, boneless	1 lb.	81.6	826

(USDA): United States Department of Agriculture

Food and Description	Measure or Quantity	Protein (grams)	Calories
(Oscar Mayer) *Jubilee,* boneless	½-lb. slice	43.1	315
(Oscar Mayer) *Jubilee,* special trim, as purchased	1 oz.	5.1	41
(Oscar Mayer) *Jubilee,* special trim, cooked	1 oz.	6.0	37
(Oscar Mayer) steak	1 slice (8 per lb.)	10.8	74
(Swift)	1¾-oz. slice (5″ × 2¼″ × ¼″)	9.0	111
(Swift) *Hostess*	1 oz. (4-lb. can)	5.6	41
(Wilson)	1 oz.	4.6	48
(Wilson) *Tender Made*	1 oz.	4.9	44
Chopped or minced, canned:			
(USDA)	1 oz.	3.9	65
(Armour Star)	1 oz.	4.2	84
(Hormel)	1 oz. (8-lb. can)	3.9	90
(Oscar Mayer)	1-oz. slice	4.5	65
Deviled, canned:			
(USDA)	1 oz.	3.9	100
(USDA)	1 T. (.5 oz.)	1.8	46
(Armour Star)	1 oz.	4.6	79
(Hormel)	1 oz. (3-oz. can)	4.3	73
(Underwood)	4½-oz. can	17.4	438
(Underwood)	1 T. (.5 oz.)	1.8	46
Freeze-dry, diced, canned (Wilson) *Campsite:*			
Dry	¾-oz. can	12.2	105
*Reconstituted	1 oz.	8.5	71
Chopped, spiced or unspiced, canned:			
(USDA)	1 oz.	4.3	83
Chopped (USDA)	1 cup (4.8 oz.)	20.4	400
Diced (USDA)	1 cup (5 oz.)	21.2	415
(Hormel)	1 oz. (5-lb. can)	4.2	78
HAM & CHEESE (Oscar Mayer):			
Loaf	1 oz.	4.8	69
Roll	1 oz.	4.8	64
Spread	1 oz.	4.8	75
HAM CROQUETTE, home recipe (USDA)	4 oz.	18.5	285

(USDA): United States Department of Agriculture
*Prepared as Package Directs

Food and Description	Measure or Quantity	Protein (grams)	Calories
HAM DINNER:			
Frozen:			
(Banquet)	10-oz. dinner	16.8	369
(Morton)	10½-oz. dinner	20.9	447
(Swanson)	10¼-oz. dinner	22.7	366
Mix, au gratin (Jeno's) *Add 'n Heat*	35-oz. pkg.	74.4	1627
HAM SPREAD, salad (Oscar Mayer)	1 oz.	2.6	60
HAMBURGER (see also **BEEF,** Ground):			
Big Mac (see **BIG MAC**)			
Regular (McDonald's)	1 piece (3.4 oz.)	13.0	251
Regular, cheese (McDonald's)	1 piece (3.9 oz.)	16.1	310
¼ pound (McDonald's)	1 piece (5.5 oz.)	26.5	416
¼ pound, cheese (McDonald's)	1 piece (6.6 oz.)	31.5	523
Freeze dry, canned (Wilson) *Campsite:*			
Dry	3¼-oz. can	50.4	526
*Reconstituted	4 oz.	25.3	262
HAWAIIAN DINNER MIX (Hunt's) *Skillet*	1-lb. pkg.	12.9	760
HAWAIIAN-STYLE VEGETABLES, frozen (Birds Eye)	⅓ of 10-oz. pkg.	.9	92
HAWS, SCARLET, raw (USDA):			
Whole	1 lb. (weighed with core)	7.3	316
Flesh & skin	4 oz.	2.3	99
HAZELNUT (see **FILBERT**)			
HEADCHEESE:			
(USDA)	1 oz.	4.4	76
(Oscar Mayer)	1 slice (8 per ½ lb.)	4.5	52

(USDA): United States Department of Agriculture
*Prepared as Package Directs

Food and Description	Measure or Quantity	Protein (grams)	Calories
HEART (USDA):			
Beef:			
Lean, raw	1 lb.	77.6	490
Lean, braised	4 oz.	35.5	213
Lean, braised, chopped or diced	1 cup (5.1 oz.)	45.4	273
Lean with visible fat, raw	1 lb.	69.9	1148
Lean with visible fat, braised	4 oz.	29.3	422
Calf, raw	1 lb.	68.0	562
Calf, braised	4 oz.	31.5	236
Chicken, raw	1 lb.	84.4	608
Chicken, simmered	1 heart (5 grams)	1.3	9
Chicken, simmered, chopped or diced	1 cup (5.1 oz.)	36.7	251
Hog, raw	1 lb.	76.2	513
Hog, braised	4 oz.	34.9	221
Lamb, raw	1 lb.	76.2	735
Lamb, braised	4 oz.	33.5	295
Turkey, raw	1 lb.	73.5	776
Turkey, simmered	4 oz.	25.6	245
Turkey, simmered, chopped or diced	1 cup (5.1 oz.)	32.8	313
HEARTLAND, cereal (Pet)	1 oz.	3.1	122
HERRING:			
Raw (USDA):			
Atlantic, whole	1 lb. (weighed whole)	40.0	407
Atlantic, meat only	4 oz.	19.6	200
Pacific, meat only	4 oz.	19.8	111
Canned:			
Plain, solids & liq. (USDA)	4 oz.	22.6	236
Plain, solids & liq (USDA)	15-oz. can	84.6	884
Bismark, drained (Vita)	5-oz. jar	26.3	273
Cocktail, drained (Vita)	8-oz. jar	30.2	342
In cream sauce (Vita)	8-oz. jar	24.3	397
In tomato sauce, solids & liq. (USDA)	4 oz.	17.9	200
In wine sauce, drained (Vita)	8-oz. jar	34.6	401
Lunch, drained (Vita)	8-oz. jar	37.9	483
Matjis, drained (Vita)	8 oz.	30.4	304
Party snacks, drained (Vita)	8-oz. jar	34.6	401

(USDA): United States Department of Agriculture

[179]

Food and Description	Measure or Quantity	Protein (grams)	Calories
Tastee Bits, drained (Vita)	8-oz. jar	25.4	361
Pickled, Bismarck type (USDA)	4 oz.	23.1	253
Salted or brined (USDA)	4 oz.	21.5	247
Smoked (USDA):			
Bloaters	4 oz.	22.2	222
Hard	4 oz.	41.8	340
Kippered	4 oz.	25.2	239
HICKORY NUT (USDA):			
Whole	1 lb. (weighed in shell)	21.0	1068
Shelled	4 oz.	15.0	763
HO-HO (Hostess)	1-oz. cake	1.0	133
HOMINY GRITS:			
Dry:			
Degermed (USDA)	½ cup (2.8 oz.)	6.8	282
Creamy (Pocono)	1 oz.	1.1	101
Degermed (Albers)	½ cup (3 oz.)	8.0	296
Instant (Quaker)	.8-oz. packet	1.9	78
Cooked:			
Degermed (USDA)	1 cup (8.6 oz.)	2.9	125
(Albers)	1 cup	2.9	122
(Aunt Jemima/Quaker)	⅔ cup	2.4	100
HONEY, strained:			
(USDA)	½ cup (5.7 oz.)	.5	496
(USDA)	1 T. (.7 oz.)	<.1	61
HONEYCOMB, cereal (Post)	1⅓ cups (1 oz.)	1.6	108
HONEYDEW, fresh (USDA):			
Whole	1 lb. (weighed whole)	2.3	94
Wedge	2″ × 7″ wedge (5.3 oz.)	.8	31
Flesh only	4 oz.	.9	37
Flesh only, diced	1 cup (5.9 oz.)	1.3	55

(USDA): United States Department of Agriculture

Food and Description	Measure or Quantity	Protein (grams)	Calories
HORSERADISH:			
Raw (USDA):			
Whole	1 lb. (weighed unpared)	10.6	288
Pared	1 oz.	.9	25
Dehydrated (Heinz)	1 T.	.9	25
Prepared:			
(USDA)	1 oz.	.4	11
(Kraft)	1 oz.	.4	3
Cream style (Kraft)	1 oz.	.4	9
Oil style (Kraft)	1 oz.	.4	20
HOTCHAS (General Mills)	15 pieces (.5 oz.)	1.2	69
*HOT DOG BEAN SOUP, canned (Campbell)	1 cup	8.6	154
HUSH PUPPIES (Borden)	1 piece (.8 oz.)	1.5	58
HYACINTH BEAN (USDA):			
Young pod, raw:			
Whole	1 lb. (weighed untrimmed)	11.2	140
Trimmed	4 oz.	3.2	40
Dry seeds	4 oz.	25.2	383

I

Food and Description	Measure or Quantity	Protein (grams)	Calories
ICE CREAM and FROZEN CUSTARD (see also listing by flavor or brand name, e.g., **CHOCOLATE ICE CREAM** or *DREAMSICLE*, and **FROZEN DESSERT**)			
Sweetened:			
10% fat (USDA) regular ice cream	3-fl.-oz. container (1.8 oz.)	2.2	97
10% fat (USDA) regular ice cream	1 cup (4.7 oz.)	6.0	257

(USDA): United States Department of Agriculture
*Prepared as Package Directs

Food and Description	Measure or Quantity	Protein (grams)	Calories
10% fat (USDA) frozen custard or French ice cream	3-fl.-oz. container (1.8 oz.)	2.2	97
10% fat (USDA) frozen custard or French ice cream	1 cup (4.7 oz.)	6.0	257
12% fat (USDA) ice cream	2½-oz. slice (⅛ of qt. brick)	2.8	147
12% fat (USDA) ice cream	3½-fl.-oz. container (2.2 oz.)	2.5	128
12% fat (USDA) ice cream	1 cup (5 oz.)	5.7	294
16% fat (USDA) rich ice cream	1 cup (5 oz.)	3.8	329
Candy, 11% fat (Dean)	¼ pt. (2.8 oz.)	3.4	180
Chocolate, 11.7% fat (Dean)	¼ pt. (2.8 oz.)	3.6	176
*Fruit, 10.4% fat (Dean)	¼ pt. (2.8 oz.)	2.8	168
Nut, 14% fat (Dean)	¼ pt. (2.8 oz.)	3.7	188
Strawberry, 10.5% fat (Dean)	¼ pt. (2.8 oz.)	2.8	168
Vanilla, 10.1% fat (Dean)	¼ pt. (2.6 oz.)	3.3	148
Vanilla, 12% fat (Dean)	¼ pt. (2.8 oz.)	3.4	170
ICE CREAM BAR, chocolate-coated (Sealtest)	2½-fl.-oz. bar (1.7 oz.)	1.6	149
ICE CREAM CONE, cone only:			
(USDA)	1 piece (5 grams)	.5	19
(Comet)	1 piece (4 grams)	.3	19
Assorted colors (Comet)	1 piece (4 grams)	.3	19
Rolled sugar (Comet)	1 piece (.4 oz.)	1.0	49
ICE CREAM CUP, cup only:			
(Comet)	1 piece (5 grams)	.4	20
Assorted colors (Comet)	1 piece (5 grams)	.4	20
Pilot (Comet)	1 piece (4 grams)	.3	19
ICE CREAM SANDWICH (Sealtest)	3 fl. oz. (2.2 oz.)	3.1	173
ICE MILK:			
Hardened (USDA)	1 cup (4.6 oz.)	6.3	199
Soft-serve (USDA)	1 cup (6.2 oz.)	8.4	266

(USDA): United States Department of Agriculture
*Prepared as Package Directs

Food and Description	Measure or Quantity	Protein (grams)	Calories
Any flavor (Borden) 2.5% fat	¼ pt. (2.3 oz.)	2.8	93
Any flavor (Borden) 3.25% fat	¼ pt. (2.4 oz.)	2.6	97
Any flavor (Borden) *Lite Line*	¼ pt.	3.8	99
(Dean) 5% fat	¼ pt. (2.5 oz.)	3.2	113
Count Calorie (Dean) 2.1% fat	¼ pt. (2.4 oz.)	3.4	78
Light n' Lively (Sealtest):			
Banana	¼ pt. (2.4 oz.)	2.7	103
Banana strawberry twirl	¼ pt. (2.5 oz.)	2.4	111
Buttered almond	¼ pt. (2.4 oz.)	3.4	117
Caramel nut	¼ pt. (2.4 oz.)	3.4	120
Cherry pineapple	¼ pt. (2.4 oz.)	2.5	98
Chocolate	¼ pt. (2.4 oz.)	2.8	105
Coffee	¼ pt. (2.4 oz.)	3.0	102
Lemon	¼ pt. (2.4 oz.)	3.0	103
Lemon chiffon	¼ pt. (2.4 oz.)	2.4	119
Orange pineapple	¼ pt. (2.4 oz.)	2.7	102
Peach	¼ pt. (2.4 oz.)	2.5	103
Raspberry	¼ pt. (2.4 oz.)	2.9	99
Strawberry	¼ pt. (2.4 oz.)	2.6	100
Strawberry royale	¼ pt. (2.4 oz.)	2.7	112
Toffee	¼ pt. (2.4 oz.)	2.9	110
Toffee crunch	¼ pt. (2.4 oz.)	2.8	119
Vanilla	¼ pt. (2.3 oz.)	3.0	102
Vanilla fudge royale	¼ pt. (2.5 oz.)	3.0	113
ICE MILK BAR, chocolate-coated (Sealtest)	2½-fl.-oz. bar (1.8 oz.)	2.0	132
ICE STICK, Twin Pops (Sealtest)	3 fl. oz. (3.1 oz.)	Tr.	70
ICES (see **LIME ICE**)			
ICING (see **CAKE ICING**)			
INCONNU or SHEEFISH, raw:			
Whole (USDA)	1 lb. (weighed whole)	56.9	417
Meat only (USDA)	4 oz.	22.6	166
INDIAN PUDDING, New England (B & M)	½ cup (4 oz.)	2.7	120

(USDA): United States Department of Agriculture

Food and Description	Measure or Quantity	Protein (grams)	Calories
INSTANT BREAKFAST (see individual brand name or company listings)			
IRISH WHISKEY (See **DISTILLED LIQUOR**)			
ITALIAN DINNER, frozen:			
(Banquet)	11-oz. dinner	19.7	446
(Swanson)	13½-oz. dinner	17.2	448
ITALIAN-STYLE VEGETABLES frozen (Birds Eye)	⅓ of 10-oz. pkg.	2.6	98

J

Food and Description	Measure or Quantity	Protein (grams)	Calories
JACKFRUIT, fresh (USDA):			
Whole	1 lb. (weighed with seeds & skin)	1.7	124
Flesh only	4 oz.	1.5	111
JACK MACKEREL, raw, meat only (USDA)	4 oz.	24.5	162
JAM, sweetened (see also individual listings by flavor):			
(USDA)	1 oz.	.2	77
(USDA)	1 T. (.7 oz.)	.1	54
JAPANESE-STYLE VEGETABLES, frozen (Birds Eye)	⅓ of 10-oz. pkg.	1.8	105
JELLY, sweetened (see also individual listings by flavor):			
(USDA)	1 oz.	<.1	77
(USDA)	1 T. (.6 oz.)	<.1	49
All flavors (Bama)	1 T. (.7 oz.)	.1	51
All flavors (Kraft)	1 oz.	<.1	74
All flavors (Smucker's)	1 T. (.7 oz.)	0.	49

(USDA): United States Department of Agriculture

Food and Description	Measure or Quantity	Protein (grams)	Calories
JERUSALEM ARTICHOKE (USDA):			
Unpared	1 lb. (weighed with skin)	7.2	207
Pared	4 oz.	2.6	75
JORDAN ALMOND (see **CANDY**)			
JUICE (see individual flavors)			
JUJUBE or CHINESE DATE (USDA):			
Fresh, whole	1 lb. (weighed with seeds)	5.1	443
Fresh, flesh only	4 oz.	1.4	119
Dried, whole	1 lb. (weighed with seeds)	14.9	1159
Dried, flesh only	1 oz.	1.0	81
JUNIOR FOOD (see **BABY FOOD**)			
JUNIORS (Tastykake):			
Chocolate	2¾-oz. pkg.	6.1	397
Chocolate devil food	2¾-oz. pkg.	4.0	284
Coconut	2¾-oz. pkg.	5.5	415
Coconut devil food	2¾-oz. pkg.	3.8	318
Jelly square	3¼-oz. pkg.	5.7	429
Koffee Kake	2½-oz. pkg.	8.6	395
Lemon	2¾-oz. pkg.	5.2	422

K

KABOOM, cereal (General Mills)	1 cup (1 oz.)	1.6	109
KALE:			
Raw, leaves only (USDA)	1 lb. (weighed untrimmed)	17.4	154
Raw, leaves including stems (USDA)	1 lb. (weighed trimmed)	14.1	128
Boiled, leaves only (USDA)	4 oz.	5.1	44

(USDA): United States Department of Agriculture

Food and Description	Measure or Quantity	Protein (grams)	Calories
Boiled, including stems (USDA)	½ cup (1.9 oz.)	1.8	15
Frozen:			
Not thawed (USDA)	4 oz.	3.6	36
Boiled, drained (USDA)	½ cup (3.2 oz.)	2.8	29
Chopped (Birds Eye)	½ cup (3.3 oz.)	3.0	29

KASHA (see **BUCKWHEAT,** Groats)

KETCHUP (see **CATSUP**)

KIDNEY (USDA):			
Beef, raw	4 oz.	17.5	147
Beef, braised	4 oz.	37.4	286
Beef, braised, ¼″ slices	1 cup (4.9 oz.)	46.2	353
Calf, raw	4 oz.	18.8	128
Hog, raw	4 oz.	18.5	120
Lamb, raw	4 oz.	19.1	119

KIELBASA:			
(Eckrich):			
Links	1 piece	8.0	184
Skinless	1 oz.	3.5	94
Sticks	1 oz.	3.0	100
(Oscar Mayer)	6-oz. link	22.1	530

KINGFISH, raw (USDA):			
Whole	1 lb. (weighed whole)	36.5	210
Meat only	4 oz.	20.8	119

KING VITAMAN, cereal (Quaker)	¾ cup (1 oz.)	1.1	118

KIPPERS (see **HERRING**)

KIX, cereal (General Mills)	1½ cups (1 oz.)	2.5	112

KNOCKWURST:			
(USDA)	1 oz.	4.0	79
Chubbies, all meat (Oscar Mayer)	1 link (2.4 oz.)	7.5	210

(USDA): United States Department of Agriculture

Food and Description	Measure or Quantity	Protein (grams)	Calories
KOHLRABI (USDA):			
Raw, whole	1 lb. (weighed with skin, without leaves)	6.6	96
Raw, diced	1 cup (4.9 oz.)	2.8	40
Boiled, drained	1 cup (5.5 oz.)	2.6	37
***KOOL-AID** (General Foods)	1 cup (9.3 oz.)	Tr.	98
KOOL-POPS (General Foods)	1 bar (1.5 oz.)	Tr.	32
KOTTBULLAR, canned (Hormel)	1 oz. (1-lb. can)	3.3	48
KRIMPETS (Tastykake):			
Apple spice	.9-oz. cake	2.0	135
Butterscotch	.9-oz. cake	1.7	123
Chocolate	.9-oz. cake	1.9	119
Jelly	.9-oz. cake	1.6	103
Lemon	.9-oz. cake	1.5	113
Orange	.9-oz. cake	1.6	114
KUMQUAT, fresh (USDA):			
Whole	1 lb. (weighed with seeds)	3.8	274
Flesh & skin	4 oz.	1.0	74

L

Food and Description	Measure or Quantity	Protein (grams)	Calories
LAKE HERRING, raw (USDA):			
Whole	1 lb. (weighed whole)	41.8	226
Meat only	4 oz.	20.1	109
LAKE TROUT, raw (USDA):			
Drawn	1 lb. (weighed with head, fins & bone)	30.7	282
Meat only	4 oz.	20.8	191

(USDA): United States Department of Agriculture
*Prepared as Package Directs

[187]

Food and Description	Measure or Quantity	Protein (grams)	Calories
LAKE TROUT or SISCOWET, raw (USDA):			
Less than 6.5 lb. whole	1 lb. (weighed whole)	24.0	404
Less than 6.5 lb. whole	4 oz. (meat only)	16.2	273
More than 6.5 lb. whole	1 lb. (weighed whole)	12.9	856
More than 6.5 lb. whole	4 oz. (meat only)	9.0	594
LAMB, choice grade (USDA):			
Chop, broiled:			
Loin. One 5-oz. chop (weighed with bone before cooking) will give you:			
Lean & fat	2.8 oz.	17.2	280
Lean only	2.3 oz.	18.3	122
Rib. One 5-oz. chop (weighed with bone before cooking) will give you:			
Lean & fat	2.9 oz.	16.5	334
Lean only	2 oz.	15.2	118
Fat, separable, cooked	1 oz.	1.8	201
Leg:			
Raw, lean & fat	1 lb. (weighed with bone)	67.7	845
Roasted, lean & fat	4 oz.	28.7	316
Roasted, lean only	4 oz.	32.5	211
Shoulder:			
Raw, lean & fat	1 lb. (weighed with bone)	58.9	1082
Roasted, lean & fat	4 oz.	24.6	383
Roasted, lean only	4 oz.	30.4	232
LAMB'S-QUARTERS (USDA):			
Raw, trimmed	1 lb.	19.1	195
Boiled, drained	4 oz.	3.6	36
LAMB STEW, canned (B & M)	1 cup (8.1 oz.)	16.7	192
LARD (USDA)	1 T.	0.	117
LASAGNE:			
Canned (Chef Boy-Ar-Dee)	⅕ of 40-oz. can	10.9	279

(USDA): United States Department of Agriculture

Food and Description	Measure or Quantity	Protein (grams)	Calories
Canned (Nalley's)	8 oz.	9.5	213
Frozen (Buitoni)	½ of 56-oz. pkg.	13.7	234
Frozen (Buitoni)	½ of 15-oz. pkg.	14.5	255
Frozen (Celeste)	¼ of 2-lb. pkg.	21.5	413
*Mix, dinner (Chef Boy-Ar-Dee)	8¾-oz. pkg.	12.6	273
Mix (Golden Grain) *Stir-N-Serv*	⅕ of 7-oz. pkg.	5.0	151
*Mix, dinner (Jeno's) *Add 'n Heat*	30-oz. pkg.	116.5	1599
Mix (Hunt's) *Skillet*	1-lb. 2-oz. pkg.	40.1	812
Seasoning mix (Lawry's)	1.1-oz. pkg.	1.5	86
LEEKS, raw (USDA):			
Whole	1 lb. (weighed untrimmed)	5.2	123
Trimmed	4 oz.	2.5	59
LEMON, fresh (USDA);			
Fruit, including peel	1 lb. (weighed whole)	5.4	90
Fruit, including peel	2⅛" lemon (3.8 oz., seeds removed)	1.3	22
Peeled fruit	1 med. lemon (2⅛")	.8	20
LEMONADE:			
Chilled (Sealtest)	½ cup (4.4 oz.)	Tr.	55
Frozen, concentrate, sweetened:			
(USDA)	6-fl.-oz. can (7.7 oz.)	.4	427
*Diluted with 4⅓ parts water (USDA)	½ cup (4.4 oz.)	.1	55
*(Minute Maid)	½ cup (4.2 oz.)	<.1	49
*(Snow Crop)	½ cup (4.2 oz.)	<.1	49
*Low calorie (Weight Watchers)	6 fl. oz. (5.8 oz.)	<.1	8
LEMON CAKE MIX:			
*(Duncan Hines)	1/12 of cake (2.7 oz.)	3.0	202
*(Pillsbury)	1/12 of cake	3.0	210
*Blueberry (Pillsbury) *Bundt*	1/12 of cake	4.0	310

(USDA): United States Department of Agriculture
*Prepared as Package Directs

Food and Description	Measure or Quantity	Protein (grams)	Calories
*Chiffon (Betty Crocker)	1/16 of cake	3.3	151
*Layer (Betty Crocker)	1/12 of cake	2.8	202
*Pudding cake (Betty Crocker)	1/6 of cake	2.3	227
*Streusel (Pillsbury)	1/12 of cake	4.0	350
LEMON DRINK, chilled			
(Sealtest)	6 fl. oz. (6.5 oz.)	Tr.	91
LEMON JUICE:			
Fresh:			
(USDA)	1 cup (8.6 oz.)	1.2	61
(USDA)	1 T. (.5 oz.)	<.1	4
(Sunkist)	1 lemon (3.9 oz.)	Tr.	11
(Sunkist)	1 T. (.5 oz.)	Tr.	4
Canned, unsweetened:			
(USDA)	1 cup (8.6 oz.)	1.0	56
(USDA)	1 T. (.5 oz.)	<.1	3
Plastic container:			
(USDA)	1/4 cup (2 oz.)	.2	13
(ReaLemon)	1 T. (.5 oz.)	<.1	3
Frozen, unsweetened:			
Concentrate (USDA)	1/2 cup (5.1 oz.)	3.4	169
Single strength (USDA)	1/2 cup (4.3 oz.)	.5	27
Full strength, already reconstituted (Minute Maid)	1/2 cup (4.2 oz.)	.5	27
Full strength, already reconstituted (Snow Crop)	1/2 cup (4.2 oz.)	.5	27
LEMON-LIMEADE, sweetened, concentrate, frozen:			
*(Minute Maid)	1/2 cup (4.2 oz.)	<.1	50
*(Snow Crop)	1/2 cup (4.2 oz.)	<.1	50
LEMON PEEL:			
Raw (USDA)	1 oz.	4.3	
Candied (USDA)	1 oz.	.1	90
LEMON PIE:			
(Hostess)	4½-oz. pie	3.6	447
(Mrs. Smith's) frozen	1/6 of 8" pie (4.2 oz.)	2.5	340
(Tastykake)	4-oz. pie	4.7	366

(USDA): United States Department of Agriculture
*Prepared as Package Directs

Food and Description	Measure or Quantity	Protein (grams)	Calories
Chiffon, home recipe (USDA)	⅙ of 9″ pie (3.8 oz.)	7.6	338
Cream, frozen:			
(Banquet)	2½-oz. serving	1.5	179
(Morton)	⅙ of 16-oz. pie	1.5	194
(Mrs. Smith's)	⅙ of 8″ pie (2.8 oz.)	1.2	227
Krunch (Mrs. Smith's)	⅙ of 8″ pie (4.3 oz.)	3.0	383
Krunch (Mrs. Smith's)	⅛ of 10″ pie (5.5 oz.)	3.6	464
Meringue, home recipe, 1 crust (USDA)	⅙ of 9″ pie (4.9 oz.)	5.2	357
Meringue, frozen (Mrs. Smith's)	⅙ of 8″ pie (3.7 oz.)	2.3	261
Meringue, frozen (Mrs. Smith's)	⅛ of 10″ pie (5.2 oz.)	3.1	356
Tart, frozen (Pepperidge Farm)	1 pie tart (3 oz.)	2.7	317
LEMON PIE FILLING, canned:			
(Comstock)	⅙ of 8″ pie (3.5 oz.)	.2	144
(Lucky Leaf)	8 oz.	.6	412
Wilderness	22-oz. can	1.9	1104
LEMON PIE FILLING MIX (see **LEMON PUDDING or PIE MIX**)			
LEMON PUDDING, canned:			
(Betty Crocker)	½ cup	.4	198
(Hunt's)	5-oz. can	<.1	175
(Thank You)	½ cup (4.5 oz.)	.4	183
LEMON PUDDING or PIE MIX:			
Regular:			
*(Jell-O)	½ cup (5.1 oz.)	2.0	178
*(Royal)	⅛ of 9″ pie (including crust, 4.6 oz.)	2.8	224
Instant:			
*(Jell-O)	½ cup (5.3 oz.)	4.3	178
*(Royal)	½ cup (5.1 oz.)	4.4	178
LEMON RENNET MIX:			
Powder:			
Dry (Junket)	1 oz.	.3	116
*(Junket)	4 oz.	3.6	109

(USDA): United States Department of Agriculture
*Prepared as Package Directs

Food and Description	Measure or Quantity	Protein (grams)	Calories
Tablet:			
Dry (Junket)	1 tablet (9 grams)	Tr.	1
*& sugar (Junket)	4 oz.	3.7	101
LEMON TURNOVER, frozen			
(Pepperidge Farm)	1 turnover (3.3 oz.)	3.1	341
LENTIL:			
Whole:			
Dry:			
(USDA)	½ lb.	56.0	771
(USDA)	1 oz.	7.0	96
(USDA)	1 cup (6.7 oz.)	47.2	649
Cooked, drained (USDA)	½ cup (3.6 oz.)	7.9	107
Split, dry, without seed coat			
(USDA)	½ lb.	56.0	782
***LENTIL SOUP:**			
*Canned (Manischewitz)	1 cup	6.7	166
Mix (Lipton) *Cup-a-Soup*	1.2-oz. pkg.	6.9	123
LETTUCE (USDA):			
Bibb, untrimmed	1 lb. (weighed untrimmed)	4.0	47
Bibb, untrimmed	7.8-oz. head (4″ dia.)	2.0	23
Boston, untrimmed	1 lb. (weighed untrimmed)	4.0	47
Boston, untrimmed	7.8-oz. head (4″ dia.)	2.0	23
Butterhead varieties (see Bibb)			
Cos (see Romaine)			
Dark green (see Romaine)			
Grand Rapids	1 lb. (weighed untrimmed)	3.8	52
Grand Rapids	2 large leaves (1.8 oz.)	.6	9
Great Lakes, untrimmed	1 lb. (weighed untrimmed)	3.9	56
Great Lakes, trimmed	1-lb. head (4¾″ dia., weighed trimmed)	4.1	59
Iceberg:			
Untrimmed	1 lb. (weighed untrimmed)	3.9	56

(USDA): United States Department of Agriculture
*Prepared as Package Directs

Food and Description	Measure or Quantity	Protein (grams)	Calories
Trimmed	1 lb. head (4¾" dia., weighed trimmed)	4.1	59
Leaves	1 cup (2.3 oz.)	.6	9
Chopped	1 cup (2 oz.)	.5	8
Chunks	1 cup (2.6 oz.)	.7	10
Looseleaf varieties (see Salad Bowl)			
New York	1 lb. (weighed untrimmed)	3.9	56
New York	1-lb. head (4¾" dia., weighed trimmed)	4.1	59
Romaine:			
Untrimmed	1 lb. (weighed untrimmed)	3.8	52
Shredded & broken into pieces	½ cup (.8 oz.)	.3	4
Salad Bowl:			
Untrimmed	1 lb. (weighed untrimmed)	3.8	52
Trimmed	2 large leaves (1.8 oz.)	.6	9
Simpson:			
Untrimmed	1 lb. (weighed untrimmed)	3.8	52
Trimmed	2 large leaves (1.8 oz.)	.6	9
White Paris (see Romaine)			
LIFE, cereal (Quaker)	⅔ cup (1 oz.)	5.1	107
LIMA BEAN (see **BEAN, LIMA**)			
LIME, fresh, whole:			
(USDA)	1 lb. (weighed with skin & seeds)	2.7	107
(USDA)	1 med. (2.4 oz., 2" dia.)	.4	15
LIMEADE, concentrate, sweetened, frozen:			
(USDA)	6-fl.-oz. can (7.7 oz.)	.4	408
*Diluted with 4⅓ parts (USDA)	½ cup (4.4 oz.)	Tr.	51

(USDA): United States Department of Agriculture
*Prepared as Package Directs

Food and Description	Measure or Quantity	Protein (grams)	Calories
*(Minute Maid)	½ cup (4.2 oz.)	.1	50
*(Snow Crop)	½ cup (4.2 oz.)	.1	50
LIME ICE, home recipe (USDA)	8 oz. (by wt.)	.9	177
LIME JUICE:			
Fresh (USDA)	1 cup (8.7 oz.)	.7	64
Canned or bottled, unsweetened:			
(USDA)	1 fl. oz. (1.1 oz.)	<.1	8
(USDA)	1 cup (8.7 oz.)	.7	64
Plastic container, *ReaLime*	1 T. (.5 oz.)	<.1	2
LIME PIE, Key lime, cream, frozen (Banquet)	2½-oz. serving	1.8	204
LIME PIE FILLING MIX, Key lime (Royal)	⅛ of 9″ pie (including crust, 4.6 oz.)	2.8	222
LINGCOD, raw (USDA):			
Whole	1 lb. (weighed whole)	27.6	130
Meat only	4 oz.	20.3	95
LITCHI NUT (USDA):			
Fresh:			
Whole	4 oz. (weighed in shell with seeds)	.6	44
Flesh only	4 oz.	1.0	73
Dried:			
Whole	4 oz. (weighed in shell with seeds)	2.0	145
Flesh only	2 oz.	2.2	157
LIVER:			
Beef, raw (USDA)	1 lb.	90.3	635
Beef, fried (USDA)	4 oz.	29.9	260
Beef, fried (USDA)	6½″ × 2⅜″ × ⅜″ slice (3 oz.)	22.4	195
Calf, raw (USDA)	1 lb.	87.1	635
Calf, fried (USDA)	4 oz.	33.5	296

(USDA): United States Department of Agriculture
*Prepared as Package Directs

Food and Description	Measure or Quantity	Protein (grams)	Calories
Calf, fried (USDA)	6½″ × 2⅜″ × ⅜″ slice (3 oz.)	25.1	222
Chicken, raw (USDA)	1 lb.	89.4	585
Chicken, raw, frozen (Swanson)	8-oz. pkg.	40.6	240
Chicken, simmered (USDA)	4 oz.	30.1	187
Chicken, simmered (USDA)	2″ × 2″ × ⅝″ liver (.9 oz.)	6.6	41
Goose, raw (USDA)	1 lb.	74.8	826
Hog, raw (USDA)	1 lb.	93.4	594
Hog, fried (USDA)	4 oz.	33.9	273
Hog, fried (USDA)	6½″ × 2⅜″ × ⅜″ slice (3 oz.)	25.4	205
Lamb, raw (USDA)	1 lb.	95.3	617
Lamb, broiled (USDA)	4 oz.	36.6	296
Lamb, broiled (USDA)	6½″ × 2⅜″ × ⅜″ slice (3 oz.)	27.5	222
Turkey, raw (USDA)	1 lb.	96.2	626
Turkey, simmered (USDA)	4 oz.	31.6	197
Turkey, simmered, chopped (USDA)	1 cup (4.9 oz.)	39.1	244

LIVER PATE (see **PATE**)

LIVER SAUSAGE or LIVERWURST:

Fresh (USDA)	1 oz.	4.6	87
Sliced (Oscar Mayer)	.9-oz. slice (10 slices to 9 oz.)	3.4	95
Ring (Oscar Mayer)	1 oz.	3.7	86
Smoked (USDA)	1 oz.	4.2	90

LIVERWURST SPREAD

(Underwood)	1 T. (.5 oz.)	2.1	45

LOBSTER (USDA):

Raw:

Whole	1 lb. (weighed whole)	19.9	107
Meat only	4 oz.	19.2	103
Cooked, meat only	4 oz.	21.2	108
Cooked, meat only	1 cup (½″ cubes, 5.1 oz.)	27.1	138
Canned, meat only	4 oz.	21.2	108

(USDA): United States Department of Agriculture

Food and Description	Measure or Quantity	Protein (grams)	Calories
Frozen, South African rock lobster tail:			
3 in 8-oz. pkg.	1 tail	20.4	87
4 in 8-oz. pkg.	1 tail	15.4	65
5 in 8-oz. pkg.	1 tail	11.9	51
LOBSTER NEWBURG:			
Home recipe made with butter, egg yolks, sherry & cream (USDA)	1 cup (8.8 oz.)	46.2	485
Frozen (Stouffer's)	11½-oz. pkg.	28.0	671
LOBSTER PASTE, canned (USDA)	1 oz.	5.9	51
LOBSTER SALAD, home recipe (USDA)	4 oz.	11.5	125
LOBSTER SOUP MIX, bisque (Lipton) *Cup-a-Soup*	.8-oz. pkg.	2.1	104
LOGANBERRY (USDA):			
Fresh:			
Untrimmed	1 lb. (weighed with caps)	4.3	267
Trimmed	1 cup (5.1 oz.)	1.4	89
Canned, solids & liq.:			
Water pack	4 oz.	.8	45
Juice pack	4 oz.	.8	61
Light syrup	4 oz.	.8	79
Heavy syrup	4 oz.	.7	101
Extra heavy syrup	4 oz.	.7	122
LOG CABIN, syrup	1 T. (.7 oz.)	Tr.	46
LONGAN (USDA):			
Fresh:			
Whole	1 lb. (weighed with shell & seeds)	2.4	147
Flesh only	4 oz.	1.1	69

(USDA): United States Department of Agriculture

Food and Description	Measure or Quantity	Protein (grams)	Calories
Dried:			
Whole	1 lb. (weighed with shell & seeds)	8.0	467
Flesh	4 oz.	5.6	324
LOQUAT, fresh (USDA):			
Whole	1 lb. (weighed with seeds)	1.4	168
Flesh only	4 oz.	.5	54
LUCKY CHARMS, cereal (General Mills)	1 cup (1 oz.)	2.1	110
LUNCHEON MEAT (see also individual listings, e.g., **BOLOGNA**):			
Banquet (Eckrich)	1 slice	2.5	54
All meat (Oscar Mayer)	1-oz. slice	3.7	98
Bar-B-Q Loaf (Oscar Mayer)	1-oz. slice	4.3	48
Cocktail loaf (Oscar Mayer)	1-oz. slice	3.4	62
Gourmet loaf (Eckrich)	1-oz. slice	4.0	38
Gourmet loaf (Eckrich) Smorgas Pac	1 slice	3.0	27
Honey loaf, regular or Smorgas Pac (Eckrich)	1 slice	4.0	42
Ham & cheese (see **HAM & CHEESE**)			
Honey loaf (Oscar Mayer)	1-oz. slice	5.1	40
Jellied:			
Beef loaf (Oscar Mayer)	1-oz. slice	6.8	41
Corned beef loaf (Oscar Mayer)	1-oz. slice	6.8	39
Luncheon roll, sausage, all meat (Oscar Mayer)	.8-oz. slice	3.9	27
Luxury Loaf (Oscar Mayer)	1-oz. slice (8 per ½ lb.)	4.8	40
Meat loaf (USDA)	1 oz.	4.5	57
Minced roll sausage, all meat (Oscar Mayer)	.8-oz. slice	3.2	54
Old fashioned loaf (Eckrich) regular or Smorgas Pac	1 slice	3.5	76

(USDA): United States Department of Agriculture

Food and Description	Measure or Quantity	Protein (grams)	Calories
Old fashioned loaf (Oscar Mayer)	1-oz. slice	4.3	62
Olive loaf (Oscar Mayer)	1-oz. slice	3.4	62
Peppered loaf (Oscar Mayer)	1-oz. slice	4.5	46
Pickle & pimento:			
(Hormel)	1 oz. (6-lb. can)	3.7	81
(Oscar Mayer)	1-oz. slice	3.4	62
Pickle loaf, regular or Smorgas Pac (Eckrich)	1 slice	3.0	86
Picnic loaf (Oscar Mayer)	1-oz. slice	4.3	64
Plain loaf (Oscar Mayer)	1-oz. slice	4.0	75
Pure beef (Oscar Mayer)	1-oz. slice	4.3	75
Spiced (Hormel)	1 oz.	3.9	70
LUNG, raw (USDA):			
Beef	1 lb.	79.8	435
Calf	1 lb.	76.2	481
Lamb	1 lb.	87.5	467

M

Food and Description	Measure or Quantity	Protein (grams)	Calories
MACADAMIA NUT (USDA):			
Whole	1 lb. (weighed in shell)	11.0	972
Shelled	4 oz.	8.8	784
Canned (Royal Hawaiian)	¼ cup (2 oz.)	4.4	394

MACARONI. Plain macaroni products are essentially the same in protein value on the same weight basis. The longer they are cooked, the more water is absorbed and this affects the nutritive values.

Food and Description	Measure or Quantity	Protein (grams)	Calories
Dry:			
(USDA)	1 oz.	3.5	105
Elbow type (USDA)	1 cup (4.8 oz.)	17.0	502
1-inch pieces (USDA)	1 cup (3.8 oz.)	13.8	406
2-inch pieces (USDA)	1 cup (3 oz.)	10.8	317
Cooked (USDA):			
8-10 minutes, firm	4 oz.	5.7	168
8-10 minutes, firm	1 cup (4.6 oz.)	6.5	192
14-20 minutes, tender	4 oz.	3.9	126

(USDA): United States Department of Agriculture

Food and Description	Measure or Quantity	Protein (grams)	Calories
14-20 minutes, tender	1 cup (4.9 oz.)	4.8	155
20% Protein, dry (Buitoni)	1 oz.	5.7	101
MACARONI & BEEF:			
Canned, tiny meatballs & sauce (Buitoni)	4 oz.	5.4	111
Canned, in tomato sauce (Franco-American)	1 cup	12.3	225
Frozen:			
(Banquet) buffet	2 lb. pkg.	52.5	1092
In tomato sauce (Kraft)	11½-oz. pkg.	19.9	447
With tomatoes (Stouffer's)	11½-oz. pkg.	20.8	410
(Swanson)	11¼-oz. dinner	16.0	302
MACARONI & CHEESE:			
Home recipe, baked (USDA)	1 cup (7.1 oz.)	16.8	430
Canned:			
(USDA)	1 cup (8.5 oz.)	9.4	228
(Franco-American)	1 cup	8.4	219
(Heinz)	8¼-oz. can	9.0	231
Frozen:			
(Banquet) cooking bag	8-oz. bag	11.8	279
(Banquet) entrée	8-oz. pkg.	11.8	279
(Banquet) entrée	20-oz. pkg.	31.8	742
(Kraft)	12½-oz. pkg.	25.8	612
(Morton) casserole	8-oz. pkg.	13.6	295
(Morton) casserole	20-oz. pkg.	34.0	737
(Stouffer's)	12-oz. pkg.	21.4	477
*MACARONI & CHEESE MIX:			
Cheddar sauce (Betty Crocker)	1 cup	11.9	325
Dinner (Golden Grain)	¼ of 7¼-oz. pkg.	8.0	202
MACARONI DINNER:			
& beef, frozen (Banquet)	12-oz. dinner	12.6	394
& beef, frozen (Morton)	11-oz. dinner	9.4	287
& cheese:			
*(Chef Boy-Ar-Dee)	4½-oz. pkg.	9.0	201
*(Kraft)	4 oz.	6.6	203
*(Kraft) deluxe	4 oz.	8.3	202
Frozen (Banquet)	12-oz. dinner	13.3	326
Frozen (Morton)	12¾-oz. dinner	14.5	384
Frozen (Swanson)	12¾-oz. dinner	12.6	367

(USDA): United States Department of Agriculture
*Prepared as Package Directs

Food and Description	Measure or Quantity	Protein (grams)	Calories
Creole, with mushrooms			
(Heinz)	8¾-oz. can	5.3	169
*Italian-style (Kraft)	4 oz.	4.1	119
*Mexican-style (Kraft)	4 oz.	4.3	126
*Monte Bello (Betty Crocker)	1 cup	19.9	350
MACARONI ENTREE, shells in			
meat sauce (Buitoni)	4 oz.	5.9	117
MACARONI SALAD, canned			
(Nalley's)	4 oz.	3.4	203
MACKEREL (USDA):			
Atlantic:			
Raw:			
Whole	1 lb. (weighed whole)	46.5	468
Meat only	4 oz.	21.5	217
Broiled with butter or margarine	8½″ × 2½″ × ½″ fillet (3.7 oz.)	22.9	248
Canned, solids & liq.	4 oz.	21.9	208
Canned, solids & liq.	15-oz. can	82.0	778
Pacific:			
Raw:			
Dressed	1 lb. (weighed with bones & skin)	71.5	519
Meat only	4 oz.	24.8	180
Canned, solids & liq.	4 oz.	23.9	204
Salted	4 oz.	21.0	346
Smoked	4 oz.	27.0	248
MACKEREL, JACK (see **JACK MACKEREL**)			
MALT, dry (USDA)	1 oz.	3.7	104
MALTED MILK MIX:			
Dry powder, "unfortified" (USDA)	(3 heaping tsps.) 1 oz.	4.2	116

(USDA): United States Department of Agriculture
*Prepared as Package Directs

Food and Description	Measure or Quantity	Protein (grams)	Calories
*Prepared with whole milk (USDA)	1 cup (8.3 oz.)	11.0	244
Chocolate, instant (Borden)	2 heaping tsp. (.7 oz.)	1.3	77
Chocolate (Carnation)	3 heaping tsps. (.7 oz.)	1.2	85
Chocolate (Horlicks)	3 heaping tsps. (1.1 oz.)	1.9	124
Chocolate, dry (Kraft)	2 heaping tsps. (.4 oz.)	1.0	51
*Chocolate (Kraft)	1 cup (8.7 oz.)	9.9	241
Natural, instant (Borden)	2 heaping tsps. (.7 oz.)	2.6	80
Natural (Carnation)	3 heaping tsps. (.7 oz.)	2.5	88
Natural (Horlicks)	3 heaping tsps. (1.1 oz.)	4.4	127
Natural, dry (Kraft)	2 heaping tsps. (.4 oz.)	1.6	52
*Natural (Kraft)	1 cup (8.6 oz.)	10.8	240
MALTEX, cereal	1 oz.	3.3	109
MALT EXTRACT, dried (USDA)	1 oz.	1.7	104
MALT LIQUOR, *Champale*, 6.25% alcohol	12 fl. oz. (12.6 oz.)	.7	173
MALT-O-MEAL, cereal:			
Chocolate	¾ cup	3.0	102
Quick	¾ cup	2.9	102
MAMEY or MAMMEE APPLE, fresh (USDA):			
Whole	1 lb. (weighed with skin & seeds)	1.4	143
Flesh only	4 oz.	.6	58
MANDARIN ORANGE, Canned:			
Light syrup (Del Monte)	½ cup (4.5 oz.)	.6	77

(USDA): United States Department of Agriculture
*Prepared as Package Directs

Food and Description	Measure or Quantity	Protein (grams)	Calories
Low calorie, solids & liq. (Diet Delight)	½ cup (4.3 oz.)	.6	31
Low calorie, solids & liq. (S and W) *Nutradiet,* unsweetened	4 oz.	.5	31
MANDARIN ORANGE, fresh (see **TANGERINE**)			
MANGO, fresh (USDA)			
Whole	1 lb. (weighed with seeds & skin)	2.1	201
Whole	1 med. (7.1 oz.)	.9	88
Fresh only, diced or sliced	½ cup (2.9 oz.)	.6	54
MANGO DRINK:			
(Yoo-Hoo) High Protein	6 fl. oz. (6.4 oz.)	6.0	100
& pineapple (Alegre)	6 fl. oz. (6.6 oz.)	Tr.	140
MANICOTTI, frozen:			
Without sauce (Buitoni)	4 oz.	12.0	218
With sauce (Buitoni)	4 oz.	8.3	150
With sauce, dinner (Celeste)	2 manicotti (13 oz.)	22.4	428
MAPLE RENNET MIX:			
Powder:			
Dry (Junket)	1 oz.	.2	117
*(Junket)	4 oz.	3.7	109
Tablet:			
Dry (Junket)	1 tablet (< 1 gram)	Tr.	1
*& sugar (Junket)	4 oz.	3.7	101
MARBLE CAKE MIX:			
Dry (USDA)	1 oz.	1.4	120
*Prepared with eggs, boiled white icing (USDA)	4 oz.	5.0	375
*Layer (Betty Crocker)	¹⁄₁₂ of cake	3.1	206
MARGARINE:			
(USDA)	1 lb.	2.7	3266
(USDA)	4 oz. (1 stick)	.7	816
(USDA)	1 cup or 1 tub (8 oz.)	1.4	1633
(USDA)	1 T. (⅛ of stick, .5 oz.)	< .1	101

(USDA): United States Department of Agriculture
*Prepared as Package Directs

Food and Description	Measure or Quantity	Protein (grams)	Calories
(USDA)	1 pat (1″ × ⅓″ × 1″, 5 grams)	Tr.	36
MARGARINE, IMITATION, diet:			
(Fleischmann's)	1 T. (.5 oz.)	0.	50
(Imperial)	1 T. (.5 oz.)	0.	49
(Mazola)	1 cup (8 oz.)	0.	805
(Mazola)	1 T. (.5 oz.)	0.	51
(Parkay)	1 T. (.5 oz.)	0.	55
MARGARINE, WHIPPED:			
(USDA)	1 stick or ½ cup (2.7 oz.)	.5	547
(Blue Bonnet)	1 T. (9 grams)	Tr.	67
(Imperial)	1 T. (10 grams)	<.1	64
(Miracle) cottonseed-soybean	1 T. (9 grams)	<.1	67
(Parkay) cup	1 T. (9 grams)	<.1	67
MARINADE MIX:			
(Adolph's) chicken	1-oz. pkg.	.6	66
(Adolph's) meat	.8-oz. pkg.	.2	38
*(Durkee) meat	6 T. (.9-oz. pkg.)	.2	69
(Lawry's) beef	1.6-oz. pkg.	1.8	69
(Lawry's) lemon pepper	2.7-oz. pkg.	2.7	159
MARMALADE: Sweetened:			
(USDA)	1 T. (.7 oz.)	.1	51
(Bama)	1 T. (.7 oz.)	.1	54
(Kraft)	1 oz.	.1	78
(Smucker's)	1 T. (.7 oz.)	<.1	53
Low calorie:			
(Kraft)	1 oz.	<.1	35
(S and W) *Nutradiet*	1 T. (.5 oz.)	<.1	11
(Slenderella)	1 T. (.6 oz.)	Tr.	22
MARMALADE PLUM (see SAPOTES)			
MASA HARINA (Quaker)	2 tortillas (6″ dia.)	3.2	139
MASA TRIGO (Quaker)	2 tortillas (6″ dia.)	3.0	150

(USDA): United States Department of Agriculture
*Prepared as Package Directs

Food and Description	Measure or Quantity	Protein (grams)	Calories
MATZO:			
Regular (Manischewitz)	1 matzo (1.1 oz.)	2.9	114
American (Manischewitz)	1 matzo (1 oz.)	3.1	121
Diet-10's (Goodman's)	1 small square (⅑ of matzo, 3 grams)	.3	12
Diet-10's (Goodman's)	1 matzo (1 oz.)	3.1	109
Diet-thins (Manischewitz)	1 matzo (1 oz.)	3.1	113
Egg (Manischewitz)	1 matzo (1.2 oz.)	4.0	133
Egg'n Onion (Manischewitz)	1 matzo (1 oz.)	3.0	116
Onion Tams (Manischewitz)	1 piece (2 grams)	.2	13
Tam Tams (Manischewitz)	1 piece (3 grams)	.2	14
Tasteas (Manischewitz)	1 matzo (1 oz.)	2.5	119
Tea (Goodman's)	1 matzo (.6 oz.)	2.0	70
Tea (Goodman's) Midgetea	1 matzo (.4 oz.)	1.1	40
Thin tea (Manischewitz)	1 matzo (1 oz.)	2.9	114
Unsalted (Goodman's)	1 matzo (1 oz.)	3.1	109
Whole wheat (Manischewitz)	1 matzo (1.2 oz.)	5.2	124
MATZO MEAL (Manischewitz)	1 cup (4.1 oz.)	11.1	438
MAYONNAISE:			
(USDA)	1 cup (7.8 oz.)	2.4	1587
(USDA)	1 T. (.5 oz.)	.2	101
(Bama)	1 T. (.5 oz.)	.2	95
(Bennett's)	1 T. (.5 oz.)	.2	113
(Best Foods) *Real*	1 T. (.5 oz.)	.2	102
(Hellmann's) *Real*	1 T. (.5 oz.)	.2	102
(Kraft)	1 T. (.5 oz.)	.2	102
(Kraft) *Salad Bowl*	1 T. (.5 oz.)	.2	102
(Nalley's)	1 oz.	.4	214
Saffola	1 cup (7.2 oz.)	2.3	1451
Saffola	1 T. (.5 oz.)	.1	92
MAYPO, cereal, dry, any flavor:			
Instant	1 oz.	3.8	107
1-minute	1 oz.	3.8	105
MEAL (see **CORNMEAL** or **CRACKER MEAL** or **MATZO MEAL**)			

(USDA): United States Department of Agriculture

Food and Description	Measure or Quantity	Protein (grams)	Calories

MEAT EXTENDER (see **TEXTURED VEGETABLE PROTEIN**)

MEATBALL:

Food and Description	Measure or Quantity	Protein (grams)	Calories
In sauce, canned (Prince)	1 can (3.7 oz.)	8.9	171
Stew, canned (Chef Boy-Ar-Dee)	¼ of 30-oz. can	9.4	179
Stew, canned (Morton House)	1 cup (8 oz.)	13.8	295
With gravy, canned (Chef Boy-Ar-Dee)	¼ of 15¼-oz. can	7.2	118
With gravy & whipped potato, frozen (Swanson)	9¼-oz. pkg.	17.0	330

MEAT LOAF DINNER, frozen:

With tomato sauce, mashed potatoes & peas (USDA)	12 oz.	27.2	445
(Banquet)	11-oz. dinner	20.9	412
(Kraft)	5 oz.	19.1	332
(Morton)	11-oz. dinner	18.7	390
(Morton) 3-course	15½-oz. dinner	26.4	656
(Swanson)	10¾-oz. dinner	19.0	419
(Swanson) 3-course	16½-oz. dinner	28.6	544
In brown gravy (Morton House)	4⅙-oz. serving	8.9	178
In tomato sauce (Morton House)	4⅙-oz. serving	8.7	188

MEAT LOAF ENTREE:

Canned:			
In brown gravy (Morton House)	⅓ of 12½-oz. can	8.9	178
In tomato sauce (Morton House)	⅓ of 12½-oz. can	8.7	188
Frozen:			
(Banquet)	5-oz. bag	13.8	281
In tomato sauce (Swanson)	9-oz. pkg.	19.9	316

MEAT LOAF SEASONING MIX:

(Contadina)	3¾-oz. pkg.	14.7	363
(Lawry's)	3½-oz. pkg.	12.8	333

MEAT, POTTED:

(Armour Star)	3-oz. can	12.2	181

(USDA): United States Department of Agriculture

[205]

Food and Description	Measure or Quantity	Protein (grams)	Calories
(Hormel)	3-oz. can	10.8	158
(Van Camp)	½ cup (3.9 oz.)	19.2	272
MEAT TENDERIZER:			
Unseasoned (Adolph's)	1 tsp. (5 grams)	Tr.	2
Unseasoned (French's)	1 tsp. (5 grams)	.2	2
Seasoned (Adolph's)	1 tsp. (5 grams)	Tr.	2
Seasoned (French's)	1 tsp. (5 grams)	.2	2
MELBA TOAST:			
Garlic (Keebler)	1 piece (2 grams)	.3	9
Garlic, round (Old London)	1 piece (2 grams)	.3	10
Onion (Keebler)	1 piece (2 grams)	.3	9
Onion, round (Old London)	1 piece (2 grams)	.3	10
Plain (Keebler)	1 piece (2 grams)	.3	9
Pumpernickel (Old London)	1 piece (5 grams)	.7	17
Rye (Old London)	1 piece (4 grams)	.6	14
Rye, unsalted (Old London)	1 piece (5 grams)	.7	18
Sesame (Keebler)	1 piece (2 grams)	.3	11
Sesame, round (Old London)	1 piece (2 grams)	.4	11
Wheat (Old London)	1 piece (4 grams)	.6	15
Wheat, unsalted (Old London)	1 piece (5 grams)	.7	18
White:			
(Keebler)	1 piece (4 grams)	.6	16
Round (Old London)	1 piece (2 grams)	.3	10
Slice (Old London)	1 piece (5 grams)	.7	17
Slice, unsalted (Old London)	1 piece (4 grams)	.7	18
MELLORINE (Sealtest)	¼ pt. (2.3 oz.)	2.3	132
MELON (see individual listings, e.g., **CANTALOUPE, WATERMELON**)			
MELON BALL, in syrup, frozen, (USDA)	½ cup (4.1 oz.)	.7	72
MENHADEN, Atlantic, canned, solids & liq. (USDA)	4 oz.	21.2	195
MEXICAN DINNER:			
Frozen:			
(Banquet) combination	12-oz. dinner	22.1	571

(USDA): United States Department of Agriculture

Food and Description	Measure or Quantity	Protein (grams)	Calories
(Banquet) Mexican style	16-oz. dinner	21.3	608
(Morton)	14-oz. dinner	19.8	409
(Swanson)	16¼-oz. dinner	26.7	658
(Swanson) 3-course	18-oz. dinner	21.9	613
Mix (Hunt's) *Skillet*	1-lb. 2-oz. pkg.	19.5	699
MEXICAN-STYLE VEGETABLES, frozen (Bird's Eye)	⅓ of 10-oz. pkg.	4.4	144
MILK, CONDENSED, sweetened, canned:			
(USDA)	1 cup (10.8 oz.)	24.8	982
Dime Brand	1 fl. oz. (1.3 oz.)	3.0	125
Eagle Brand	1 T. (.7 oz.)	1.4	60
Magnolia Brand	1 fl. oz. (1.3 oz.)	2.9	125
MILK, DRY:			
Whole:			
(USDA) packed	1 cup (5.1 oz.)	38.3	728
(USDA) spooned	1 cup (4.3 oz.)	31.9	607
Nonfat, instant:			
⅞ cup makes 1 qt. (USDA)	⅞ cup (3.2 oz.)	32.7	330
1⅓ cups make 1 qt. (USDA)	1⅓ cups (3.2 oz.)	32.3	327
(Carnation)	1 cup (2.4 oz.)	22.4	244
*(Carnation)	1 cup (8.6 oz.)	7.5	81
*(Sanalac)	1 cup	8.0	80
(Weight Watchers)	1 packet (3 grams)	1.0	10
Chocolate (Carnation)	1 cup (2.4 oz.)	9.2	260
*Chocolate (Carnation)	1 cup (8.6 oz.)	4.6	129
MILK, EVAPORATED, canned:			
Regular:			
Unsweetened (USDA)	1 cup (8.9 oz.)	17.6	345
(Borden)	14.5-oz. can	28.8	563
(Carnation)	1 cup (8.9 oz.)	17.6	348
Skimmed:			
(Carnation)	1 cup (9 oz.)	19.1	192
Sunshine (Defiance Milk)	1 cup (8.9 oz.)	19.7	200
MILK, FRESH:			
Whole:			
3.5% fat (USDA)	1 cup (8.6 oz.)	8.5	159

(USDA): United States Department of Agriculture
*Prepared as Package Directs

Food and Description	Measure or Quantity	Protein (grams)	Calories
3.7% fat (USDA)	1 cup (8.5 oz.)	8.4	159
3.25% fat, homogenized			
(Borden)	1 cup (8.6 oz.)	8.4	152
3.25% fat (Sealtest)	1 cup (8.6 oz.)	7.6	144
3.5% fat (Dean)	1 cup (8.6 oz.)	7.5	151
3.5% fat (Sealtest)	1 cup (8.6 oz.)	7.7	151
3.7% fat (Sealtest)	1 cup (8.6 oz.)	7.8	157
Multivitamin (Sealtest)	1 cup (8.6 oz.)	7.7	151
Skim:			
(USDA)	1 cup (8.6 oz.)	8.8	88
2% fat with 1-2% nonfat			
milk solids added (USDA)	1 cup (8.7 oz.)	10.3	145
(Borden)	1 cup (8.6 oz.)	9.7	99
(Dean) .5% fat	1 cup (8.2 oz.)	8.3	91
(Dean) 1% fat	1 cup (8.7 oz.)	8.4	103
(Dean) 2% fat	1 cup (8.7 oz.)	8.8	133
(Sealtest)	1 cup (8.6 oz.)	7.9	79
Diet (Sealtest)	1 cup (8.6 oz.)	9.7	103
Light n' Lively, lowfat			
(Sealtest)	1 cup (8.6 oz.)	9.5	114
Lite-line, fortified, low fat			
(Borden)	1 cup (8.6 oz.)	10.1	119
Pro-Line, 2% fat (Borden)	1 cup (8.6 oz.)	10.1	140
Skim-Line, fortified (Borden)	1 cup (8.6 oz.)	9.7	99
Vita Lure, 2% fat (Sealtest)	1 cup (8.6 oz.)	9.5	137
Buttermilk, cultured, fresh:			
(USDA)	1 cup (8.6 oz.)	8.8	88
0.1% fat (Borden)	1 cup (8.6 oz.)	8.8	88
1.0% fat (Borden)	1 cup (8.6 oz.)	8.8	107
3.5% fat (Borden)	1 cup (8.6 oz.)	8.5	159
(Dean)	1 cup (8.6 oz.)	8.3	95
Golden Nugget (Sealtest)	1 cup (8.6 oz.)	8.5	92
Light n' Lively (Sealtest)	1 cup (8.6 oz.)	8.8	95
Lowfat (Sealtest)	1 cup (8.6 oz.)	8.1	114
Skimmilk (Sealtest)	1 cup (8.6 oz.)	8.1	71
Buttermilk, cultured, dried			
(USDA)	1 cup (4.2 oz.)	41.2	464
Chocolate milk drink, fresh:			
With whole milk:			
(USDA)	1 cup (8.8 oz.)	8.5	212
3.4% fat (Sealtest)	1 cup (8.6 oz.)	7.5	207
3.5% fat (Dean)	1 cup (8.8 oz.)	8.2	212

(USDA): United States Department of Agriculture

Food and Description	Measure or Quantity	Protein (grams)	Calories
With skim milk, 2% added butterfat (USDA)	1 cup (8.8 oz.)	8.2	190
With skim milk:			
0.5% fat (Sealtest)	1 cup (8.6 oz.)	7.7	146
1% fat (Dean)	1 cup (8.9 oz.)	7.8	166
1% fat (Sealtest)	1 cup (8.6 oz.)	7.7	158
2% fat (Sealtest)	1 cup (8.6 oz.)	7.6	178
MILK, GOAT, whole (USDA)	1 cup (8.6 oz.)	7.8	163
MILK, HUMAN (USDA)	1 oz. (by wt.)	.3	22
MILK, REINDEER (USDA)	1 oz. (by wt.)	3.1	66
MILK SHAKE (McDonald's):			
Chocolate	1 serving (9.5 oz.)	11.3	318
Strawberry	1 serving (9.4 oz.)	9.9	313
Vanilla	1 serving (9.7 oz.)	11.2	324
MILLET, whole-grain (USDA)	1 lb.	44.9	1483
MINCEMEAT:			
(Wilderness)	22-oz. can	4.8	1291
Condensed (None Such)	9-oz. pkg.	4.1	950
Ready-to-use (None Such)	½ cup (5.3 oz.)	1.2	345
With brandy & rum (None Such)	½ cup (5.2 oz.)	1.0	338
MINCE PIE:			
Home recipe, 2 crust (USDA)	⅙ of 9″ pie (5.6 oz.)	4.0	428
(Tastykake)	4-oz. pie	3.7	373
Frozen:			
(Banquet)	5-oz. serving	4.8	401
(Morton)	⅙ of 24-oz. pie	2.7	297
(Mrs. Smith's)	⅙ of 8″ pie (4.2 oz.)	2.0	339
(Mrs. Smith's)	⅛ of 10″ pie (5.6 oz.)	2.8	450
MINESTRONE SOUP:			
Canned:			
Condensed (USDA)	8 oz. (by wt.)	9.1	197
*Prepared with equal volume water (USDA)	1 cup (8.6 oz.)	4.9	105
*(Campbell)	1 cup	3.8	82

(USDA): United States Department of Agriculture
*Prepared as Package Directs

[209]

Food and Description	Measure or Quantity	Protein (grams)	Calories
MINT MIST PIE, frozen (Kraft)	¼ of 13-oz. pie	4.1	325
MISO, cereal & soybeans (USDA)	4 oz.	11.9	194
*MOCHA NUT PUDDING MIX, instant (Royal)	½ cup (5.1 oz.)	5.2	199
MOCHA PIE, frozen (Kraft)	3 oz.	5.4	336
MOR (Wilson) canned luncheon meat	3 oz.	12.8	266
MORTADELLA, sausage (USDA)	1 oz.	5.8	89
MUFFIN:			
Blueberry, home recipe (USDA)	3″ muffin (1.4 oz.)	2.9	112
Blueberry, frozen (Morton)	1.6-oz. muffin	2.4	116
Blueberry, frozen (Mrs. Smith's)	.9-oz. muffin	2.0	116
Bran:			
Home recipe (USDA)	3″ muffin (1.4 oz.)	3.1	104
(Thomas') with raisins	1.9-oz. muffin	3.4	170
Corn:			
Home recipe, prepared with whole-ground cornmeal (USDA)	2⅜″ muffin (1.4 oz.)	2.9	115
Home recipe, prepared with degermed cornmeal (USDA)	2⅜″ muffin (1.4 oz.)	2.8	126
(Morton) frozen	1.7-oz. muffin	2.8	132
(Mrs. Smith's) frozen	1-oz. muffin	3.0	157
(Thomas')	2-oz. muffin	4.3	194
English:			
(Arnold)	2.2-oz. muffin	5.6	145
(Newly Weds)	2.5-oz. muffin	7.2	167
(Thomas')	2.1-oz. muffin	5.0	140
(Wonder)	2-oz. muffin	4.4	133
Golden Egg Toasting (Arnold)	2.2-oz. muffin	5.5	162
Plain, home recipe (USDA)	3″ muffin (1.4 oz.)	3.1	118
Scone, *Raisin Round* (Wonder)	2-oz. piece	4.4	147
Sour dough (Wonder)	2-oz. muffin	4.5	133
Wheat berry (Wonder)	2-oz. muffin	4.5	136

(USDA): United States Department of Agriculture
*Prepared as Package Directs

Food and Description	Measure or Quantity	Protein (grams)	Calories
MUFFIN MIX:			
*Apple cinnamon (Betty Crocker)	2¾" muffin	2.2	159
*Banana nut (Betty Crocker)	2¾" muffin	2.7	167
*Blueberry (Betty Crocker)	2¾" muffin	2.0	118
*Blueberry (Duncan Hines)	1 muffin (1.2 oz.)	1.4	89
*Butter pecan (Betty Crocker)	2¾" muffin	2.2	159
Corn:			
With enriched flour (USDA)	1 oz.	1.8	118
*Prepared with egg & milk (USDA)	2⅜" muffin (1.4 oz.)	2.0	92
With cake flour & nonfat dry milk (USDA)	1 oz.	1.8	116
*Prepared with egg & water (USDA)	2⅜" muffin (1.4 oz.)	1.8	119
(Albers)	1 oz.	1.4	118
*(Betty Crocker)	2¾" muffin	3.4	156
*(Dromedary)	2½" muffin (1.4 oz.)	2.4	144
*(Flako)	1 med. muffin (¹⁄₁₂ of pkg.)	2.7	133
*Date nut (Betty Crocker)	2¾" muffin	2.5	152
*Honey bran (Betty Crocker)	2¾" muffin	2.7	154
*Lemon (Betty Crocker)	1 muffin	2.1	149
*Oatmeal (Betty Crocker)	2¾" muffin	3.1	166
*Orange (Betty Crocker)	1 muffin	2.3	153
*Spice (Betty Crocker)	1 muffin	2.1	151
MULLET, raw (USDA):			
Whole	1 lb. (weighed whole)	47.1	351
Meat only	4 oz.	22.2	166
MUNG BEAN SPROUT (see **BEAN SPROUT**)			
MUSHROOM:			
Raw (USDA):			
Whole	½ lb. (weighed untrimmed)	6.0	62
Trimmed, slices	½ cup (1.2 oz.)	.9	10

(USDA): United States Department of Agriculture
*Prepared as Package Directs

. [211]

Food and Description	Measure or Quantity	Protein (grams)	Calories
Trimmed, slices (Shady Oak)	½ cup (1.3 oz.)	.9	10
Canned, solids & liq.:			
(USDA)	½ cup (4.3 oz.)	2.3	21
Sliced, chopped or whole,			
broiled in butter (B in B)	6-oz. can	3.2	50
Whole or sliced (Green			
Giant)	4-oz. can	1.7	24
Solids & liq. (Shady Oak)	4-oz. can	2.0	19
Dried (HEW/FAO)	1 oz.	7.2	72
Frozen, whole, in butter sauce,			
Sauté	6-oz. pkg.	2.7	90
MUSHROOM, CHINESE:			
Dried (HEW/FAO)	1 oz.	2.8	81
Dried, soaked, drained			
(HEW/FAO)	1 oz.	.6	12
MUSHROOM SOUP, canned:			
*Barley (Manischewitz)	8 fl. oz.	1.7	72
Cream of:			
Condensed (USDA)	8 oz. (by wt.)	4.3	252
Prepared with equal volume			
water (USDA)	1 cup (8.5 oz.)	2.4	134
Prepared with equal volume			
milk (USDA)	1 cup (8.6 oz.)	6.9	216
*(Campbell)	1 cup	1.9	131
*(Heinz)	1 cup (8.5 oz.)	2.4	124
(Heinz) *Great American*	1 cup (8¾ oz.)	4.5	131
*Dietetic (Claybourne)	8 oz.	2.0	79
*Dietetic (Slim-ette)	8 oz. (by wt.)	1.2	60
*Golden (Campbell)	1 cup	3.3	80
Low sodium (Campbell)	7¼-oz. can	1.3	124
MUSHROOM SOUP MIX:			
(Wyler's)	1 oz.	2.9	93
*Beef flavor (Lipton)	1 cup	1.8	39
Cream of (Lipton) *Cup-a-Soup*	1 pkg. (.7 oz.)	1.8	86
Cream of (Wyler's)	1 pkg. (.7 oz.)	1.8	78
MUSKELLUNGE, raw (USDA):			
Whole	1 lb. (weighed whole)	44.9	242
Meat only	4 oz.	22.9	124

(HEW/FAO): Health, Education, and Welfare/
Food and Agricultural Organization
(USDA): United States Department of Agriculture
*Prepared as Package Directs

Food and Description	Measure or Quantity	Protein (grams)	Calories
MUSKMELON (see **CANTALOUPE, CASABA** or **HONEYDEW**)			
MUSKRAT, roasted (USDA)	4 oz.	30.8	174
MUSSEL (USDA):			
Raw, Atlantic & Pacific, with liq., in shell	1 lb. (weighed in shell)	27.2	153
Raw, Atlantic & Pacific, meat only	4 oz.	16.3	108
Canned, Pacific, drained	4 oz.	20.6	129
MUSTARD, prepared:			
Brown:			
(USDA)	1 tsp. (9 grams)	.5	8
Spicy (French's)	1 tsp.	.3	6
(Heinz)	1 tsp.	.4	8
German style (Kraft)	1 oz.	1.8	30
Grey Poupon	1 tsp. (5 grams)	.3	5
Horseradish (Best Foods)	1 tsp. (6 grams)	.3	4
Horseradish (French's)	1 tsp.	.2	6
Horseradish (Kraft)	1 oz.	1.8	29
Medford (French's)	1 tsp.	.2	5
Onion (French's)	1 tsp.	.2	7
Ring Star (French's)	1 tsp.	.2	4
Salad:			
(French's)	1 tsp.	.2	4
Glass or squeeze pack (Kraft)	1 oz.	1.5	23
Plastic squeeze bottle (Kraft)	1 oz.	1.5	23
Yellow:			
(USDA)	1 tsp. (9 grams)	.4	7
(Heinz)	1 tsp.	.3	5
MUSTARD GREEN:			
Raw, whole (USDA)	1 lb. (weighed untrimmed)	9.5	98
Boiled, drained (USDA)	1 cup (7.8 oz.)	4.9	51
Frozen:			
Not thawed (USDA)	4 oz.	2.6	23
Boiled, drained (USDA)	½ cup (3.8 oz.)	2.4	21
Chopped (Bird's Eye)	½ cup (3.3 oz.)	2.2	19

(USDA): United States Department of Agriculture

[213]

Food and Description	Measure or Quantity	Protein (grams)	Calories
MUSTARD SPINACH (USDA):			
Raw	1 lb.	10.0	100
Boiled, drained	4 oz.	1.9	18

N

NATTO, fermented soybean (USDA)	4 oz.	19.2	189
NATURAL CEREAL:			
100% (Quaker)	¼ cup (1 oz.)	3.4	140
100%, with fruit (Quaker)	¼ cup (1 oz.)	2.8	136
NEAR BEER (see **BEER, NEAR**)			
NEAPOLITAN CREAM PIE, frozen:			
(Banquet)	2½-oz. serving	2.0	188
(Morton)	⅙ of 16-oz. pie	1.7	198
(Mrs. Smith's)	⅙ of 8″ pie (2.8 oz.)	1.3	244
NECTARINE, fresh (USDA):			
Whole	1 lb. (weighed with pits)	2.5	267
Flesh only	4 oz.	.7	73
NEW ZEALAND SPINACH (USDA):			
Raw	1 lb.	10.0	86
Boiled, drained	4 oz.	1.9	15
NOODLE. Plain noodle products are essentially the same in protein value on the same weight basis. The longer they are cooked, the more water is absorbed and this affects the nutritive values (USDA):			
Dry, 1½″ strips	1 cup (2.6 oz.)	9.3	283
Dry, 1½″ strips	1 oz.	3.6	110
Cooked	1 cup (5.6 oz.)	6.6	200
Cooked	1 oz.	1.2	35

(USDA): United States Department of Agriculture

Food and Description	Measure or Quantity	Protein (grams)	Calories
NOODLE & BEEF:			
Canned (Heinz)	8½-oz. can	10.1	171
Canned (Nalley's)	8 oz.	9.9	152
Frozen (Banquet) buffet	2-lb. pkg.	50.0	735
NOODLE, CHOW MEIN,			
canned:			
(USDA)	1 cup (1.6 oz.)	5.9	220
(Hung's)	1 oz.	2.6	148
NOODLE DINNER:			
Cantong dinner mix (Betty			
Crocker)	1 cup	19.1	403
*Romanoff, mix (Kraft)	8 oz.	13.6	431
*Stroganoff dinner mix (Betty			
Crocker)	1 cup	24.2	500
*With cheese, mix (Kraft)	8 oz.	11.6	408
*With chicken, mix (Kraft)	8 oz.	10.2	245
With chicken, frozen (Swanson)	11-oz. dinner	12.8	370
NOODLE MIX:			
*Almondine (Betty Crocker)	½ cup	5.7	213
*Almondine, *Noodle-Roni*	4 oz.	5.0	143
*Au gratin, *Noodle-Roni*	4 oz.	5.3	129
Egg Noodles Plus (Pennsylvania			
Dutch Brand):			
Beef sauce	½ cup	4.7	137
Butter sauce	½ cup	4.4	151
Cheese sauce	½ cup	5.4	147
Chicken sauce	½ cup	4.6	143
Mushroom sauce	½ cup	5.0	137
Onion sauce	½ cup	4.7	140
*Italiano (Betty Crocker)	½ cup	6.9	207
*Parmesano, *Noodle Roni*	⅕ of 6-oz. pkg.	5.0	130
*Romanoff (Betty Crocker)	½ cup	7.1	241
*Romanoff, *Noodle Roni*	4 oz.	6.1	179
NOODLE SOUP:			
Beef (see **BEEF SOUP**)			
Chicken (see **CHICKEN SOUP**)			
*Canned, with ground beef			
(Campbell)	1 cup	4.1	91

(USDA): United States Department of Agriculture
*Prepared as Package Directs

[215]

Food and Description	Measure or Quantity	Protein (grams)	Calories
NUT, mixed (see individual kinds):			
Dry roasted:			
(Flavor House)	1 oz.	6.0	172
(Planters)	1 oz.	5.6	176
(Skippy)	1 oz.	6.3	168
Oil roasted:			
With peanuts (Planters)	1 oz.	5.6	176
Without peanuts (Planters)	1 oz.	4.8	178
NUT LOAF (see **BREAD, CANNED**)			
NUTRIMATO (Mott's)	4 oz.	.8	55

O

Food and Description	Measure or Quantity	Protein (grams)	Calories
OAT FLAKES, cereal (Post)	⅔ cup (1 oz.)	5.1	107
OATMEAL:			
Instant:			
(H-O)	1 cup (2.3 oz.)	11.3	258
(H-O)	1 T. (4 grams)	.7	16
(Quaker)	1-oz. packet (¾ cup cooked)	3.8	107
(Ralston)	4 T. (1 oz.)	4.1	103
Sweet & Mellow (H-O)	(1.4-oz.) packet	4.8	149
*With apple & cinnamon (Quaker)	1⅛-oz. packet (¾ cup cooked)	2.9	119
With dates & caramel (H-O)	(1.4-oz.) packet	4.2	147
With maple & brown sugar (Quaker)	1⅝-oz. packet (¾ cup cooked)	3.9	177
With raisins & spice (H-O)	(1.6-oz.) packet	4.7	167
With raisins & spice (Quaker)	1½-oz. packet (¾ cup cooked)	3.6	154
Quick:			
Dry:			
(H-O)	1 cup (2.3 oz.)	11.7	265
(H-O)	1 T. (4 grams)	.7	17
(Ralston)	⅓ cup (1 oz.)	4.0	113

*Prepared as Package Directs

Food and Description	Measure or Quantity	Protein (grams)	Calories
Cooked:			
*(Albers)	1 cup	5.4	148
*(Quaker)	⅔ cup (1 oz. dry)	4.1	107
Regular:			
Dry:			
(USDA)	1 cup (2.5 oz.)	10.2	281
(USDA)	1 T. (4 grams)	.6	18
(H-O) old fashioned	1 cup (2.3 oz.)	11.5	280
(H-O) old fashioned	1 T. (4 grams)	.7	18
(Ralston)	⅓ cup (1 oz.)	4.1	128
Cooked:			
*(USDA)	1 cup (8.5 oz.)	4.8	132
*(Albers) old fashioned	1 cup	5.4	148
*(Quaker) old fashioned	⅔ cup (1 oz. dry)	4.1	107
OCEAN PERCH:			
Atlantic:			
Raw, whole (USDA)	1 lb. (weighed whole)	25.3	124
Fried, dipped in egg, milk & bread crumbs (USDA)	4 oz.	21.5	257
Frozen, breaded, fried, reheated (USDA)	4 oz.	21.4	362
Pacific, raw:			
Whole (USDA)	1 lb. (weighed whole)	23.3	116
Meat only (USDA)	4 oz.	21.5	108
Frozen (Groton)	⅓ of 1-lb. pkg.	27.0	133
OCEAN PERCH MEALS,			
frozen:			
(Banquet)	8¾-oz. dinner	19.1	434
(Weight Watchers)	18-oz. dinner	44.4	307
& broccoli (Weight Watchers)	9½-oz. luncheon	20.3	185
OCTOPUS, raw, meat only (USDA)	4 oz.	17.4	83
OIL, salad or cooking (USDA)	1 T. (5 oz.)	0.	124
OKRA:			
Raw, whole (USDA)	1 lb. (weighed untrimmed)	9.4	140

(USDA): United States Department of Agriculture
*Prepared as Package Directs

OKRA (Continued)

Food and Description	Measure or Quantity	Protein (grams)	Calories
Boiled, drained (USDA):			
Whole	½ cup (3.1 oz.)	1.8	26
Pods	8 pods (3″ × ⅝″, 3 oz.)	1.7	25
Slices	½ cup (2.8 oz.)	1.6	23
Frozen:			
Cut & pods, not thawed (USDA)	4 oz.	2.6	44
Cut, boiled, drained (USDA)	½ cup (3.2 oz.)	2.0	35
Whole, boiled, drained (USDA)	½ cup (2.4 oz.)	1.5	26
Cut (Birds Eye)	½ cup (3.3 oz.)	2.5	36
Whole (Birds Eye)	½ cup (2.5 oz.)	1.9	27

OLEOMARGARINE (see **MARGARINE**)

OLIVE:

Greek style, salt-cured, oil-coated:			
Pitted (USDA)	1 oz.	.6	96
With pits, drained (USDA)	4 oz.	2.0	308
Green, pitted & drained:			
(USDA)	1 oz.	.4	33
(USDA)	4 med. or 3 extra large or 2 giant (.6 oz.)	.2	19
(USDA)	1 olive (¹³⁄₁₆″ × 1¹⁄₁₆″, 6 grams)	<.1	6
Ripe, by variety, pitted & drained:			
Ascalano, any size (USDA)	1 oz.	.3	37
Manzanilla, any size (USDA)	1 oz.	.3	37
Mission, any size (USDA)	1 oz.	.3	52
Mission (USDA)	3 small or 2 large (.4 oz.)	.1	18
Mission, slices (USDA)	½ cup (2.2 oz.)	.7	114
Sevillano, any size (USDA)	1 oz.	.3	26
*1·2·3, dessert (Jell-O)	⅔ cup (5.3 oz.)	1.5	135

(USDA): United States Department of Agriculture
*Prepared as Package Directs

Food and Description	Measure or Quantity	Protein (grams)	Calories
ONION:			
Raw (USDA):			
Whole	1 lb. (weighed untrimmed)	6.2	157
Whole	2½″ onion (3.9 oz.)	1.5	38
Chopped	½ cup (3 oz.)	1.3	33
Chopped	1 T. (.4 oz.)	.1	4
Grated	1 T. (.5 oz.)	.1	5
Ground	1 T. (.5 oz.)	.2	6
Slices	½ cup (2 oz.)	.8	21
Boiled (USDA):			
Whole	½ cup (3.7 oz.)	1.2	30
Halves or pieces	½ cup (3.2 oz.)	1.1	26
Pearl onions	½ cup (3.2 oz.)	1.1	27
Canned, boiled, solids & liq. (Durkee) O & C	¼ of 16-oz. jar	1.2	32
Canned, in cream sauce (Durkee) O & C	¼ of 15½-oz. can	2.0	88
Dehydrated:			
Flakes (USDA)	1 cup (2.3 oz.)	5.6	224
Flakes (USDA)	1 tsp. (1 gram)	.1	5
Flakes (Gilroy)	1 tsp. (2 grams)	.1	6
Powder (Gilroy)	1 tsp. (2 grams)	.2	9
Frozen:			
Chopped (Birds Eye)	¼ cup (1 oz.)	.4	11
Small, whole (Birds Eye)	½ cup (4 oz.)	1.3	51
Small, with cream sauce (Birds Eye)	⅓ of 9-oz. pkg.	2.3	132
Small, with cream sauce (Green Giant)	⅓ of 10-oz. pkg.	2.1	44
French-fried rings:			
Canned (Durkee) O & C	3-oz. can	5.1	534
Frozen (Birds Eye)	2 oz.	2.6	166
Frozen, batter-fried (Mrs. Paul's)	9-oz. pkg.	12.6	567
Frozen, breaded (Mrs. Paul's)	9-oz. pkg.	12.1	762
ONION BOUILLON:			
(Herb-Ox)	1 cube (4 grams)	.6	10
(Herb-Ox) instant	1 packet (5 grams)	.9	15
(Steero)	1 cube (4 grams)	.1	7

(USDA): United States Department of Agriculture

[219]

Food and Description	Measure or Quantity	Protein (grams)	Calories
(Wyler's)	1 cube (4 grams)	.6	10
ONION, GREEN, raw (USDA):			
Whole	1 lb. (weighed untrimmed)	6.5	157
Bulb & entire top	1 oz.	.4	10
Bulb & white portion of top	3 small onions (.9 oz.)	.3	11
Slices, bulb & white portion of top	½ cup (1.8 oz.)	.6	22
Tops only	1 oz.	.5	8
ONION SOUP:			
Condensed (USDA)	8 oz. (by wt.)	9.8	122
*Prepared with equal volume water (USDA)	1 cup (8.5 oz.)	5.3	65
*(Campbell)	1 cup	3.4	41
(Hormel)	15-oz. can	14.0	144
ONION SOUP MIX:			
Dry (USDA)	1½-oz. pkg.	6.0	151
*Prepared (USDA)	1 cup (8.1 oz.)	1.4	34
*(Lipton)	1 cup	1.2	34
(Lipton) *Cup-a-Soup*	1 pkg. (.4 oz.)	1.1	30
*(Wyler's)	1 cup	1.1	37
ONION, WELSH, raw (USDA):			
Whole	1 lb. (weighed untrimmed)	5.6	100
Trimmed	4 oz.	2.2	39
OPOSSUM, roasted, meat only (USDA)	4 oz.	34.2	251
ORANGE, fresh:			
All varieties:			
Whole (USDA)	1 lb. (weighed with rind & seeds)	3.3	162
Whole (USDA)	small orange (2½" dia., 5.3 oz.)	1.1	54
Whole (USDA)	med. orange (3" dia., 5.5 oz.)	1.6	77

(USDA): United States Department of Agriculture
*Prepared as Package Directs

Food and Description	Measure or Quantity	Protein (grams)	Calories
Whole (USDA)	large orange (3⅝" dia., 8.4 oz.)	2.4	116
Diced or sliced, drained (USDA)	1 cup (7.7 oz.)	2.2	107
Sections (USDA)	1 cup (8.5 oz.)	2.4	118
Sections, sweetened, chilled, bottled (Kraft)	4 oz.	.8	61
Sections, unsweetened, chilled, bottled (Kraft)	4 oz.	.6	35
California Navel:			
Whole (USDA)	1 lb. (weighed with rind & seeds)	4.0	157
Whole (USDA)	2⅘" orange (6.3 oz.)	1.6	62
Sections (USDA)	1 cup (8.5 oz.)	3.1	123
California Navel or Valencia:			
Unpeeled, wedge or slice (Sunkist)	⅙ orange (1.1 oz.)	Tr.	10
Peeled, cut bite-size (Sunkist)	½ cup (6.3-oz. orange)	1.0	62
California Valencia (USDA):			
Whole	1 lb. (weighed with rind & seeds)	4.1	174
Fruit including peel	2⅝" orange (6.3 oz.)	2.3	72
Sections	1 cup (8.5 oz.)	2.9	123
Florida, all varieties (USDA):			
Whole	1 lb. (weighed with rind & seeds)	2.3	158
Whole	3" orange (5.5 oz.)	1.1	73
Sections	1 cup (8.5 oz.)	1.7	113
ORANGEADE:			
Chilled (Sealtest)	½ cup (4.4 oz.)	.1	64
Frozen:			
*(Minute Maid)	½ cup (4.2 oz.)	.1	63
*(Snow Crop)	½ cup (4.2 oz.)	.1	63

(USDA): United States Department of Agriculture
*Prepared as Package Directs

Food and Description	Measure or Quantity	Protein (grams)	Calories
ORANGE-APRICOT JUICE DRINK, canned:			
(USDA) 40% fruit juices	1 cup (8.8 oz.)	.7	124
(Del Monte)	1 cup (8.6 oz.)	.5	118
ORANGE CAKE MIX:			
*Chiffon (Betty Crocker)	1/16 of cake	2.7	135
*(Betty Crocker) layer	1/12 of cake	2.8	201
*(Duncan Hines)	1/12 of cake (2.7 oz.)	3.0	201
ORANGE CREAM BAR			
(Sealtest)	2½-fl.-oz. bar	1.2	103
ORANGE DRINK:			
Chilled (Sealtest)	6 fl. oz. (6.5 oz.)	.2	87
Canned (Alegre)	6 fl. oz. (6.6 oz.)	.2	102
Canned (Del Monte)	6 fl. oz. (6.5 oz.)	.2	77
Canned (Hi-C)	6 fl. oz. (6.3 oz.)	.8	98
*Crystals (Wagner)	6 fl. oz.	0.	90
ORANGE-GRAPEFRUIT JUICE:			
Bottled, chilled (Kraft)	½ cup (4.5 oz.)	.8	60
Canned:			
Sweetened (USDA)	½ cup (4.4 oz.)	.6	63
Sweetened (Del Monte)	½ cup (4.3 oz.)	.8	54
Sweetened (Stokely-Van Camp)	½ cup (4.4 oz.)	.6	63
Unsweetened (USDA)	½ cup (4.3 oz.)	.7	53
Unsweetened (Stokely-Van Camp)	½ cup (4.4 oz.)	.8	54
Frozen, concentrate, unsweetened:			
(USDA)	6-fl.-oz. can (7.4 oz.)	4.4	330
*Diluted with 3 parts water (USDA)	½ cup (4.4 oz.)	.7	55
*Minute Maid)	½ cup (4.2 oz.)	.8	51
*(Snow Crop)	½ cup (4.2 oz.)	.8	51
ORANGE ICE (Sealtest)	¼ pt. (3.2 oz.)	Tr.	130

(USDA): United States Department of Agriculture
*Prepared as Package Directs

Food and Description	Measure or Quantity	Protein (grams)	Calories
ORANGE JUICE:			
Fresh:			
All varieties (USDA)	½ cup (4.4 oz.)	.9	56
California Navel (USDA)	½ cup (4.4 oz.)	1.2	60
California Valencia (USDA)	½ cup (4.4 oz.)	1.2	58
California Navel or Valencia			
(Sunkist)	½ cup (4.4 oz.)	1.0	52
Florida, early or mid-season			
(USDA)	½ cup (4.4 oz.)	.6	50
Florida Temple (USDA)	½ cup (4.4 oz.)	.6	67
Florida Valencia (USDA)	½ cup (4.4 oz.)	.7	56
Chilled (Kraft)	½ cup (4.4 oz.)	.8	60
Chilled (Minute Maid)	½ cup (4.4 oz.)	.8	55
Chilled (Sealtest)	½ cup (4.3 oz.)	1.1	64
Canned or bottled, sweetened:			
(USDA)	½ cup (4.4 oz.)	.9	66
(Del Monte)	½ cup (4.3 oz.)	.8	58
(Heinz)	5½-fl.-oz. can	.8	91
(Stokely-Van Camp)	½ cup (4.4 oz.)	.8	65
Canned or bottled,			
unsweetened:			
(USDA)	½ cup (4.4 oz.)	1.0	60
(Del Monte)	½ cup (4.3 oz.)	.8	46
(Heinz)	5½-fl.-oz. can	.9	71
*Reconstituted (Kraft)	½ cup (4.4 oz.)	.9	64
(Stokely-Van Camp)	½ cup (4.4 oz.)	1.0	60
Canned, concentrate,			
unsweetened:			
(USDA)	4 oz.	1.0	56
*Diluted with 5 parts water			
(USDA)	½ cup (4.4 oz.)	1.0	57
Dehydrated, crystals:			
(USDA)	4-oz. can	5.7	431
*Reconstituted (USDA)	½ cup (4.4 oz.)	.7	57
Frozen, concentrate: (USDA)	6-fl. oz. can (7.5 oz.)	4.9	337
*Diluted with 3 parts water			
(USDA)	½ cup (4.4 oz.)	.9	56
*(Birds Eye)	½ cup (4 oz.)	.9	51
*(Lake Hamilton)	½ cup (4.4 oz.)	1.0	58
*(Minute Maid)	½ cup (4.2 oz.)	.9	60
*(Nature's Best)	½ cup (4.4 oz.)	1.0	58
*(Snow Crop)	½ cup (4.2 oz.)	.9	60

(USDA): United States Department of Agriculture
*Prepared as Package Directs

Food and Description	Measure or Quantity	Protein (grams)	Calories
ORANGE, MANDARIN (see **TANGERINE**)			
ORANGE PEEL:			
Raw (USDA)	1 oz.	.4	
Candied (USDA)	1 oz.	.1	90
ORANGE-PINEAPPLE DRINK, canned (Hi-C)	6 fl. oz. (6.3 oz.)	.6	98
ORANGE-PINEAPPLE JUICE, bottled, chilled (Kraft)	½ cup (4.4 oz.)	.6	64
ORANGE-PINEAPPLE PIE, (Tastykake)	4-oz. pie	4.1	374
ORANGE PLUS (Bird's Eye)	½ cup (4.4 oz.)	.6	67
ORANGE PUDDING (Royal) *Creamerino*	5-oz. container	3.4	225
ORANGE RENNET MIX:			
Powder:			
Dry (Junket)	1 oz.	.2	116
*(Junket)	4 oz.	3.7	108
Tablet:			
Dry (Junket)	1 tablet (< 1 gram)	Tr.	1
*& sugar (Junket)	4 oz.	3.7	101
ORANGE SHERBET (see **SHERBET**)			
ORIENTAL DINNER, (Hunt's) *Skillet*	1-lb. 1-oz. pkg.	24.1	675
OVALTINE, dry:			
Chocolate	1 oz.	1.5	111
Malt	1 oz.	2.3	110

(USDA): United States Department of Agriculture
*Prepared as Package Directs

Food and Description	Measure or Quantity	Protein (grams)	Calories
OYSTER:			
Raw:			
Eastern, meat only:			
(USDA)	1 lb. (weighed with shell & liq.)	3.8	30
(USDA)	12 oysters (weighed in shell, 4 lbs.)	15.2	120
(USDA)	4 oz.	9.5	75
(USDA)	19–31 small or 13–19 med. oysters (1 cup, 8.5 oz.)	20.2	158
Pacific & Western, meat only:			
(USDA)	4 oz.	12.0	103
(USDA)	6–9 small or 4–6 med. oysters (1 cup, 8.5 oz.)	25.4	218
Canned, solids & liq. (USDA)	4 oz.	9.6	86
Canned, solids & liq. (Bumble Bee)	4 oz.	9.6	86
Fried, dipped in egg, milk & bread crumbs (USDA)	4 oz.	9.8	271
OYSTER CRACKER (see CRACKER))			
OYSTER STEW:			
Home recipe (USDA):			
1 part oysters to 1 part milk by volume	1 cup (6-8 oysters, 8.5 oz.)	16.6	245
1 part oysters to 2 parts milk by volume	1 cup (8.5 oz.)	12.5	233
1 part oysters to 3 parts milk by volume	1 cup (8.5 oz.)	11.8	206
*Canned (Campbell)	1 cup	6.0	142
Frozen (USDA):			
Condensed	8 oz. (by wt.)	10.4	231
*Prepared with equal volume water	1 cup (8.5 oz.)	5.5	122

(USDA): United States Department of Agriculture
*Prepared as Package Directs

Food and Description	Measure or Quantity	Protein (grams)	Calories
*Prepared with equal volume milk	1 cup (8.5 oz.)	10.1	202

P

PANCAKE:
Home recipe, wheat (USDA)	4″ pancake (1 oz.)	1.9	62
Frozen, breakfast, with link sausage (Swanson)	6-oz. breakfast	20.2	463

PANCAKE & WAFFLE MIX (see also **PANCAKE & WAFFLE MIX, DIETETIC**):
Buckwheat:			
(USDA)	1 oz.	3.0	93
(USDA)	1 cup (4.8 oz.)	14.2	443
*Prepared with egg & milk			
(USDA)	4″ pancake (1 oz.)	1.8	54
*(Aunt Jemima)	4″ pancake (1.2 oz.)	2.3	61
Buttermilk:			
(USDA)	1 oz.	2.4	101
(USDA)	1 cup (4.8 oz.)	11.6	481
*Prepared with milk (USDA)	4″ pancake (1 oz.)	1.6	55
*Prepared with milk & egg			
(USDA)	4″ pancake (1 oz.)	1.9	61
*(Aunt Jemima)	4″ pancake (1 oz.)	2.9	84
*(Duncan Hines)	4″ pancake (2 oz.)	4.0	109
*(Pillsbury) *Hungry Jack*	4″ pancake	2.3	80
*(Pillsbury) *Hungry Jack*, complete	4″ pancake	1.7	73
Flapjack (Albers)	1 cup	13.5	459
Plain:			
(USDA)	1 oz.	2.4	101
(USDA)	1 cup (4.8 oz.)	11.6	481
*Prepared with milk (USDA)	4″ pancake (1 oz.)	1.6	55
*Prepared with milk & egg (USDA)	4″ pancake (1 oz.)	1.9	61
*Prepared with milk & egg (USDA)	6″ × ½″ pancake (7 T. batter)	5.3	164
*(Aunt Jemima) Complete	4″ pancake (1.2 oz.)	1.6	60
*(Aunt Jemima) Easy Pour	4″ pancake (1.2 oz.)	2.6	78

(USDA): United States Department of Agriculture
*Prepared as Package Directs

Food and Description	Measure or Quantity	Protein (grams)	Calories
*(Aunt Jemima) Original	4″ pancake (1 oz.)	2.0	60
*(Pillsbury) extra light	4″ pancake	2.3	90
*(Pillsbury) *Hungry Jack,* complete	4″ pancake	1.7	73
*(Pillsbury) *Hungry Jack,* extra light	4″ pancake	2.3	73
*PANCAKE & WAFFLE MIX, DIETETIC, buttermilk or plain (Tillie Lewis)	4″ pancake	2.0	43
PANCREAS, raw (USDA):			
Beef, lean only	4 oz.	20.0	160
Beef, medium-fat	4 oz.	15.3	321
Calf	4 oz.	21.8	183
Hog or hog sweetbread	4 oz.	16.7	274
PAPAW, fresh (USDA):			
Whole	1 lb. (weighed with rind & seeds)	17.7	289
Flesh only	4 oz.	5.9	96
PAPAYA, fresh:			
Whole (USDA)	1 lb. (weighed with skin & seeds)	1.8	119
Flesh only (USDA)	4 oz.	.7	44
Cubed (USDA)	1 cup (6.4 oz.)	1.1	71
Juice, fresh (HEW/FAO)	4 oz. (by wt.)	Tr.	78
PARISIAN-STYLE VEGETABLES, frozen (Bird's Eye)	⅓ of 10-oz. pkg.	1.3	90
PARSLEY, fresh (USDA):			
Whole	½ lb.	8.2	100
Chopped	1 T. (4 grams)	.1	2
PARSNIP (USDA)			
Raw, whole	1 lb. (weighed unpared)	6.6	293
Boiled, drained, cut in pieces	½ cup (3.7 oz.)	1.6	70

(HEW/FAO): Health, Education and Welfare/ Food and Agricultural Organization
(USDA): United States Department of Agriculture
*Prepared as Package Directs

[227]

Food and Description	Measure or Quantity	Protein (grams)	Calories
PARV-A-ZERT (SugarLo)	⅓ pt. (3.5 oz.)	.4	131
PASSION FRUIT:			
Giant (HEW/FAO):			
Whole	1 lb. (weighed with seeds, aril and shell	1.8	53
Pulp	4 oz.	.8	23
Juice	4 oz. (by wt.)	1.0	50
Purple			
Whole (USDA)	1 lb. (weighed with shell)	5.2	212
Pulp and seeds (USDA)	4 oz.	2.5	102
Juice, fresh (HEW/FAO)	4 oz. (by wt.)	.7	48
Yellow (HEW/FAO)			
Juice, fresh	4 oz. (by wt.)	.7	83
Juice, canned, sweetened	4 oz. (by wt.)	.9	198
PASTINAS, dry (USDA):			
Carrot	1 oz.	3.4	105
Egg	1 oz.	3.7	109
Spinach	1 oz.	3.5	104
PASTRY SHELL (see also PIECRUST):			
Home recipe, baked (USDA)	1 shell (1.5 oz.)	2.6	212
(Stella D'oro)	1 shell (1 oz.)	2.0	143
Pot, bland (Stella D'oro)	1 shell (1.6 oz.)	3.1	205
Pot pie, bland (Keebler)	4″ shell (1.7 oz.)	3.6	236
Tart, sweet (Keebler)	3″ shell (1 oz.)	2.4	158
Frozen (Pepperidge Farm)	1 shell (1.8 oz.)	2.5	232
Frozen, tart (Pepperidge Farm)	1 pie tart (3 oz.)	2.3	276
PATE, canned:			
De foie gras (USDA)	1 oz.	3.2	131
De foie gras (USDA)	1 T. (.5 oz.)	1.7	69
Liver (Hormel)	1 oz.	4.2	78
Liver (Sell's)	1 T. (.5 oz.)	2.1	45
PDQ:			
Chocolate	1 T. (.6 oz.)	.7	64
Egg Nog	2 heaping tsps. (1 oz.)	.1	114
Strawberry	1 T. (.5 oz.)	Tr.	61

(USDA): United States Department of Agriculture

Food and Description	Measure or Quantity	Protein (grams)	Calories
PEA, GREEN:			
Raw (USDA):			
In pod	1 lb. (weighed in pod)	10.9	145
Shelled	1 lb.	28.6	381
Shelled	½ cup (2.4 oz.)	4.3	58
Boiled, drained (USDA)	½ cup (2.9 oz.)	4.4	58
Canned, regular pack:			
Drained solids (Del Monte)	½ cup (3 oz.)	2.9	43
Seasoned, drained solids (Del Monte)	½ cup (3 oz.)	4.1	52
Alaska, Early or June, solids & liq. (USDA)	½ cup (4.4 oz.)	4.3	82
Alaska, Early or June, drained solids (USDA)	½ cup (3 oz.)	4.0	76
Alaska, Early or June, drained liq. (USDA)	4 oz.	1.5	29
Alaska, drained solids (Butter Kernel)	½ cup (4.1 oz.)	4.0	69
Early, solids & liq., *April Showers*	½ of 8½-oz. can	3.6	65
Early, solids & liq., *Le Sueur*	½ of 8.5-oz. can	3.3	58
Early, solids & liq. (Stokely-Van Camp)	½ cup (4.1 oz.)	4.0	76
Early June, with onions (Green Giant)	¼ of 17-oz. can	3.6	67
Sweet, solids & liq. (USDA)	½ cup (4.4 oz.)	4.2	71
Sweet, drained solids (USDA)	½ cup (3 oz.)	4.0	69
Sweet, drained liq. (USDA)	4 oz.	1.5	25
Sweet, drained solids (Butter Kernel)	½ cup (4.1 oz.)	4.0	60
Sweet, solids & liq. (Green Giant)	½ of 8.5-oz. can	3.6	59
Sweet, honey pod, solids & liq. (Stokely-Van Camp)	½ cup (4 oz.)	3.9	65
Canned, dietetic pack:			
Alaska, Early or June, solids & liq. (USDA)	4 oz.	4.1	62
Alaska, Early or June, drained solids (USDA)	4 oz.	5.4	88
Alaska, Early or June, drained liq. (USDA)	4 oz.	1.6	25
Sweet, solids & liq. (USDA)	4 oz.	3.7	53

(USDA): United States Department of Agriculture

Food and Description	Measure or Quantity	Protein (grams)	Calories
Sweet, drained solids (USDA)	4 oz.	5.0	82
Sweet, drained liq. (USDA)	4 oz.	1.5	20
Solids & liq. (Diet Delight)	½ cup (4.4 oz.)	3.4	52
Sweet, solids & liq. (Blue Boy)	4 oz.	3.6	49
Sweet, solids & liq. (Tillie Lewis)	½ cup (4.4 oz.)	4.1	54
Unseasoned (S and W) *Nutradiet*	4 oz.	3.0	40
Frozen:			
Not thawed (USDA)	1 cup (2.5 oz.)	3.9	53
Boiled, drained (USDA)	½ cup (3 oz.)	4.3	57
Sweet (Birds Eye)	½ cup (3.3 oz.)	5.0	70
Tender tiny (Birds Eye)	½ cup (3.3 oz.)	5.1	70
In butter sauce (Green Giant)	⅓ of 10-oz. pkg.	3.3	78
In butter sauce, *Le Sueur*	⅓ of 10-oz. pkg.	3.8	86
With cream sauce (Birds Eye)	⅓ of 8-oz. pkg.	4.3	125
With cream sauce (Green Giant)	⅓ of 10-oz. pkg.	3.8	66
With sliced mushrooms (Birds Eye)	⅓ of 10-oz. pkg.	4.7	66

PEA, MATURE SEED, dry:

Raw:			
Whole (USDA)	1 lb.	109.3	1542
Whole (USDA)	1 cup (7.1 oz.)	48.2	680
Split, without seed coat (USDA)	1 lb.	109.8	1579
Split, without seed coat (USDA)	1 cup (7.2 oz.)	49.1	706
Cooked, split, without seed coat, drained (USDA)	½ cup (3.4 oz.)	7.8	112

PEA POD, edible-podded or Chinese (USDA):

Raw	1 lb. (weighed untrimmed)	14.7	228
Boiled, drained	4 oz.	3.3	49

PEA & CARROT:

Canned, regular pack, drained (Del Monte)	½ cup (3 oz.)	2.1	38

(USDA): United States Department of Agriculture

Food and Description	Measure or Quantity	Protein (grams)	Calories
Canned, dietetic pack, solids & liq.:			
(Blue Boy)	4 oz.	2.2	26
(Diet Delight)	½ cup (4.2 oz.)	2.9	47
(S and W) *Nutradiet*	4 oz.	2.2	36
Frozen:			
Not thawed (USDA)	4 oz.	3.7	62
Boiled, drained (USDA)	½ cup (3.1 oz.)	2.8	46
(Birds Eye)	½ cup (3.3 oz.)	3.0	55
In cream sauce (Green Giant)	⅓ of 10-oz. pkg.	3.3	60
PEA & CELERY, frozen (Birds Eye)	½ cup (3.3 oz.)	3.6	55
PEA & ONION, frozen:			
(Birds Eye)	½ cup (3.3 oz.)	4.2	67
In butter sauce (Green Giant)	⅓ of 9-oz. pkg.	3.0	75
In cream sauce (Green Giant)	⅓ of 9-oz. pkg.	2.6	67
PEA & POTATO, with cream sauce, frozen (Birds Eye)	⅓ of 8-oz. pkg.	3.5	131
PEA SOUP, green;			
Canned, condensed:			
(USDA)	8 oz. (by wt.)	10.4	240
*Prepared with equal volume water (USDA)	1 cup (8.6 oz.)	5.6	130
*Prepared with equal volume milk (USDA)	1 cup (8.6 oz.)	10.3	208
*(Campbell)	1 cup	7.8	131
Canned, low sodium:			
(Campbell)	7½-oz. can	7.5	140
*(Claybourne)	8 oz.	5.1	98
Dry mix:			
(USDA)	1 oz.	6.4	103
*(USDA)	1 cup (8.5 oz.)	7.5	121
*(Lipton)	1 cup	8.2	138
(Lipton) *Cup-a-Soup*	1.2-oz. pkg.	6.9	127
Frozen, condensed:			
With ham (USDA)	8 oz. (by wt.)	17.2	256
*With ham prepared with equal volume water (USDA)	8 oz. (by wt.)	8.6	129

(USDA): United States Department of Agriculture
*Prepared as Package Directs

[231]

Food and Description	Measure or Quantity	Protein (grams)	Calories
PEA SOUP, SPLIT:			
Canned, regular pack:			
Condensed (USDA)	8 oz. (by wt.)	15.9	268
*Prepared with equal volume			
water (USDA)	1 cup (8.6 oz.)	8.6	145
*(Manischewitz)	1 cup	3.7	133
*With ham (Campbell)	1 cup	11.6	160
*With ham (Heinz)	1 cup (8¾ oz.)	9.6	153
With smoked ham (Heinz)			
Great American	1 cup (9 oz.)	11.0	186
*Canned, dietetic (Slim-ette)	8 oz. (by wt.)	2.3	50
Canned, low sodium (Tillie			
Lewis)	1 cup (8 oz.)	10.7	152
PEACH:			
Fresh without skin (USDA):			
Whole, peeled of thin skins	1 lb. (weighed		
	unpeeled)	2.4	150
Whole, pared	1 lb. (weighed		
	unpeeled)	2.1	131
Whole	4-oz. peach (2½″		
	dia.)	.6	38
Diced	½ cup (4.7 oz.)	.8	51
Slices	½ cup (3 oz.)	.5	32
Canned, regular pack, solids & liq.:			
Juice pack (USDA)	4 oz.	.7	51
Light syrup (USDA)	4 oz.	.5	66
Heavy syrup (USDA)	2 med. halves & 2 T. syrup (4.1 oz.)	.5	91
Heavy syrup, halves (USDA)	½ cup (4.5 oz.)	.5	100
Heavy syrup, slices (USDA)	½ cup (4.4 oz.)	.5	98
Extra heavy syrup (USDA)	4 oz.	.5	110
Heavy syrup:			
(Del Monte) cling	½ cup (4.6 oz.)	.6	95
(Del Monte) freestone	½ cup (4.6 oz.)	.5	108
(Hunt's)	½ cup (4.5 oz.)	.5	96
(Stokely-Van Camp)	½ cup (4 oz.)	.5	89
Spiced (Del Monte)	½ cup (4.5 oz.)	.5	94

(USDA): United States Department of Agriculture
*Prepared as Package Directs

Food and Description	Measure or Quantity	Protein (grams)	Calories
Canned, dietetic or unsweetened pack:			
Water pack, solids & liq.			
(USDA)	½ cup (4.3 oz.)	.5	38
(Blue Boy) slices, solids & liq.	4 oz.	.4	32
(Diet Delight) cling, halves or slices	½ cup (4.4 oz.)	.6	61
(Diet Delight) freestone, halves or slices	½ cup (4.4 oz.)	.8	64
(S and W) *Nutradiet*, cling, halves, low calorie, undrained	2 halves (3.5 oz.)	.4	25
(S and W) *Nutradiet*, cling, halves, unsweetened	2 halves (3.5 oz.)	.5	28
(S and W) *Nutradiet*, cling, slices, low calorie, undrained	4 oz.	.5	28
(S and W) *Nutradiet*, cling, slices, unsweetened	4 oz.	.6	27
(S and W) *Nutradiet*, freestone, halves, low calorie, undrained	4 halves (3.5 oz.)	.5	26
(S and W) *Nutradiet*, freestone, slices, low calorie, solids & liq.	4 oz.	.6	27
(Tillie Lewis) cling, solids & liq.	¼ of 1-lb. can	.5	43
(Tillie Lewis) Elberta, solids & liq.	¼ of 1-lb. can	.6	43
Dehydrated, sulfured, nugget or pieces:			
Uncooked (USDA)	1 oz.	1.4	96
Cooked with added sugar, solids & liq. (USDA)	½ cup (5–6 halves & 3 T. liq., 5.4 oz.)	1.7	184
Dried:			
Uncooked (USDA)	1 lb.	14.1	1188
Uncooked (USDA)	½ cup (3.1 oz.)	2.7	231
Uncooked (Del Monte)	½ cup (3.1 oz.)	2.8	199

(USDA): United States Department of Agriculture

[233]

Food and Description	Measure or Quantity	Protein (grams)	Calories
Cooked, unsweetened (USDA)	½ cup (5–6 halves & 3 T. liq., 4.8 oz.)	1.4	111
Cooked, with added sugar (USDA)	½ cup (5–6 halves & 3 T. liq., 5.4 oz.)	1.4	181
Frozen: Not thawed, slices, sweetened:			
(USDA)	12-oz. pkg.	1.4	299
(USDA)	16-oz. pkg.	1.8	400
(USDA)	½ cup (4.2 oz.)	.5	104
Quick thaw (Birds Eye)	½ cup (5 oz.)	.7	87
PEACH BUTTER (Smucker's)	1 T. (.7 oz.)	.2	45
PEACH DUMPLING, frozen (Pepperidge Farm)	1 dumpling (3.3 oz.)	2.4	293
PEACH FRUIT DRINK (Alegre)	6 fl. oz. (6.6 oz.)	< .2	93
PEACH NECTAR, canned:			
(USDA)	1 cup (8.8 oz.)	.5	120
(Del Monte)	1 cup (8.7 oz.)	1.0	140
PEACH PIE:			
Home recipe, 2 crust (USDA)	⅙ of 9″ pie (5.6 oz.)	4.0	403
(Hostess)	4½-oz. pie	3.8	415
(Tastykake)	4-oz. pie	3.5	360
Frozen:			
(Banquet)	5-oz. serving	3.5	320
(Morton)	⅙ of 24-oz. pie	2.2	282
(Mrs. Smith's)	⅙ of 8″ pie (4.2 oz.)	2.0	301
(Mrs. Smith's)	⅛ of 10″ pie	2.5	399
(Mrs. Smith's) natural juice	⅙ of 8″ pie (4.2 oz.)	2.5	329

(USDA): United States Department of Agriculture

Food and Description	Measure or Quantity	Protein (grams)	Calories
PEACH PIE FILLING:			
(Comstock)	½ cup (5.1 oz.)	.4	175
(Lucky Leaf)	8 oz.	.6	300
(Wilderness)	21-oz. can	2.4	679
PEACH PRESERVE:			
Sweetened:			
(Bama)	1 T. (.7 oz.)	.1	51
(Smucker's)	1 T. (.7 oz.)	<.1	50
Low calorie or dietetic:			
(Kraft)	1 oz.	<.1	35
(Louis Sherry)	1 T. (.6 oz.)	Tr.	6
(Tillie Lewis)	1 T. (.8 oz.)	.2	11
PEACH TURNOVER, frozen			
(Pepperidge Farm)	1 turnover (3.3 oz.)	2.8	323
PEANUT:			
Raw (USDA):			
In shell	1 lb. (weighed in shell)	86.1	1868
With skins	1 oz.	7.4	160
Without skins, whole	1 oz.	7.5	161
Boiled (USDA)	1 oz.	4.4	107
Roasted:			
Whole (USDA)	1 lb. (weighed in shell)	79.6	1769
With skins (USDA)	1 oz.	7.4	165
Without skins (USDA)	1 oz.	7.4	166
Halves (USDA)	½ cup (2.5 oz.)	18.7	421
Chopped (USDA)	½ cup (2.4 oz.)	17.9	404
Chopped (USDA)	1 T. (9 grams)	2.3	53
Dry (Flavor House)	1 oz.	7.4	166
Dry (Franklin)	1 oz.	8.3	163
Dry (Frito-Lay)	1 oz.	7.0	168
Dry (Planters)	1 oz. (jar)	7.1	170
Dry (Skippy)	1 oz.	8.0	167
Oil (Planters) cocktail	15¢ bag (1¼ oz.)	8.5	224
Oil (Planters) cocktail	1 oz. (can)	6.8	179
Oil (Skippy)	1 oz.	7.5	178
(Nabisco) *Nab*	¾-oz. pkg.	5.5	134
(Nabisco) *Nab*	1¼-oz. pkg.	9.2	223

(USDA): United States Department of Agriculture

[235]

Food and Description	Measure or Quantity	Protein (grams)	Calories
(Nabisco) *Nab*	1⅞-oz. pkg.	13.9	334
Toasted (Tom Houston)	2 T. (1.1 oz.)	7.8	176
Spanish, dry roasted (Planters)	1 oz. (jar)	7.9	175
Spanish, oil roasted, *Freshnut*	1 oz.	6.5	170
Spanish, oil roasted (Planters)	1 oz. (can)	7.6	182
PEANUT BUTTER:			
(USDA):			
Small amounts of fat added	½ cup (4.4 oz.)	35.0	732
Small amounts of fat added	1 T. (.6 oz.)	4.4	93
Small amounts of fat &			
sweetener added	½ cup (4.4 oz.)	32.1	733
Small amounts of fat &			
sweetener added	1 T. (.6 oz.)	4.1	93
Moderate amounts of fat &			
sweetener added	½ cup (4.4 oz.)	31.8	742
Moderate amounts of fat &			
sweetener added	1 T. (.6 oz.)	4.0	94
(Bama) crunchy	1 T. (.6 oz.)	4.3	100
(Bama) smooth	1 T. (.6 oz.)	4.0	103
(Jif)	1 T. (.6 oz.)	3.8	95
(Peter Pan)	1 T. (.5 oz.)	4.4	93
(Planters)	1 T. (.6 oz.)	4.0	100
(Skippy) chunk	1 T. (.6 oz.)	4.6	96
(Skippy) creamy	1 T. (.6 oz.)	4.7	96
(Smucker's) creamy or crunchy	1 T. (.5 oz.)	4.0	85
(Smucker's) old fashioned	1 T. (.5 oz.)	4.4	86
& jelly (Smucker's) *Goober*	1 T. (.5 oz.)	1.9	64
Imitation (Bama) *Skyway*	1 T. (.6 oz.)	3.0	96
With crackers (see **CRACKERS**)			
PEANUT SPREAD:			
(USDA)	1 oz.	5.8	170
Diet (Peter Pan)	1 T. (.5 oz.)	4.6	100
PEAR:			
Fresh:			
Whole (USDA)	1 lb. (weighed with stems & core)	2.9	252
Whole (USDA)	6.4-oz. pear (3″ × 2½″)	1.2	101

(USDA): United States Department of Agriculture

Food and Description	Measure or Quantity	Protein (grams)	Calories
Quartered (USDA)	½ cup (3.4 oz.)	.7	59
Slices, including skin (USDA)	½ cup (2.9 oz.)	.6	50
Canned, regular pack, solids & liq.:			
Juice pack (USDA)	4 oz.	.3	52
Light syrup (USDA)	4 oz.	.2	69
Heavy syrup, halves (USDA)	½ cup (4 oz.)	.2	87
Heavy syrup, halves (USDA)	2 med. halves & 2 T. syrup (4.1 oz.)	.2	89
Extra heavy syrup (USDA)	4 oz.	.2	104
Heavy syrup (Del Monte)	½ cup (4 oz.)	.4	84
Heavy syrup (Hunt's)	½ cup (4.5 oz.)	.3	90
(Stokely-Van Camp)	½ cup (4 oz.)	.2	87
Canned, unsweetened or low calorie:			
Water pack, solids & liq. (USDA)	½ cup (4.3 oz.)	.2	39
Solids & liq. (Blue Boy) Bartlett	4 oz.	.2	33
Solids & liq. (Diet Delight) halves or quarters	½ cup (4.4 oz.)	.4	66
(S and W) *Nutradiet*, halves, low calorie, undrained	4 halves (3.5 oz.)	.3	28
(S and W) *Nutradiet*, halves, unsweetened	2 halves (3.5 oz.)	.3	30
(S and W) *Nutradiet*, quartered, low calorie, undrained	4 oz.	.2	29
(S and W) *Nutradiet*, quartered, unsweetened	4 oz.	.2	31
Solids & liq. (Tillie Lewis) Bartlett	¼ of 1-lb. can	.3	44
Dried:			
(USDA)	1 lb.	14.1	1216
Uncooked (Del Monte)	½ cup (2.8 oz.)	1.8	178
Cooked without added sugar, solids & liq. (USDA)	4 oz.	1.7	143
Cooked with added sugar, solids & liq. (USDA)	4 oz.	1.5	171
PEAR, CANDIED (USDA)	1 oz.	.4	86

(USDA): United States Department of Agriculture

Food and Description	Measure or Quantity	Protein (grams)	Calories
PEAR NECTAR:			
Sweetened (USDA)	1 cup (8.5 oz.)	.7	125
Sweetened (Del Monte)	1 cup (8.7 oz.)	.5	140
(S and W) *Nutradiet*	4 oz. (by wt.)	.5	34
PEAR PRESERVE (Bama)	1 T. (.7 oz.)	.1	51
PECAN:			
In shell (USDA)	1 lb. (weighed in shell)	22.1	1652
Shelled (USDA):			
Whole	1 lb.	41.7	3116
Halves	½ cup (1.9 oz.)	5.0	371
Halves	12–14 halves (.5 oz.)	1.3	96
Chopped	½ cup (1.8 oz.)	4.8	357
Chopped	1 T. (7 grams)	.6	48
Dry-roasted (Flavor House)	1 oz.	2.6	195
Dry-roasted (Planters)	1 oz.	2.6	206
PECAN PIE:			
Home recipe, 1 crust (USDA)	⅙ of 9″ pie (4.9 oz.)	7.0	577
Frozen (Morton)	⅙ of 16-oz. pie	3.0	275
Frozen (Mrs. Smith's)	⅙ of 8″ pie (4 oz.)	4.2	430
Frozen (Mrs. Smith's)	⅛ of 10″ pie (5.6 oz.)	4.5	470
PEP, cereal (Kellogg's)	¾ cup (1 oz.)	2.8	103
PEPPER:			
Black:			
Seasoned (French's)	1 tsp. (3 grams)	.3	7
Seasoned (Lawry's)	1 pkg. (1.6 oz.)	4.4	158
Seasoned (Lawry's)	1 tsp. (2 grams)	.2	8
Lemon (Durkee)	1 tsp. (3 grams)	Tr.	1
& lemon seasoning (French's)	1 tsp. (4 grams)	Tr.	5
PEPPER, HOT CHILI:			
Green (USDA):			
Raw, whole	4 oz.	1.1	31
Raw, without seeds	4 oz.	1.5	42

(USDA): United States Department of Agriculture

Food and Description	Measure or Quantity	Protein (grams)	Calories
Canned, chili sauce	1 oz.	.2	6
Canned, pods, without seeds, solids & liq.	4 oz.	1.0	28
Red:			
Raw, whole (USDA)	4 oz. (weighed with seeds)	4.2	105
Raw, trimmed, pods only (USDA)	4 oz.	1.9	54
Canned, chili sauce (USDA)	1 oz.	.3	6
Canned, solids & liq. (Del Monte)	¼ cup	.3	11
Dried:			
Pods (USDA)	1 oz.	3.7	91
Pods (Chili Products)	1 oz.	3.8	88
Powder with added seasoning (USDA)	1 T. (.5 oz.)	2.1	51
*PEPPER POT SOUP, canned (Campbell)	1 cup	6.9	94
PEPPER, STUFFED:			
Home recipe, with beef & crumbs (USDA)	2¾" × 2½" pepper with 1⅛ cups stuffing (6.5 oz.)	24.0	314
Frozen, with veal (Weight Watchers)	12-oz. dinner	27.2	224
PEPPER, SWEET:			
Green:			
Raw, whole (USDA)	1 lb. (weighed untrimmed)	4.5	82
Raw, without stem & seeds: (USDA)	1 med. pepper (2.6 oz.)	.7	13
Chopped (USDA)	½ cup (2.6 oz.)	.9	16
Slices (USDA)	½ cup (1.4 oz.)	.5	9
Strips (USDA)	½ cup (1.7 oz.)	.6	11
Boiled, drained (USDA)	1 med. pepper (2.6 oz.)	.7	12
Boiled, strips, drained (USDA)	½ cup (2.4 oz.)	.7	13

(USDA): United States Department of Agriculture
*Prepared as Package Directs

Food and Description	Measure or Quantity	Protein (grams)	Calories
Red (USDA):			
Raw, whole	1 lb. (weighed with stem & seeds)	5.1	112
Raw, without stem & seeds	1 med. pepper (2.2 oz.)	.8	19
PEPPERMINT PIE, pink, frozen (Kraft)	¼ of 13-oz. pie (3.2 oz.)	4.0	328
PEPPERONI (Hormel)	1 oz.	5.7	140
PERCH:			
Raw (USDA):			
White, whole	1 lb. (weighed whole)	31.5	193
White, meat only	4 oz.	21.9	134
Yellow, whole	1 lb. (weighed whole)	34.5	161
Yellow, meat only	4 oz.	22.1	103
Frozen, breaded (Gorton)	⅓ of 11-oz. pkg.	16.0	113
PERSIMMON (USDA):			
Japanese or Kaki, fresh:			
With seeds	1 lb. (weighed with skin, calyx & seeds)	2.6	286
With seeds	4.4-oz. persimmon	.7	79
Seedless	1 lb. (weighed with skin & calyx)	2.7	293
Seedless	4.4-oz. persimmon	.7	81
Native, fresh, whole	1 lb. (weighed with seeds & calyx)	3.0	472
Native, fresh, flesh only	4 oz.	.9	144
PETTIJOHNS, rolled whole wheat (Quaker)	⅔ cup (1 oz. dry)	2.6	97

(USDA): United States Department of Agriculture
*Prepared as Package Directs

Food and Description	Measure or Quantity	Protein (grams)	Calories
PHEASANT, raw (USDA):			
Ready-to-cook	1 lb. (weighed with bones)	95.9	596
Meat & skin	4 oz.	28.0	172
Meat only	4 oz.	26.8	184
Giblets	2 oz.	11.8	79
PICKEREL, chain, raw (USDA):			
Whole	1 lb. (weighed whole)	43.3	194
Meat only	4 oz.	21.2	95
PICKLE:			
Chowchow (see **CHOW CHOW**)			
Cucumber, fresh or bread & butter:			
(USDA)	½ cup (3 oz.)	.8	62
(USDA)	3 slices (¼″ × 1½″, .7 oz.)	.2	15
(Aunt Jane's)	4 slices or sticks (1 oz.)	.3	21
(Del Monte)	3 med. pieces (.9 oz.)	.2	4
(Fanning's)	14-fl.-oz. bottle	2.6	200
(Heinz)	3 slices	.1	20
Dill:			
(USDA)	4″ × 1¾″ pickle (4.8 oz.)	.9	15
(USDA)	3¾″ × 1¼″ pickle (2.3 oz.)	.5	7
(Aunt Jane's)	1 pickle (2 oz.)	.4	6
(Del Monte)	1 large pickle (3.5 oz.)	.4	7
(Heinz)	4″ pickle	.4	7
Processed (Heinz)	3″ pickle	.1	1
(Smucker's) baby, fresh pack	2¾″ pickle (.8 oz.)	.2	3
Dill, candied sticks (Smucker's)	4″ pickle (.8 oz.)	.2	46
Dill, hamburger (Heinz)	3 slices	Tr.	1
Dill, hamburger (Smucker's)	3 slices (.4 oz.)	<.1	2
Hot, mixed (Smucker's)	4 pieces (.7 oz.)	.2	5
Hot peppers (Smucker's)	4″ pepper (1 oz.)	.4	10
Kosher dill (Smucker's)	3½″ pickle (.5 oz.)	.4	8

(USDA): United States Department of Agriculture

[241]

Food and Description	Measure or Quantity	Protein (grams)	Calories
Sour:			
Cucumber (USDA)	1¾″ × 4″ pickle		
	(4.8 oz.)	.7	14
(Aunt Jane's)	1 pickle (2 oz.)	.3	6
Cucumber (Del Monte)	1 large pickle		
	(3.5 oz.)	.3	10
(Heinz)	2″ pickle	.1	1
Sweet:			
Cucumber (USDA):			
Whole	1 oz.	.2	41
Whole, gherkin	2½″ × ¾″		
	pickle (.5 oz.)	.1	22
Chopped	½ cup (2.6 oz.)	.5	108
Chopped	1 T. (9 grams)	<.1	13
(Aunt Jane's)	1 pickle (1.5 oz.)	.3	62
(Del Monte)	1 med. pickle (.4 oz.)	<.1	14
(Smucker's)	2½″ pickle (.4 oz.)	.1	17
Candied (Borden)	1 pickle (1.5 oz.)	.3	72
Candied, midgets (Smucker's)	2″ pickle (9		
	grams)	<.1	14
Cherry (Del Monte)	½ cup (1.9 oz.)	.8	30
Chips, fresh pack (Smucker's)	1 piece (5 grams)	<.1	10
Gherkin (Heinz)	2″ pickle	.1	16
Mixed (Heinz)	3 pieces	.1	23
Mixed (Smucker's)	1 piece (8 grams)	<.1	12
Sticks, fresh pack (Smucker's)	4″ stick (1.1 oz.)	.3	29
Wax, mild (Del Monte)	½ cup (2.1 oz.)	.6	21
PIE (see individual kinds)			
PIECRUST (see also **PASTRY SHELL**):			
Home recipe, baked 9″ pie (USDA)	1 crust (6.3 oz.)	11.0	900
Frozen:			
(Mrs. Smith's)	8″ shell (5 oz.)	8.0	692
(Mrs. Smith's) old fashioned	9″ shell (6 oz.)	9.0	867
(Mrs. Smith's)	10″ shell (8 oz.)	12.0	1177
PIECRUST MIX:			
Dry, pkg. or stick (USDA)	10-oz. pkg. (2 crusts)	20.4	1482

(USDA): United States Department of Agriculture

Food and Description	Measure or Quantity	Protein (grams)	Calories
*Prepared with water, baked (USDA)	4 oz.	7.3	526
*Double crust (Betty Crocker)	⅙ of 2 crusts	3.3	302
*2 crust (Pillsbury) stick	⅙ of 2 crusts	4.0	290
*Graham cracker (Betty Crocker)	⅙ of crust	1.7	159
*(Flako)	⅙ of 9″ shell (.8 oz.)	1.6	116

PIE FILLING (see individual kinds)

PIGEON (see SQUAB)

PIGEONPEA (USDA):

Raw, immature seeds in pods	1 lb.	12.7	207
Dry seeds	1 lb.	92.5	1551

PIGNOLIA (see PINE NUT)

PIGS FEET, pickled:

(USDA)	4 oz.	18.9	226
(Hormel)	1-pt. can (8.6 oz.)	26.4	442

PIKE, raw (USDA):

Blue, whole	1 lb. (weighed whole)	38.1	180
Blue, meat only	4 oz.	21.7	102
Northern, whole	1 lb. (weighed whole)	21.6	104
Northern, meat only	4 oz.	20.8	100
Walleye, whole	1 lb. (weighed whole)	49.9	240
Walleye, meat only	4 oz.	21.9	105

PILI NUT (USDA):

In shell	1 lb. (weighed in shell)	9.3	543
Shelled	4 oz.	12.9	759

PILLSBURY INSTANT BREAKFAST:

Chocolate or chocolate malt	1 pouch	6.0	130

(USDA): United States Department of Agriculture
*Prepared as Package Directs

Food and Description	Measure or Quantity	Protein (grams)	Calories
Strawberry or vanilla	1 pouch	5.0	130
PIMIENTO, canned:			
Solids & liq. (USDA)	1 med. pod (1.3 oz.)	.3	10
Solids & liq. (Stokely-Van Camp)	½ cup (4.1 oz.)	1.0	31
Whole pods, slices, pieces (Dromedary)	¼ cup (1.7 oz.)	.3	8
PINCH of HERBS:			
(Lawry's)	1 pkg. (2.2 oz.)	8.0	212
(Lawry's)	1 tsp. (3 grams)	.3	9
PINEAPPLE:			
Fresh:			
Whole (USDA)	1 lb. (weighed untrimmed)	.9	123
Diced (USDA)	½ cup (2.8 oz.)	.3	41
Slices (USDA)	¾″ × 3½″ slice (3 oz.)	.3	44
(Del Monte)	½ cup (2.5 oz.)	.3	42
Chunks (Dole)	½ cup (3.5 oz.)	.4	52
Canned, regular pack, solids & liq.:			
Juice pack:			
(USDA)	4 oz.	.5	66
(Del Monte)	½ cup (4.9 oz.)	.7	79
Chunks or crushed (Dole)	½ of 8-oz. can	.4	64
Light syrup (USDA)	4 oz.	.3	67
Heavy syrup:			
Crushed (USDA)	½ cup (4.6 oz.)	.4	97
Slices (USDA)	½ cup (4.9 oz.)	.4	103
Slices (USDA)	2 small or 1 large slice & 2 T. syrup (4.3 oz.)	.4	90
Tidbits (USDA)	½ cup (4.6 oz.)	.4	95
(Del Monte)	½ cup (5 oz.)	.7	100
Chunks (Dole)	½ of 8-oz. can	.3	84
Slices, chunks & crushed (Stokely-Van Camp)	½ cup (4 oz.)	.4	84
Extra heavy syrup (USDA)	4 oz.	.3	102

(USDA): United States Department of Agriculture

Food and Description	Measure or Quantity	Protein (grams)	Calories
Canned, unsweetened, low calorie or dietetic, solids & liq.:			
Water pack, except crushed (USDA)	4 oz.	.3	44
Chunks:			
(Diet Delight)	½ cup (4.4 oz.)	.5	65
(S and W) *Nutradiet*	4 oz.	< .1	56
Crushed (Diet Delight)	½ cup (4.4 oz.)	.6	77
Slices:			
(Diet Delight)	½ cup (4.4 oz.)	.5	65
(Dole)	2 med. slices & 2½ T. juice (3.7 oz.)	.4	66
(S and W) *Nutradiet,* unsweetened	2½ slices (3.5 oz.)	.4	69
Tidbits:			
(Diet Delight)	½ cup (4.4 oz.)	.5	65
(S and W) *Nutradiet,* unsweetened	4 oz.	.5	78
(Tillie Lewis)	½ cup (4.4 oz.)	.5	76
Frozen:			
Chunks, sweetened, not thawed (USDA)	½ cup (4.3 oz.)	.5	105
Chunks in heavy syrup (Dole)	11 chunks & 2½ T. syrup (4 oz.)	.5	87
PINEAPPLE-APRICOT JUICE DRINK (Del Monte)	1 cup (8.6 oz.)	.7	132
PINEAPPLE CAKE MIX:			
*(Betty Crocker) layer	1/12 of cake	2.8	199
*(Duncan Hines)	1/12 of cake (2.7 oz.)	3.0	201
*(Pillsbury)	1/12 of cake	3.0	210
PINEAPPLE, CANDIED (USDA)	1 oz.	.2	90
PINEAPPLE-CHERRY JUICE DRINK, canned (Del Monte) *Merry*	½ cup (4.3 oz.)	.8	52

Food and Description	Measure or Quantity	Protein (grams)	Calories
PINEAPPLE & GRAPEFRUIT JUICE DRINK, canned:			
(USDA) 40% fruit juices	½ cup (4.4 oz.)	.2	68
(Del Monte)	½ cup (4.3 oz.)	.2	62
(Del Monte) pink	½ cup (4.3 oz.)	.1	64
(Dole) regular or pink	6-fl.-oz. can	.3	91
(Dole) regular or pink	½ cup (4.4 oz.)	.2	67
PINEAPPLE JUICE:			
Canned, unsweetened:			
(USDA)	½ cup (4.4 oz.)	.5	68
(Del Monte)	½ cup (4.3 oz.)	.5	60
(Dole)	6-fl.-oz. can	.7	93
(Dole)	½ cup (4.4 oz.)	.5	69
(Heinz)	5½-fl.-oz. can	.7	101
(S and W) *Nutradiet*	4 oz. (by wt.)	.4	67
(Stokely-Van Camp)	½ cup (4 oz.)	.5	63
Frozen, concentrate:			
Unsweetened, undiluted			
(USDA)	6-fl.-oz. can (7.6 oz.)	2.8	387
*Unsweetened, diluted with			
3 parts water (USDA)	½ cup (4.4 oz.)	.5	64
Unsweetened, undiluted			
(Dole)	6-fl.-oz. can	2.8	389
PINEAPPLE & ORANGE JUICE DRINK, canned:			
(USDA) 40% fruit juices	½ cup (4.4 oz.)	.2	67
(Del Monte)	½ cup (4.3 oz.)	.2	62
PINEAPPLE-PEAR JUICE DRINK, canned (Del Monte)	½ cup (4.3 oz.)	.2	68
PINEAPPLE PIE:			
Home recipe:			
(USDA) 2 crust	⅙ of 9" pie (5.6 oz.)	3.5	400
Chiffon, 1 crust (USDA)	⅙ of 9" pie (3.8 oz.)	7.1	311
Custard, 1 crust (USDA)	⅙ of 9" pie (5.4 oz.)	6.1	334
(Hostess)	4½-oz. pie	3.8	415
(Tastykake)	4-oz. pie	4.3	389
With cheese (Tastykake)	4-oz. pie	3.9	436
Frozen (Mrs. Smith's)	⅙ of 8" pie (4.2 oz.)	1.8	300

(USDA): United States Department of Agriculture
*Prepared as Package Directs

Food and Description	Measure or Quantity	Protein (grams)	Calories
Frozen, with cheese (Mrs. Smith's)	⅙ of 8″ pie (4 oz.)	6.8	263
Frozen, with cheese (Mrs. Smith's)	⅛ of 10″ pie (5.5 oz.)	8.8	339
PINEAPPLE PIE FILLING, canned:			
(Comstock)	½ cup (5.4 oz.)	.3	140
(Lucky Leaf)	8 oz.	.6	240
PINEAPPLE PRESERVE:			
Sweetened (Bama)	1 T. (.7 oz.)	.1	54
Sweetened (Smucker's)	1 T. (.7 oz.)	Tr.	52
Low calorie (Tillie Lewis)	1 T. (.5 oz.)	Tr.	10
***PINEAPPLE PUDDING MIX, CREAM,** instant (Jell-O)	½ cup (5.3 oz.)	4.3	178
PINEAPPLE SOFT DRINK (Yoo-Hoo) High Protein	6 fl. oz. (6.4 oz.)	6.0	100
PINE NUT (USDA):			
Pignolias, shelled	4 oz.	35.3	626
Piñon, whole	4 oz. (weighed in shell)	8.6	418
Piñon, shelled	4 oz.	14.7	720
PINK PANTHER FLAKES, cereal (Post)	⅓ cup (1 oz.)	1.0	112
PISTACHIO NUT:			
In shell (USDA)	4 oz. (weighed in shell)	11.0	337
In shell (USDA)	½ cup (2.3 oz.)	6.4	197
Shelled (USDA)	½ cup (2.2 oz.)	12.0	368
Shelled (USDA)	1 T. (8 grams)	1.5	46
Dry-roasted (Flavor House)	1 oz.	5.5	168
***PISTACHIO NUT PUDDING MIX,** instant (Royal)	½ cup (5.1 oz.)	4.5	179

(USDA): United States Department of Agriculture
*Prepared as Package Directs

[247]

Food and Description	Measure or Quantity	Protein (grams)	Calories
PITANGA, fresh (USDA):			
Whole	1 lb. (weighed whole)	2.9	187
Flesh only	4 oz.	.9	58
PIZZA PIE:			
Home recipe, with cheese topping:			
(USDA)	4 oz.	13.6	268
(USDA)	5½" sector (⅛ of 14" pie, 2.6 oz.)	9.0	177
Home recipe, with sausage topping:			
(USDA)	4 oz.	8.8	265
(USDA)	5½" sector (⅛ of 14" pie, 2.6 oz.)	5.8	176
Chilled, partially baked (USDA)	4 oz.	8.8	236
Chilled, baked (USDA)	4 oz.	10.4	278
Frozen:			
Partially baked (USDA)	4 oz.	10.1	260
Baked (USDA)	4 oz.	10.8	278
Baked (USDA)	⅛ of 14" pie (2.6 oz.)	7.1	184
(Celeste) *Bambino,* deluxe	10-oz. pie	27.0	634
With cheese (Buitoni)	4 oz.	10.7	270
With cheese (Celeste)	20-oz. pie	65.6	1284
With cheese (Chef Boy-Ar-Dee)	⅙ of 12½" pie (2.1 oz.)	5.4	131
With cheese, little (Chef Boy-Ar-Dee)	1 pie (2½ oz.)	6.3	162
With cheese (Jeno's)	13-oz. pie	47.5	881
With cheese (Jeno's) Serv-A-Slice	1 slice (1.7 oz.)	6.1	116
With cheese (Jeno's) snack tray	½-oz. pizza	1.5	30
With cheese (Kraft)	14-oz. pie	42.1	826
With cheese, (Kraft) *Pee Wee*	2½-oz. pie	7.7	169
With hamburger (Jeno's)	13½-oz. pie	48.5	922
With pepperoni (Buitoni)	4 oz.	9.8	288
With pepperoni (Chef Boy-Ar-Dee)	⅙ of 14-oz. pie	6.1	150
With pepperoni (Jeno's)	13¼-oz. pie	46.4	1010

(USDA): United States Department of Agriculture
*Prepared as Package Directs

Food and Description	*Measure or Quantity*	*Protein (grams)*	*Calories*
With pepperoni (Jeno's) Serv-A-Slice	1 slice (1.8 oz.)	7.0	145
With pepperoni (Jeno's) snack tray	½-oz. pizza	1.6	37
With sausage (Buitoni)	4 oz.	13.5	281
With sausage (Celeste)	23-oz. pie	71.6	1600
With sausage (Celeste) *Bambino*	9-oz. pie	27.4	642
With sausage (Chef Boy-Ar-Dee)	⅙ of 13¼-oz. pie	6.0	141
With sausage, little (Chef Boy-Ar-Dee)	2½-oz. pie	7.4	168
With sausage (Jeno's)	13½-oz. pie	44.2	903
With sausage (Jeno's) Serv-A-Slice	1 slice (2 oz.)	8.1	143
With sausage (Jeno's) snack tray	½-oz. pizza	1.7	35
With sausage (Kraft)	14½-oz. pie	44.0	999
With sausage, (Kraft) *Pee Wee*	2½-oz. pizza	7.6	191
PIZZA PIE MIX:			
Regular (Jeno's)	14-oz. pie	30.2	960
Cheese (Chef Boy-Ar-Dee)	⅕ of 15½-oz. pie	6.8	187
Cheese (Jeno's)	14¾-oz. pie	34.7	1096
*Cheese (Kraft)	4 oz.	16.1	265
Pepperoni (Jeno's)	14¾-oz. pie	44.3	1263
Sausage (Chef Boy-Ar-Dee)	⅕ of 17-oz. pie	7.1	222
Sausage (Jeno's)	16½-oz. pie	50.1	1244
Sausage (Kraft)	4 oz.	17.0	274
PIZZA ROLL, Frozen:			
Cheeseburger (Jeno's) 12 to pkg.	½-oz. roll	1.6	44
Pepperoni (Jeno's) 12 to pkg.	½-oz. roll	1.5	42
Sausage (Jeno's) 12 to pkg.	½-oz. roll	1.4	42
Shrimp (Jeno's) 12 to pkg.	½-oz. roll	1.3	35
Snack tray (Jeno's):			
Hamburger	½-oz. roll	1.7	41
Pepperoni	½-oz. roll	1.6	38
Sausage	½-oz. roll	1.5	39
PIZZA SAUCE:			
Canned (Buitoni)	4 oz.	2.6	76

*Prepared as Package Directs

Food and Description	Measure or Quantity	Protein (grams)	Calories
Canned (Chef Boy-Ar-Dee)	½ of 10½-oz. can	2.8	116
Canned (Contadina)	1 cup	4.0	144
Mix (French's)	1-oz. pkg.	1.9	77
*Mix (French's)	2 T.	.6	18
PIZZA SEASONING (French's)	1 tsp. (4 grams)	.2	4
PIZZARIA MIX (Hunt's) *Skillet*	14.1-oz. pkg.	24.8	507
PLANTAIN, raw (USDA):			
Whole	1 lb. (weighed with skin)	3.6	389
Flesh only	4 oz.	1.2	135
PLUM:			
Damson, fresh (USDA):			
Whole	1 lb. (weighed with pits)	2.1	272
Flesh only	4 oz.	.6	75
Japanese & hybrid, fresh (USDA):			
Whole	1 lb. (weighed with pits)	2.1	205
Whole	2″ plum (2.1 oz.)	.3	27
Diced	½ cup (2.3 oz.)	.4	39
Halves	½ cup (3.1 oz.)	.4	42
Slices	½ cup (3 oz.)	.4	40
Prune-type, fresh (USDA):			
Whole	1 lb. (weighed with pits)	3.4	320
Halves	½ cup (2.8 oz.)	.6	60
Canned, purple, regular pack, solids & liq.:			
Light syrup (USDA)	4 oz.	.5	71
Heavy syrup (USDA)	½ cup (with pits, 4.5 oz.)	.5	106
Heavy syrup (USDA)	½ cup (without pits, 4.2 oz.)	.5	100
Heavy syrup (USDA)	3 plums without pits & 2 T. syrup (4.3 oz.)	.5	101
Heavy syrup (Del Monte)	½ cup (4.1 oz.)	.6	112
Extra heavy syrup (USDA)	4 oz.	.5	116

(USDA): United States Department of Agriculture
*Prepared as Package Directs

Food and Description	Measure or Quantity	Protein (grams)	Calories
(Stokely-Van Camp)	½ cup (4.2 oz.)	.5	100
Canned, unsweetened or low calorie, solids & liq.:			
Greengage, water pack (USDA)	4 oz.	.5	37
Purple:			
Water pack (USDA)	4 oz.	.5	52
(Diet Delight)	½ cup (4.4 oz.)	.8	80
(S and W) *Nutradiet*	4 oz.	.5	57
(Tillie Lewis)	½ of 8-oz. can	.6	61
PLUM PIE (Tastykake)	4-oz. pie	3.4	364
PLUM PRESERVE or JAM, sweetened, Damson or red (Bama)	1 T. (.7 oz.)	.1	51
PLUM PUDDING (Richardson & Robbins)	½ cup (4 oz.)	4.1	300
P.M., fruit juice drink (Motts)	½ cup	.1	62
POHA (see **GROUND-CHERRY**)			
POKE SHOOTS (USDA):			
Raw	1 lb.	11.8	104
Boiled, drained	4 oz.	2.6	23
POLISH-STYLE SAUSAGE:			
(USDA)	1 oz.	4.5	86
(Oscar Mayer) all meat	1 oz. (from 8-oz. link)	4.5	81
Kolbase (Hormel)	1 oz.	4.4	80
POLLOCK (USDA):			
Raw, drawn	1 lb. (weighed with head, tail, fins & bones)	41.6	194
Raw, meat only	4 oz.	23.1	108
Creamed, made with flour, butter & milk	4 oz.	15.8	145

(USDA): United States Department of Agriculture

Food and Description	Measure or Quantity	Protein (grams)	Calories
POLYNESIAN-STYLE DINNER, frozen (Swanson)	11¾-oz. dinner	20.6	512
POMEGRANATE, raw (USDA):			
Whole	1 lb. (weighed whole)	1.3	160
Pulp only	4 oz.	.6	71
POMPANO, raw (USDA):			
Whole	1 lb. (weighed whole)	47.8	422
Meat only	4 oz.	21.3	188
POPCORN:			
Unpopped (USDA)	1 oz.	3.4	103
Popped (USDA):			
Plain, large kernel	1 oz.	3.6	109
Plain, large kernel	1 cup (6 grams)	.8	23
Butter added	1 oz.	2.8	129
Butter added	1 cup (9 grams)	.9	41
Coconut oil	1 oz.	2.8	129
Coconut oil	1 cup (9 grams)	.9	41
Sugar-coated	1 oz.	1.7	109
Sugar-coated	1 cup (1.2 oz.)	2.1	134
(Jiffy Pop)	½ pkg. (2½ oz.)	4.9	244
(Tom Houston)	1 cup (.5 oz.)	1.5	68
Buttered (Jiffy Pop)	½ pkg. (2½ oz.)	4.9	247
Buttered (Wise)	1-oz. bag	2.3	137
Caramel-coated:			
Without peanuts (Old London)	1¾-oz. bag	1.9	195
With peanuts (Old London)	1 cup (1.3 oz.)	2.0	142
Cheese (Old London)	¾-oz. bag	1.8	110
Cheese (Wise)	⅝-oz. bag	2.0	90
Cracker Jack	¾-oz. bag	1.5	90
Cracker Jack	1⅜-oz. box	2.7	165
Seasoned (Old London)	1¼-oz. bag	2.9	175
POPOVER:			
Home recipe (USDA)	1.4-oz. popover (2¾″ dia. at top, ¼ cup batter)	3.5	90

(USDA): United States Department of Agriculture

Food and Description	Measure or Quantity	Protein (grams)	Calories
*Mix (Flako)	2.3-oz. popover (⅙ of pkg.)	7.0	163
POPSICLE, fruit flavors (Popsicle Industries)	3-fl.-oz. pop	0.	70
POP-UP (see **TOASTER CAKE**)			
PORGY, raw (USDA):			
Whole	1 lb. (weighed whole)	35.3	208
Meat only	4 oz.	21.5	127
PORK, medium fat:			
Fresh (USDA):			
All lean cuts:			
Lean only:			
Raw, diced	1 cup (8.2 oz.)	44.0	426
Raw, strips	1 cup (8.2 oz.)	43.8	428
Roasted, chopped	1 cup (5 oz.)	39.2	348
Boston butt:			
Raw	1 lb. (weighed with bone & skin)	65.9	1220
Roasted, lean & fat	4 oz.	25.5	400
Roasted, lean only	4 oz.	30.6	277
Chop:			
Broiled, lean & fat	4-oz. chop (weighed with bone)	18.6	295
Broiled, lean & fat	3-oz. chop (weighed without bone)	21.0	332
Broiled, lean only	3-oz. chop (weighed without bone)	26.0	230
Fat, separable, cooked	1 oz.	1.4	219
Ham (see also **HAM**):			
Raw	1 lb. (weighed with bone & skin)	61.3	1188
Roasted, lean & fat	4 oz.	26.1	424

(USDA): United States Department of Agriculture
*Prepared as Package Directs

Food and Description	Measure or Quantity	Protein (grams)	Calories
Roasted, lean only	4 oz.	33.7	246
Loin:			
Raw	1 lb. (weighed with bone)	61.1	1065
Broiled, lean & fat	4 oz.	28.0	443
Broiled, lean only	4 oz.	34.7	309
Roasted, lean & fat	4 oz.	27.8	411
Roasted, lean only	4 oz.	33.3	288
Picnic:			
Raw	1 lb. (weighed with bone & skin)	59.0	1083
Simmered, lean & fat	4 oz.	26.3	424
Simmered, lean only	4 oz.	32.9	240
Spareribs:			
Raw, with bone	1 lb. (weighed with bone)	39.2	976
Raw, without bone	1 lb. (weighed without bone)	65.8	1637
Braised, lean & fat	4 oz.	23.6	499
Cured, light commercial cure:			
Bacon (see **BACON**)			
Boston butt (USDA):			
Raw	1 lb. (weighed with bone & skin)	72.5	1227
Roasted, lean & fat	4 oz.	26.0	374
Roasted, lean only	4 oz.	31.5	276
Smoked Tasty Meat (Wilson)	4 oz.	15.0	288
Ham (see also **HAM**):			
Raw (USDA)	1 lb. (weighed with bone & skin)	68.3	1100
Raw, lean only, ground (USDA)	1 cup (6 oz.)	36.6	286
Roasted, lean & fat (USDA)	4 oz.	23.7	328
Roasted, lean only:			
(USDA)	4 oz.	28.7	212
Chopped (USDA)	1 cup (4.9 oz.)	34.9	258
Diced (USDA)	1 cup (5.2 oz.)	37.2	275
Ground (USDA)	1 cup (3.8 oz.)	27.6	204

(USDA): United States Department of Agriculture

Food and Description	Measure or Quantity	Protein (grams)	Calories
Fully cooked, bone-in			
(Hormel)	4 oz.	17.1	207
Fully cooked, boneless:			
(Armour Star) Parti-Style	4 oz.	21.7	167
(Hormel) *Cure 81*	4 oz.	21.5	193
(Hormel) *Curemaster*	4 oz.	21.5	133
(Wilson) Certified, rolled	4 oz.	20.3	222
(Wilson) Festival, smoked	4 oz.	20.6	191
Cured, long-cure, country-style (USDA):			
Ham, Virginia, raw	1 lb. (weighed with bone & skin)	66.7	1535
Ham, Virginia, raw	1 lb. (weighed without bone & skin)	76.7	1765
Picnic:			
Raw (USDA)	1 lb. (weighed with bone & skin)	62.5	1060
Raw (Wilson) smoked	4 oz.	14.9	279
Roasted, lean & fat (USDA)	4 oz.	25.4	366
Roasted, lean only (USDA)	4 oz.	32.2	239
Canned (Hormel)	4 oz. (3-lb. can)	18.7	206
PORK, canned, chopped luncheon meat:			
(USDA)	1 oz.	4.3	83
Chopped (USDA)	1 cup (4.8 oz.)	20.4	400
Diced (USDA)	1 cup (5 oz.)	21.2	415
(Hormel)	1 oz. (8-lb. can)	4.1	70
PORK DINNER, loin of pork, frozen (Swanson)	10-oz. dinner	25.3	460
PORK, FREEZE DRY, canned (Wilson) *Campsite:*			
Chop, dry	2-oz. can	35.5	294
*Chop, reconstituted	4 oz.	28.8	238
Patties, dry	2-oz. can	28.2	299
*Patties, reconstituted	4 oz.	23.0	244

(USDA): United States Department of Agriculture
*Prepared as Package Directs

Food and Description	Measure or Quantity	Protein (grams)	Calories
PORK & GRAVY, 90% pork, canned (USDA)	4 oz.	18.6	290
PORK LOIN (Oscar Mayer) thin sliced	1 slice (16 to 3 oz.)	1.1	10
PORK RINDS, fried, *Bakenets* (see also other brand names)	1 oz.	18.3	139
PORK ROAST, canned (Wilson) *Tender Made*	1 oz.	6.0	44
PORK SAUSAGE:			
Uncooked:			
Links or bulk (USDA)	1 oz.	2.7	141
(Armour Star)	1-oz. sausage	2.8	133
Country Style (Hormel)	1 oz.	4.5	107
Little Sizzlers (Hormel)	1 piece (.8 oz.)	2.5	105
Midget (Hormel)	1 piece (.8 oz.)	5.5	90
Smoked (Hormel)	1 oz.	4.1	97
Bulk style (Oscar Mayer)	1 oz.	2.8	127
Italian Brand (Oscar Mayer)	1 oz.	3.4	134
Little Friers (Oscar Mayer):			
6–8 per lb.	1 link (2.3 oz.)	6.5	296
14–18 per lb.	1 link (1 oz.)	2.8	129
(Wilson)	1 oz.	2.9	135
Cooked:			
Links or bulk (USDA)	1 oz.	5.1	135
Links (USDA) 16 per lb. raw	2 links (.9 oz)	4.7	124
Bulk style (Oscar Mayer)	1 oz.	6.2	99
Italian Brand (Oscar Mayer)	1 oz.	6.0	86
Little Friers (Oscar Mayer):			
6–8 per lb. raw	1 link (1 oz. cooked)	6.7	106
14–18 per lb. raw	1 oz. (.4 oz. cooked)	2.8	44
Canned, solids & liq. (USDA)	1 oz.	3.9	118
Canned, drained (USDA)	1 oz.	5.2	108
POST TOASTIES, cereal (Post)	1 cup (1 oz.)	2.0	108
POSTUM, instant	1 cup (6 oz.)	.3	16

(USDA): United States Department of Agriculture
*Prepared as Package Directs

Food and Description	Measure or Quantity	Protein (grams)	Calories
POTATO:			
Raw, whole (USDA)	1 lb. (weighed unpared)	7.7	279
Raw, pared (USDA):			
Chopped	1 cup (5.2 oz.)	3.1	112
Diced	1 cup (5.5 oz.)	3.3	119
Slices	1 cup (5.2 oz.)	3.1	113
Cooked:			
Au gratin or scalloped, without cheese (USDA)	½ cup (4.3 oz.)	3.7	127
Au gratin, with cheese (USDA)	½ cup (4.3 oz.)	6.5	177
Baked, peeled after baking (USDA)	2½" dia. potato (3.5 oz., 3 raw per lb.)	2.6	92
Boiled, peeled after boiling (USDA)	4.8-oz. potato (3 raw per lb.)	2.9	103
Boiled, peeled before boiling (USDA):			
Whole	4.3-oz. potato (3 raw per lb.)	2.3	79
Diced	½ cup (2.8 oz.)	1.5	51
Mashed	½ cup (3.7 oz.)	2.0	68
Riced	½ cup (4 oz.)	2.2	74
Slices	½ cup (2.8 oz.)	1.5	52
French-fried in deep fat (USDA)	10 pieces (2" × ½" × ½", 2 oz.)	2.5	156
French-fried (McDonald's)	1 serving (2.4 oz.)	2.9	218
Hash-browned (USDA)	½ cup (3.4 oz.)	3.0	223
Mashed, milk added (USDA)	½ cup (3.5 oz.)	2.1	64
Mashed, milk & butter added (USDA)	½ cup (3.5 oz.)	2.1	92
Pan-fried from raw (USDA)	½ cup (3.5 oz.)	3.4	228
Scalloped (see Au gratin)			
Canned:			
Solids & liq. (USDA)	1 cup (8.8 oz.)	2.8	110
White (Butter Kernel)	3-4 small potatoes (4.1 oz.)	2.3	96
Solids & liq. (Stokely-Van Camp)	½ cup (4.1 oz.)	1.2	51

(USDA): United States Department of Agriculture

Food and Description	Measure or Quantity	Protein (grams)	Calories
Whole, new, solids & liq. (Del Monte)	½ cup (2.6 oz.)	1.0	23
Whole, new, drained solids (Del Monte)	½ cup (5 oz.)	1.1	42
Whole, new, drained liq. (Del Monte)	4 oz.	1.2	11
Dehydrated, mashed:			
Flakes, without milk (USDA):			
Dry	½ cup (.8 oz.)	1.7	84
*Prepared with water, milk & fat	½ cup (3.8 oz.)	2.0	100
(Borden) *Country Store*	½ cup (1.1 oz.)	2.6	120
Granules, without milk (USDA):			
Dry	½ cup (3.5 oz.)	8.3	352
*Prepared with water, milk & butter	½ cup (3.7 oz.)	2.1	101
Granules with milk (USDA):			
Dry	½ cup (3.5 oz.)	10.9	358
*Prepared with water & fat	½ cup (3.7 oz.)	2.1	83
Frozen:			
Au gratin (Stouffer's)	11½-oz. pkg.	14.6	304
Diced for hash-browning, not thawed (USDA)	4 oz.	1.4	83
Diced, hash-browned (USDA)	4 oz.	2.3	254
French-fried:			
Not thawed (USDA)	9-oz. pkg.	7.1	434
Heated (USDA)	10 pieces (2″ × ½″ × ½″, 2 oz.)	2.1	125
Mashed, not thawed (USDA)	4 oz.	1.9	85
Mashed, heated (USDA)	4 oz.	2.0	105
Stuffed:			
Bake-A-Tata (Holloway House):			
With cheese	1 potato (5 oz.)	3.6	226
With sour cream	1 potato (5 oz.)	3.6	200
With cheese or sour cream & chives	1 potato (6 oz.)	5.1	296

POTATO CHIP:

(USDA)	1 oz.	1.5	161

(USDA): United States Department of Agriculture
*Prepared as Package Directs

Food and Description	Measure or Quantity	Protein (grams)	Calories
(USDA)	10 2" chips or 7 3" chips (.7 oz.)	1.1	114
(Lay's)	1 oz.	1.8	158
(Nalley's)	1 oz.	1.9	158
(Pringle's)	10 chips (.5 oz.)	.8	73
(Pringle's)	1 oz.	1.4	147
Ruffles	1 oz.	1.9	158
(Tom Houston)	10 chips (.7 oz.)	1.1	114
(Wise)	1-oz. bag	2.0	156
(Wonder)	1 oz.	1.5	158
Barbecue (Lay's)	1 oz.	2.3	155
Barbecue (Wise)	1-oz. bag	2.0	150
Barbecue (Wonder)	1 oz.	1.5	154
Light, blanched (Wise)	1-oz. bag	1.8	166
Onion-garlic (Wise)	1-oz. bag	2.0	154
Ridgies (Wise)	1-oz. bag	2.0	156
Sour cream & onion, regular or Ridges (Wise)	1-oz. bag	2.4	150
POTATO MIX:			
*Au gratin (Betty Crocker)	½ cup	4.6	160
Au gratin (French's)	5½-oz. pkg.	22.0	573
*Au gratin (French's)	½ cup	3.7	95
*Au gratin (Pillsbury's)	½ cup	5.0	170
*Buds (Betty Crocker)	½ cup	2.4	134
*Hash brown (Pillsbury) *Hungry Jack*	½ cup	2.0	140
Mashed, country style (French's)	2⅔-oz. pkg.	6.3	267
*Mashed, country style (French's)	½ cup	2.7	137
Mashed, granules (French's)	3¼-oz. pkg.	7.6	324
*Mashed, granules (French's)	½ cup	2.4	114
*Mashed (Pillsbury) *Hungry Jack*	½ cup	3.0	170
*Mashed, complete (Pillsbury) *Hungry Jack*	½ cup	3.0	150
*Scalloped (Betty Crocker)	½ cup	3.5	150
Scalloped (French's)	5⅝-oz. pkg.	19.7	552
*Scalloped (French's)	½ cup	3.3	109
*Scalloped (Pillsbury)	½ cup	3.0	140
Whipped (Borden)	¼ cup (.5 oz.)	1.0	52

(USDA): United States Department of Agriculture
*Prepared as Package Directs

Food and Description	Measure or Quantity	Protein (grams)	Calories
POTATO PANCAKE MIX:			
(French's)	3-oz. pkg.	6.5	284
*(French's)	3 small pancakes (¼ of pkg.)	3.1	90
POTATO SALAD:			
Home recipe, with cooked salad dressing & seasonings (USDA)	4 oz.	⁻3.1	112
Home recipe, with mayonnaise & French dressing, hard-cooked eggs & seasonings (USDA)	½ cup (4.4 oz.)	3.8	181
Canned (Nalley's)	4 oz.	2.0	178
POTATO SOUP, Cream of:			
*Canned (Campbell)	1 cup	3.3	105
Frozen (USDA):			
Condensed	8 oz. (by wt.)	6.1	197
*Prepared with equal volume water	1 cup (8.5 oz.)	3.4	106
*Prepared with equal volume milk	1 cup (8.6 oz.)	7.8	186
POTATO SOUP MIX:			
*(Lipton)	1 cup	3.6	100
(Wyler's)	1 oz.	2.6	88
POTATO STICK:			
(USDA)	1 oz.	1.8	154
(Durkee) *O & C*	1½-oz. can	2.7	233
Julienne (Wise)	1-oz. bag	2.0	156
POUND CAKE:			
Home recipe, old fashioned, equal weights flour, sugar, eggs & butter (USDA)	1.1-oz. slice (3½″ × 3″ × ½″)	1.7	142
Home recipe, traditional (USDA)	1.1-oz. slice (3½″ × 3″ × ½″)	1.9	123
All butter, Jr. (Drake's)	1 slice (1.2 oz.)	1.7	110
Plain (Drake's)	1 slice (1.6 oz.)	2.5	153

(USDA): United States Department of Agriculture
*Prepared as Package Directs

Food and Description	Measure or Quantity	Protein (grams)	Calories
Frozen (Morton)	1 oz.	1.6	117
POUND CAKE MIX:			
*(Betty Crocker)	¹⁄₁₂ of cake	2.7	210
*(Dromedary)	1" slice (2.9 oz.)	4.2	313
*(Pillsbury) *Bundt*	¹⁄₁₂ of cake	4.0	300
PRESERVE, sweetened (see also individual listings by flavor):			
(USDA)	1 oz.	.2	77
(USDA)	1 T. (.7 oz.)	.1	54
(Kraft)	1 oz.	.1	78
(Smucker's)	1 T. (.7 oz.)	Tr.	52
PRETZEL:			
(USDA)	1 oz.	2.8	111
Dutch, twisted (USDA)	1 pretzel (.6 oz.)	1.6	62
Stick, regular (USDA)	5 regular sticks (3⅛", 3 grams)	.3	12
Stick, small (USDA)	10 small sticks (2¼", 3 grams)	.3	12
Thin, twisted (USDA)	1 pretzel (6 grams)	.6	23
(Nab) *Mister Salty Veri-Thin*	75 pieces (¾-oz. pkg.)	2.0	79
(Nab) *Mister Salty Veri-Thin*	15 pieces (1½-oz. pkg.)	4.0	158
(Nab) Pretzelette	16 pieces (1¼-oz. pkg.)	3.3	132
(Nabisco) *Mister Salty,* Dutch	1 piece (.5 oz.)	1.1	51
(Nabisco) *Mister Salty,* pretzelette	1 piece (2 grams)	.2	6
(Nabisco) *Mister Salty,* 3-ring	1 piece (3 grams)	.3	12
(Nabisco) *Mister Salty Veri-Thin*	1 piece (5 grams)	.5	20
(Nabisco) *Mister Salty Veri-Thin,* sticks	1 piece (1 gram)	<.1	1
(Old London) nuggets	2-oz. bag	5.7	211
(Old London) rings	1½-oz. bag	4.3	156
(Rold Gold) rods	1 oz.	2.6	100
(Rold Gold) twists	1 oz.	2.5	100

(USDA): United States Department of Agriculture
*Prepared as Package Directs

Food and Description	Measure or Quantity	Protein (grams)	Calories
PRICKLY PEAR, fresh (USDA):			
Whole	1 lb. (weighed with rind & seeds)	1.0	84
Flesh only	4 oz.	.6	48
PRODUCT 19, cereal (Kellogg's)	1 cup (1 oz.)	2.5	107
PRUNE:			
Dried, "softenized," uncooked:			
Small (USDA)	1 prune (5 grams)	<.1	11
Medium, whole with pits, (USDA)	1 cup (6.6 oz.)	3.3	405
Medium (USDA)	1 prune (7 grams)	.1	15
Large (USDA)	1 prune (9 grams)	.2	20
Pitted, chopped (USDA)	1 cup (5.3 oz.)	3.2	382
Pitted, ground	1 cup (9.7 oz.)	5.8	699
Dried, moist-pak (Del Monte)	1 cup (8 oz.)	7.1	538
Dried, ready-to-eat (Del Monte)	1 cup (6.6 oz.)	5.4	473
Dried, "softenized," cooked unsweetened (USDA)	1 cup (17-18 med. with ⅓ cup liq., 9.5 oz.)	2.5	295
Dried, "softenized," cooked with sugar (USDA)	1 cup (16-18 prunes & ⅓ cup liq., 11.1 oz.)	2.3	504
Dehydrated (USDA):			
Nugget-type & pieces	8 oz.	7.5	780
Nugget-type & pieces, cooked with sugar, solids & liq.	1 cup (8.9 oz.)	3.0	454
Canned:			
Cooked (Sunsweet)	1 cup	1.8	300
Stewed (Del Monte)	1 cup (9.4 oz.)	2.1	316
Stewed, pitted (Del Monte)	1 cup (9.2 oz.)	2.3	288
PRUNE JUICE:			
(USDA)	½ cup (4.5 oz.)	.5	99
(Bennett's)	½ cup (4.5 oz.)	.5	99
(Del Monte)	½ cup (4.3 oz.)	.8	58
(Heinz)	5½-fl.-oz. can	1.0	119
(Mott's) Super	4 oz. (by wt.)	.3	99

(USDA): United States Department of Agriculture

Food and Description	Measure or Quantity	Protein (grams)	Calories
Real Prune	½ cup (4.5 oz.)	.4	74
(Sunsweet)	½ cup	.3	82
& apple (Sunsweet)	4 oz. (by wt.)	.2	68
With lemon (Sunsweet)	½ cup	.3	82
PRUNE WHIP, home recipe			
(USDA)	1 cup (4.8 oz.)	5.9	211
PUDDING or PUDDING MIX (see individual kinds)			
PUFF (see **CRACKER** or individual kinds of hors d'oeuvres, such as **CHICKEN PUFF**)			
PUFFA PUFFA RICE, cereal			
(Kellogg's)	1 cup (1 oz.)	1.0	120
PUFFED OAT CEREAL			
(USDA):			
Added nutrients	1 oz.	3.4	113
Sugar-coated, added nutrients	1 oz.	1.9	112
PUFFED RICE CEREAL:			
(USDA) added nutrients	1 cup (.5 oz.)	.9	60
(USDA) honey & added nutrients	½ oz.	.6	55
(USDA) honey or cocoa fat, added nutrients	½ oz.	.6	57
(Malt-O-Meal)	½ oz.	1.0	52
(Quaker)	1¼ cups (½ oz.)	.9	56
(Sunland)	½ oz.	1.0	52
(Whiffs)	½ oz.	1.0	52
PUFFED WHEAT CEREAL:			
(USDA) added nutrients	1 cup (.4 oz.)	1.8	44
Frosted with sugar & honey (USDA)	1 cup (.4 oz.)	.7	45
(Malt-O-Meal)	½ oz.	1.0	52
(Quaker)	1⅓ cups (½ oz.)	1.9	51
(Sunland)	½ oz.	1.0	52
(Whiffs)	½ oz.	1.0	52

(USDA): United States Department of Agriculture

Food and Description	Measure or Quantity	Protein (grams)	Calories
PUMPKIN:			
Fresh, whole (USDA)	1 lb. (weighed with rind & seeds)	3.2	83
Flesh only (USDA)	4 oz.	1.1	29
Canned:			
(USDA)	½ cup (4.3 oz.)	1.2	40
(Del Monte)	½ cup (4.3 oz.)	.8	39
(Stokely-Van Camp)	½ cup (4.1 oz.)	1.1	38
PUMPKIN PIE:			
Home recipe, 1 crust (USDA)	⅙ of 9″ pie (5.4 oz.)	6.1	321
(Tastykake)	4-oz. pie	5.6	368
Frozen:			
(Banquet)	5-oz. serving	6.0	306
(Morton)	⅙ of 24-oz. pie	3.4	201
(Mrs. Smith's)	⅙ of 8″ pie (4 oz.)	4.2	242
(Mrs. Smith's)	⅛ of 10″ pie (5.6 oz.)	5.4	314
PUMPKIN PIE FILLING:			
(Comstock)	1 cup (10.8 oz.)	2.2	366
(Del Monte)	1 cup (9 oz.)	.5	218
PUMPKIN SEED, dry (USDA):			
Whole	4 oz. (weighed in hull)	24.4	464
Hulled	1 oz.	8.2	157
PUNCH DRINK, canned (Hi-C)	6 fl. oz. (6.3 oz.)	.6	98
PURSLANE, including stems (USDA):			
Raw	1 lb.	7.7	95
Boiled, drained	4 oz.	1.4	17

Q

QUAIL, raw (USDA):			
Ready-to-cook	1 lb. (weighed with bones)	102.1	686
Meat & skin only	4 oz.	28.8	195
Giblets	2 oz.	12.4	100

(USDA): United States Department of Agriculture

Food and Description	Measure or Quantity	Protein (grams)	Calories
QUANGAROOS, cereal (Quaker)	1 cup (1 oz.)	1.1	112
QUIK (see individual kinds)			
QUINCE, fresh (USDA):			
Untrimmed	1 lb. (weighed with skin & seeds)	1.1	158
Flesh only	4 oz.	.5	65
QUISP, cereal (Quaker)	1⅙ cups (1 oz.)	1.3	122

R

RABBIT (USDA):			
Domesticated:			
Ready-to-cook	1 lb. (weighed with bones)	75.0	581
Raw, meat only	4 oz.	23.8	184
Stewed, meat only	4 oz.	33.2	245
Stewed, meat only, chopped or diced	1 cup (4.9 oz.)	41.0	302
Wild, ready-to-cook	1 lb. (weighed with bones)	76.0	490
Wild, raw, meat only	4 oz.	23.8	153
RACOON, roasted, meat only (USDA)	4 oz.	33.1	289
RADISH (USDA):			
Common, raw:			
Untrimmed, without tops	½ lb. (weighed untrimmed)	2.0	34
Trimmed, whole	4 small radishes (1.4 oz.)	.4	7
Trimmed, whole	1 cup (4.7 oz.)	1.3	22
Trimmed, sliced	½ cup (2 oz.)	.6	10
Oriental, raw, without tops	½ lb. (weighed unpared)	1.6	34
Oriental, raw, trimmed & pared	4 oz.	1.0	22

(USDA): United States Department of Agriculture

Food and Description	Measure or Quantity	Protein (grams)	Calories
RAISIN:			
Dried:			
Whole (USDA)	4 oz.	2.8	328
Whole (USDA)	1 pkg. (.5 oz.)	.4	40
Whole, pressed down (USDA)	½ cup (2.9 oz.)	2.0	237
Whole, pressed down (USDA)	1 T. (.4 oz.)	.2	29
Chopped (USDA)	½ cup (2.9 oz.)	2.0	234
Ground (USDA)	½ cup (4.7 oz.)	3.4	387
Cinnamon coated (Del Monte)	½ cup (2.5 oz.)	2.3	213
Seeded, Muscat (Del Monte)	½ cup (2.5 oz.)	1.9	222
Seeded, Muscat (Sun-Maid)	1 oz.	.7	76
Seedless, California Thompson (Del Monte)	½ cup (2.5 oz.)	2.3	212
Seedless, California Thompson, natural (Sun-Maid)	½ cup (3 oz.)	2.9	250
Seedless, California Thompson, natural (Sun-Maid)	1 T. (.4 oz.)	.4	31
Seedless, golden (Del Monte)	½ cup (2.5 oz.)	2.4	209
Seedless, golden (Sun-Maid)	½ cup (3 oz.)	2.7	250
Cooked, added sugar, solids & liq. (USDA)	½ cup (4.3 oz.)	1.5	260
RAISIN PIE:			
Home recipe, 2 crust (USDA)	⅙ of 9″ pie (5.6 oz.)	4.1	427
(Tastykake)	4-oz. pie	4.0	391
Frozen (Mrs. Smith's)	⅙ of 8″ pie (4.2 oz.)	2.3	322
RAISIN PIE FILLING:			
(Comstock)	½ cup (5.4 oz.)	1.0	181
(Lucky Leaf)	8 oz.	1.8	292
(Wilderness)	2.2-oz. can	5.6	773
RALSTON, regular or instant	¼ cup (1 oz.)	3.8	106
RASPBERRY:			
Black:			
Fresh:			
(USDA)	½ lb. (weighed with caps & stems)	3.3	160

(USDA): United States Department of Agriculture

Food and Description	Measure or Quantity	Protein (grams)	Calories
(USDA) without caps & stems	½ cup (2.4 oz.)	1.0	49
Canned, water pack, unsweetened, solids & liq. (USDA)	4 oz.	1.2	58
Red:			
Fresh:			
(USDA)	½ lb. (weighed with caps & stems)	2.6	126
(USDA) without caps & stems)	½ cup (2.5 oz.)	.9	41
Canned, water pack, unsweetened or low calorie:			
Solids & liq. (USDA)	4 oz.	.8	40
Solids & liq. (Blue Boy)	4 oz.	.8	48
Frozen, sweetened:			
Not thawed (USDA)	10-oz. pkg.	2.0	278
Not thawed (USDA)	½ cup (4.4 oz.)	.9	122
Quick-thaw (Birds Eye)	½ cup (5 oz.)	1.1	148
RASPBERRY CAKE MIX			
(Pillsbury) *Bundt*	¹⁄₁₂ of cake	4.0	290
RASPBERRY PIE FILLING:			
Red (Comstock)	½ cup (5.3 oz.)	.6	212
Red (Lucky Leaf)	8 oz.	1.2	324
RASPBERRY PRESERVE or JAM:			
Sweetened, black or red (Bama)	1 T. (.7 oz.)	.1	54
Low calorie or dietetic:			
(Diet Delight)	1 T. (.6 oz.)	.1	24
Black (Kraft)	1 oz.	<.1	36
(S and W) *Nutradiet*	1 T. (.5 oz.)	.1	10
Low calorie (Louis Sherry)	1 T. (.5 oz.)	Tr.	6
Low sugar, black (Slenderella)	1 T. (.6 oz.)	<.1	25
RASPBERRY RENNET MIX:			
Powder:			
Dry (Junket)	1 oz.	.2	115
*(Junket)	4 oz.	3.7	108

(USDA): United States Department of Agriculture
*Prepared as Package Directs

[267]

Food and Description	Measure or Quantity	Protein (grams)	Calories
Tablet:			
Dry (Junket)	1 tablet (< 1 gram)	Tr.	1
*& sugar (Junket)	4 oz.	3.7	101
RASPBERRY TURNOVER,			
frozen (Pepperidge Farm)	1 turnover (3.3 oz.)	3.0	337
RAVIOLI:			
Canned:			
Beef or meat:			
(Buitoni)	8 oz.	10.1	188
In brine (Buitoni)	8 oz.	12.3	212
(Chef Boy-Ar-Dee)	⅕ of 40-oz can	8.4	211
(Nalley's)	8 oz.	14.5	399
(Prince)	3.7-oz. can	4.5	136
Cheese:			
(Buitoni)	8 oz.	10.7	218
In brine (Buitoni)	8 oz.	12.3	192
(Chef Boy-Ar-Dee)	½ of 15-oz. can	9.6	262
(Prince)	3.7-oz. can	3.9	123
Chicken (Nalley's)	8 oz.	9.7	458
Frozen:			
Beef (Celeste)	7 ravioli (4 oz.)	14.3	259
Beef dinner (Celeste)	½ of 15-oz. pkg.	12.1	262
Beef (Kraft)	12½-oz. pkg.	21.6	411
Cheese (Buitoni)	4 oz.	13.8	313
Cheese (Celeste)	7 ravioli (4 oz.)	13.4	264
Cheese dinner (Celeste)	½ of 15-oz. pkg.	11.5	255
Cheese (Kraft)	12½-oz. pkg.	17.0	407
Meat, without sauce (Buitoni)	4 oz.	14.5	278
Meat, with sauce (Buitoni)	4 oz.	9.3	143
Raviolettes (Buitoni)	4 oz.	15.6	324
REDFISH (see **DRUM, RED & OCEAN PERCH,** Atlantic)			
RED & GRAY SNAPPER, raw (USDA):			
Whole	1 lb. (weighed whole)	46.7	219
Meat only	4 oz.	22.5	105

(USDA): United States Department of Agriculture
*Prepared as Package Directs

Food and Description	Measure or Quantity	Protein (grams)	Calories
REDHORSE, SILVER, raw (USDA):			
Drawn	1 lb. (weighed eviscerated)	37.6	204
Flesh only	4 oz.	20.4	111
REINDEER, raw, lean only (USDA)	4 oz.	24.7	144
RELISH:			
Barbecue (Heinz)	1 T.	.1	35
Hamburger (Del Monte)	1 T. (.9 oz.)	.2	33
Hamburger (Heinz)	1 T.	.1	15
Hot dog (Del Monte)	1 T. (.9 oz.)	.5	28
Hot dog (Heinz)	1 T.	.2	17
India (Heinz)	1 T.	.1	17
Piccalili (Heinz)	1 T.	.2	23
Sour (USDA)	½ cup (4.3 oz.)	.9	23
Sour (USDA)	1 T. (.5 oz.)	.1	3
Sweet:			
(USDA) finely chopped	½ cup (4.3 oz.)	.6	168
(USDA) finely chopped	1 T. (.5 oz.)	<.1	21
(Aunt Jane's)	1 rounded tsp. (.4 oz.)	<.1	14
(Del Monte)	1 T. (.9 oz.)	.2	36
(Heinz)	1 T.	.1	28
(Smucker's)	1 T. (.6 oz.)	.1	23
RENNIN CUSTARD PRODUCTS (see individual flavors)			
RHUBARB:			
Fresh (USDA):			
Partly trimmed	1 lb. (weighed with part leaves, ends & trimmings)	2.0	54
Trimmed	4 oz.	.7	18
Diced	½ cup (2.2 oz.)	.4	10
Cooked, sweetened, solids & liq. (USDA)	½ cup (4.2 oz.)	.6	169
Frozen:			
Not thawed (USDA)	½ cup (3.9 oz.)	.7	82

(USDA): United States Department of Agriculture

Food and Description	Measure or Quantity	Protein (grams)	Calories
Cooked, added sugar, solids & liq. (USDA)	½ cup (4.4 oz.)	.6	177
(Birds Eye)	½ cup (4 oz.)	.6	84
RHUBARB PIE, home recipe 2 crust (USDA)	⅙ of 9″ pie (5.6 oz.)	4.0	400
RICE;			
Brown:			
Raw (USDA)	½ cup (3.7 oz.)	7.8	374
Raw (USDA)	1 oz.	2.1	102
Cooked:			
(USDA)	4 oz.	2.8	135
(Carolina)	4 oz.	2.8	135
(River Brand)	4 oz.	2.8	135
(Water Maid)	4 oz.	2.8	135
Parboiled (Uncle Ben's) no added butter	⅔ cup (4.2 oz.)	2.9	133
Parboiled (Uncle Ben's) with added butter	⅔ cup (4.3 oz.)	2.9	153
Frozen, in beef stock (Green Giant)	⅓ of 12-oz. pkg.	2.8	128
White:			
Instant or Precooked:			
Dry, long-grain (USDA)	½ cup (1.9 oz.)	4.1	206
Dry, long-grain (USDA)	1 oz.	2.1	106
Dry, long-grain (Uncle Ben's Quick)	1 oz.	1.9	101
Cooked:			
Long-grain (USDA)	⅔ cup (3.3 oz.)	2.0	101
(Carolina)	⅔ cup (3.3 oz.)	2.0	101
(Minute Rice) no added butter	⅔ cup (4 oz.)	2.0	124
Long-grain (Uncle Ben's Quick) no added butter	⅔ cup (4.1 oz.)	1.9	105
Long-grain (Uncle Ben's Quick) with added butter	⅔ cup (4.1 oz.)	1.9	126
Parboiled:			
Dry, long-grain (USDA)	1 oz.	2.1	105
Dry, long-grain (Uncle Ben's Converted)	1 oz.	2.0	101

(USDA): United States Department of Agriculture

Food and Description	Measure or Quantity	Protein (grams)	Calories
Cooked:			
Long-grain (USDA)	⅔ cup (4.1 oz.)	2.5	124
(Aunt Caroline)	⅔ cup (4.1 oz.)	2.5	124
Long-grain (Uncle Ben's Converted) no added butter	⅔ cup (4.3 oz.)	2.3	121
Regular:			
Raw (USDA)	½ cup (3.5 oz.)	6.6	359
Cooked:			
(USDA)	⅔ cup (4.8 oz.)	2.7	149
Extra long-grain (Carolina)	⅔ cup (4.8 oz.)	2.7	149
Long-grain (Mahatma)	⅔ cup (4.8 oz.)	2.7	149
(River Brand) fluffy	⅔ cup (4.8 oz.)	2.7	149
(Water Maid)	⅔ cup (4.8 oz.)	2.7	149
White & wild, frozen (Green Giant)	⅓ of 12-oz. pkg.	2.5	104
Wild (see **WILD RICE**)			
RICE BRAN (USDA)	1 oz.	3.8	78
RICE CEREAL (USDA):			
With casein & other added nutrients	1 oz.	11.3	108
Wheat gluten & other added nutrients	1 oz.	5.7	109
RICE CHEX, cereal (Ralston Purina)	1⅛ cups (1 oz.)	1.3	111
RICE FLAKES, cereal, added nutrients (USDA)	1 cup (1.1 oz.)	1.8	117
RICE, FRIED:			
Seasoning mix (Durkee)	1-oz. pkg.	1.9	62
*Seasoning mix (Durkee)	2 cups (1-oz. pkg.)	8.7	430
RICE KRISPIES, cereal (Kellogg's)	1 cup (1 oz.)	2.0	108
RICE MIX:			
Beef:			
Rice-A-Roni	⅙ of 8-oz. pkg.	3.0	129
(Uncle Ben's)	6-oz. pkg.	16.2	592

(USDA): United States Department of Agriculture
*Prepared as Package Directs

[271]

Food and Description	Measure or Quantity	Protein (grams)	Calories
*(Uncle Ben's) no added butter	½ cup (4.2 oz.)	2.8	103
*(Uncle Ben's) with added butter	½ cup (4.3 oz.)	2.8	120
*(Village Inn)	½ cup	2.5	125
Brown or wild:			
(Uncle Ben's)	6-oz. pkg.	17.0	612
*(Uncle Ben's) no added butter	½ cup (4.3 oz.)	2.7	99
*(Uncle Ben's) with added butter	½ cup (4.3 oz.)	2.7	117
Chicken:			
Rice-A-Roni	⅕ of 8-oz. pkg.	3.0	153
(Uncle Ben's)	6-oz. pkg.	16.2	612
*(Uncle Ben's) no added butter	½ cup (3.6 oz.)	2.6	100
*(Uncle Ben's) with added butter	½ cup (3.8 oz.)	2.7	133
*(Village Inn)	½ cup	2.5	125
Chinese, fried, Rice-A-Roni	4 oz.	4.8	226
Curried (Uncle Ben's)	6-oz. pkg.	16.3	586
*Curried (Uncle Ben's) no added butter	½ cup (4.2 oz.)	2.8	100
*Curried (Uncle Ben's) with added butter	½ cup (4.2 oz.)	2.8	117
*Curry (Village Inn)	½ cup	2.5	125
*Drumstick (Minute Rice)	½ cup (4.1 oz.)	3.1	152
*Ham with pineapple, Rice-A-Roni	4 oz.	3.4	113
*Herb (Village Inn)	½ cup	2.5	125
*Keriyaki dinner (Betty Crocker)	½ cup	9.2	208
Long-grain & wild (Uncle Ben's)	6-oz. pkg.	18.2	590
*Long-grain & wild (Uncle Ben's) no added butter	½ cup (4 oz.)	3.0	97
*Long-grain & wild (Uncle Ben's) with added butter	½ cup (4.1 oz.)	3.0	114
*Long-grain & wild (Village Inn)	½ cup	2.5	125
*Milanese (Betty Crocker)	½ cup	4.6	172
*Oriental, dinner (Jeno's) Add 'n Heat	40-oz. pkg.	102.1	1599

*Prepared as Package Directs

Food and Description	Measure or Quantity	Protein (grams)	Calories
Pilaf (Uncle Ben's)	6-oz. pkg.	14.4	598
*Pilaf (Uncle Ben's) no added butter	½ cup (3.3 oz.)	2.3	97
*Pilaf (Uncle Ben's) with added butter	½ cup (3.5 oz.)	2.4	129
*Provence (Betty Crocker)	½ cup	3.8	184
*Rib Roast (Minute Rice)	½ cup (4.1 oz.)	3.7	149
Spanish (see also RICE, SPANISH):			
*(Minute Rice)	½ cup (5.6 oz.)	2.8	132
Rice-A-Roni	⅙ of 7½-oz. pkg.	3.0	125
(Uncle Ben's)	5½-oz. pkg.	18.4	527
*(Uncle Ben's) no added butter	½ cup (4.6 oz.)	3.8	109
*(Uncle Ben's) with added butter	½ cup (4.6 oz.)	3.8	129
*(Village Inn)	½ cup	2.5	125
*Yellow (Village Inn)	½ cup	2.5	125
RICE & PEAS with MUSHROOMS:			
Frozen (Birds Eye)	⅓ of 9-oz. pkg.	3.4	113
Frozen (Green Giant)	⅓ of 12-oz. pkg.	2.3	100
RICE PILAF, frozen (Green Giant)	⅓ of 12-oz. pkg.	2.0	101
RICE POLISH (USDA)	1 oz.	3.4	75
RICE PUDDING:			
Home recipe, with raisins (USDA)	½ cup (4.7 oz.)	4.8	193
Canned:			
(Betty Crocker)	½ cup	3.9	154
(Hunt's)	5-oz. can	3.0	240
RICE, PUFFED (see PUFFED RICE CEREAL)			
RICE, SPANISH:			
Home recipe (USDA)	4 oz.	2.0	99
Canned:			
(Heinz)	8¾-oz. can	3.3	196

(USDA): United States Department of Agriculture
*Prepared as Package Directs

[273]

Food and Description	Measure or Quantity	Protein (grams)	Calories
(Nalley's)	4 oz.	1.6	150
(Van Camp)	½ cup (3.9 oz.)	1.9	96
Frozen (Green Giant)	⅓ of 12-oz. pkg.	1.9	87
RICE, SPANISH, SEASONING MIX:			
(Durkee)	1⅝-oz. pkg.	5.4	129
(Lawry's)	1½-oz. pkg.	7.6	125
RICE VERDI, frozen (Green Giant)	⅓ of 12-oz. pkg.	2.0	122
RICE WINE (HEW/FAO):			
Chinese, 20.7% alcohol by volume	3 fl. oz.	0.	114
Japanese, 10,6% alcohol by volume	3 fl. oz.	2.6	215
Sweetened, non-alcoholic	3 fl. oz.	.7	30
ROAST 'n BOAST (General Foods):			
For beef	1½-oz. pkg.	2.7	129
For chicken	1⅜-oz. pkg.	5.5	119
For pork	1¾-oz. pkg.	6.5	145
For stew	1½-oz. pkg.	4.6	126
ROCKFISH (USDA):			
Raw, meat only	1 lb.	85.7	440
Oven-steamed, with onion	4 oz.	20.5	121
ROE (USDA):			
Raw, carp, cod, haddock, herring, pike or shad	4 oz.	27.7	147
Raw, salmon, sturgeon, or turbot	4 oz.	28.6	235
Baked or broiled with butter & lemon juice or vinegar, cod & shad	4 oz.	24.9	143
Canned, cod, haddock, or herring, solids & liq.	4 oz.	24.4	134
ROLL & BUN:			
Barbecue (Arnold)	1 bun (1.6 oz.)	3.7	132

Food and Description	Measure or Quantity	Protein (grams)	Calories
Brown & serve:			
Unbrowned (USDA)	1 oz.	2.2	85
Browned (USDA)	1 oz.	2.5	93
(Wonder)	1 roll (1 oz.)	2.2	85
Butter crescent (Pepperidge Farm)	1 roll (1.2 oz.)	2.7	127
Butter, old fashioned (Pepperidge Farm)	1 roll (.4 oz.)	1.0	36
Butterfly (Pepperidge Farm)	1 roll (.6 oz.)	1.6	58
Cinnamon nut (Pepperidge Farm)	1 bun (1 oz.)	1.3	92
Cloverleaf, home recipe (USDA)	1 roll (1.2 oz.)	2.9	119
Club (Pepperidge Farm)	1 roll (1.6 oz.)	3.9	114
Deli Twist (Arnold)	1 roll (1.2 oz.)	3.3	115
Diet Size (Arnold)	1 roll (.5 oz.)	1.1	40
Dinner (Arnold) 12 or 24 to pkg.	1 roll (¾ oz.)	1.9	71
Dinner (Pepperidge Farm)	1 roll (.7 oz.)	1.8	61
Dinner (Wonder)	1 roll (1 oz.)	2.3	85
Dutch Egg, sandwich (Arnold)	1 bun (1.7 oz.)	4.1	143
Finger:			
(Arnold) handipan	1 roll (.7 oz.)	1.7	61
Egg (Arnold) family	1 roll (.7 oz.)	1.7	62
Poppy (Pepperidge Farm)	1 roll (.7 oz.)	1.7	59
Sesame (Pepperidge Farm)	1 roll (.7 oz.)	1.7	60
Frankfurter:			
(USDA)	1 roll (1.4 oz.)	3.3	119
(Arnold)	1 roll (1.4 oz.)	3.4	121
New England (Arnold)	1 roll (1.6 oz.)	3.7	130
(Pepperidge Farm)	1 roll (1.4 oz.)	3.7	117
(Wonder)	1 bun (2 oz.)	4.8	162
French:			
Triple (Pepperidge Farm)	1 roll (3.5 oz.)	8.9	253
Twin (Pepperidge Farm)	1 roll (5.2 oz.)	12.8	389
Golden Twist (Pepperidge Farm)	1 roll (1.2 oz.)	2.3	123
Hamburger:			
(USDA)	1 roll (1.4 oz.)	3.3	119
(Pepperidge Farm)	1 roll (1.4 oz.)	3.6	112
(Wonder)	1 bun (2 oz.)	4.8	162
Hard, round or rectangular (USDA)	1 roll (1.8 oz.)	4.9	156

(USDA): United States Department of Agriculture

Food and Description	Measure or Quantity	Protein (grams)	Calories
Hearth (Pepperidge Farm)	1 roll (.8 oz.)	1.9	59
Honey, frozen (Morton)	1 serving (2.2 oz.)	3.5	170
Kaiser, brown & serve (Arnold)	1 roll (1.7 oz.)	5.1	132
Old Fashioned (Pepperidge Farm)	1 roll (.6 oz.)	1.4	57
Parker (Arnold) handipan	1 roll (.7 oz.)	1.7	63
Parkerhouse (Pepperidge Farm)	1 roll (.7 oz.)	1.7	57
Party Pan (Pepperidge Farm):			
Plain	1 roll (.4 oz.)	1.0	34
Poppy	1 roll (.4 oz.)	1.0	34
Pecan coffee (Pepperidge Farm)	1 bun (1.7 oz.)	2.6	195
Plain (USDA)	1 roll (1 oz.)	2.3	84
Raisin (USDA)	1 oz.	2.0	78
Sandwich, soft (Arnold)	1 roll (1.5 oz.)	3.9	135
Sesame crisp (Pepperidge Farm):			
Mid-west	1 roll (.8 oz.)	2.3	69
East	1 roll (.9 oz.)	2.4	73
Soft (Arnold) handipan	1 roll (.7 oz.)	1.6	58
Sweet (USDA)	1 bun (1.5 oz.)	3.7	136
Whole-wheat (USDA)	1 roll (1.3 oz.)	3.8	98
ROLL DOUGH:			
Refrigerated:			
Butterflakes (Pillsbury)	1 roll	1.5	55
Crescent (Borden)	1 roll (1.1 oz.)	2.2	104
Crescent (Pillsbury)	1 roll	1.5	95
Crescent (Pillsbury) Ballard	1 roll	2.0	95
Crescent, Italian (Pillsbury)	1 roll	1.5	90
Dinner, Crescent, buttermilk (Pillsbury) Hungry Jack	1 roll	2.0	115
Pan rolls (Pillsbury)	1 roll	2.5	75
Parkerhouse (Pillsbury)	1 roll	1.5	60
Snowflake (Pillsbury)	1 roll	2.0	70
Frozen, unraised (USDA)	1 oz.	2.1	76
Frozen, baked (USDA)	1 oz.	2.4	88
ROLL MIX:			
Dry (USDA)	1 oz.	3.2	111
*Prepared with water (USDA)	1 oz.	2.6	85
*Caramel (Pillsbury)	1 roll	3.0	250
*Cinnamon (Pillsbury)	1 roll	3.0	240

(USDA): United States Department of Agriculture
*Prepared as Package Directs

Food and Description	Measure or Quantity	Protein (grams)	Calories
*Honey (Pillsbury)	1 roll	3.0	250
*Hot roll (Pillsbury)	1 roll	3.0	95
*Orange (Pillsbury)	1 roll	3.0	240
*ROMAN MEAL CEREAL	¾ cup (1.3 oz. dry)	5.3	130
ROSE APPLE, raw (USDA):			
Whole	1 lb. (weighed with caps & seeds)	1.8	170
Flesh only	4 oz.	.7	64
RUSK:			
(USDA)	1 piece (.5 oz.)	2.0	59
Holland (Nabisco)	1 piece (.4 oz.)	1.7	49
RUTABAGA:			
Raw, without tops (USDA)	1 lb. (weighed with skin)	4.2	177
Raw, diced (USDA)	½ cup (2.5 oz.)	.8	32
Boiled, diced, drained (USDA)	½ cup (3 oz.)	.8	30
Boiled, mashed (USDA)	½ cup (4.3 oz.)	1.1	43
RYE, whole grain (USDA)	1 oz.	3.4	95
RYE FLOUR (see FLOUR)			
RYE-KRISP (see CRACKER)			
RY-KING (see BREAD)			

S

SABLEFISH, raw (USDA):			
Whole	1 lb. (weighed whole)	24.8	362
Meat only	4 oz.	14.7	215
SAFFLOWER SEED (USDA):			
Kernels, dry, in hull	½ lb. (weighed in hull)	22.1	712
Kernels, dry, hulled	1 oz.	5.4	174
Meal, partially defatted	1 oz.	11.2	101

(USDA): United States Department of Agriculture
*Prepared as Package Directs

[277]

Food and Description	Measure or Quantity	Protein (grams)	Calories
SAINT JOHN'S-BREAD FLOUR (see **FLOUR**, Carob)			
SAKE:			
Cake, 8% alcohol by wt.	1 oz. (by wt.)	4.4	62
Wine, 19.8% alcohol by volume	3 fl. oz. (3.1 oz.)	.4	116
SALAD DRESSING (see also **SALAD DRESSING, LOW CALORIE**):			
Avocado, refrigerated (Marzetti)	1 T. (.5 oz.)	.2	80
(Bama)	1 T. (.5 oz.)	.1	59
Bennett's	1 T. (.5 oz.)	<.1	51
Blendaise (Marzetti)	1 T. (.5 oz.)	.1	60
Bleu or blue cheese:			
(USDA)	1 oz.	1.4	143
(USDA)	1 T. (.5 oz.)	.7	76
(Bernstein's) Danish	1 T. (.5 oz.)	.4	44
(Kraft) Imperial	1 T. (.5 oz.)	.5	68
(Kraft) refrigerated	1 T. (.5 oz.)	.8	74
(Kraft) Roka	1 T. (.5 oz.)	.9	55
(Lawry's)	1 T. (.5 oz.)	.5	57
(Marzetti)	1 T. (.5 oz.)	.4	71
(Marzetti) refrigerated	1 T. (.5 oz.)	.5	82
(Wish-Bone) chunky	1 T. (.5 oz.)	.3	74
Boiled, home recipe (USDA)	1 T. (.6 oz.)	.7	26
Caesar (Kraft) Golden	1 T. (.5 oz.)	.4	63
Caesar (Kraft) Imperial	1 T. (.5 oz.)	.6	77
Caesar (Lawry's)	1 T. (.5 oz.)	.3	70
Canadian (Lawry's)	1 T. (.5 oz.)	.4	72
Coleslaw (Bernstein's)	1 T. (.5 oz.)	1.0	60
Coleslaw (Kraft)	1 T. (.5 oz.)	.2	62
French:			
Home recipe (USDA)	1 T. (.6 oz.)	<.1	101
Commercial (USDA)	1 T. (.6 oz.)	.1	66
(Bennett's)	1 T. (.5 oz.)	.1	56
(Bernstein's) regular or New Orleans	1 T. (.5 oz.)	.2	56
(Hellman's) Family	1 T. (.6 oz.)	<.1	65
(Kraft)	1 T. (.5 oz.)	<.1	65
(Kraft) Casino	1 T. (.5 oz.)	<.1	60
(Kraft) Catalina	1 T. (.5 oz.)	<.1	60

(USDA): United States Department of Agriculture

Food and Description	Measure or Quantity	Protein (grams)	Calories
(Kraft) *Miracle*	1 T. (.5 oz.)	<.1	57
(Lawry's)	1 T. (.5 oz.)	.2	60
(Lawry's) San Francisco	1 T. (.5 oz.)	.3	53
(Marzetti) Blue, refrigerated	1 T. (.5 oz.)	.5	70
(Marzetti) Country	1 T. (.5 oz.)	.3	72
(Nalley's)	.5 oz.	.3	55
(Wish-Bone) Deluxe	1 T. (.6 oz.)	.1	59
(Wish-Bone) Garlic	1 T. (.6 oz.)	Tr.	66
Fruit (Kraft)	1 T. (.5 oz.)	<.1	52
Garlic:			
French (Hellmann's) *Old Homestead*	1 T. (.6 oz.)	.1	68
(Marzetti) creamy	1 T. (.5 oz.)	<.1	73
(Marzetti) creamy, refrigerated	1 T. (.5 oz.)	.1	88
German Style (Marzetti)	1 T. (.5 oz.)	<.1	55
Green Goddess:			
(Bernstein's)	1 T. (.5 oz.)	.4	45
(Kraft)	1 T. (.5 oz.)	<.1	75
(Kraft) Imperial	1 T. (.5 oz.)	<.1	85
(Lawry's)	1 T. (.5 oz.)	.3	59
(Wish-Bone)	1 T. (.5 oz.)	.1	68
Green onion (Kraft)	1 T. (.5 oz.)	<.1	71
Hawaiian (Lawry's)	1 T. (.6 oz.)	1.2	77
Herb & garlic (Kraft)	1 T. (.5 oz.)	<.1	87
Italian:			
(USDA)	1 T. (.5 oz.)	<.1	83
(Bernstein's)	1 T. (.5 oz.)	.2	62
(Kraft)	1 T. (.5 oz.)	<.1	88
(Lawry's)	1 T. (.5 oz.)	<.1	80
(Lawry's) with cheese	1 T. (.5 oz.)	.2	60
(Marzetti) creamy	1 T. (.5 oz.)	<.1	73
(Marzetti) Sunny	1 T. (.5 oz.)	<.1	81
Mayonnaise-type salad dressing (USDA)	1 T. (.5 oz.)	.2	65
Miracle Whip (Kraft)	1 T. (.5 oz.)	<.1	69
Oil & vinegar (Kraft)	1 T. (.5 oz.)	<.1	65
Onion, California (Wish-Bone)	1 T. (.5 oz.)	.1	76
Potato salad (Marzetti)	1 T. (.5 oz.)	.1	62
Ranch Style (Marzetti)	1 T. (.5 oz.)	<.1	66
Red wine vinegar & oil (Lawry's)	1 T. (.6 oz.)	0.	61
Rich 'n' Tangy (Dutch Pantry)	1 T. (.6 oz.)	Tr.	68

(USDA): United States Department of Agriculture

Food and Description	Measure or Quantity	Protein (grams)	Calories
Romano Caesar (Marzetti)	1 T. (.5 oz.)	.3	69
Roquefort:			
(USDA)	1 T. (.5 oz.)	.7	76
(Bernstein's)	1 T. (.5 oz.)	.7	50
(Kraft) refrigerated	1 T. (.5 oz.)	.6	56
(Kraft) refrigerated, Imperial	1 T. (.5 oz.)	.5	68
(Marzetti) refrigerated	1 T. (.5 oz.)	.6	80
Royal Scandia (Bernstein's)	1 T. (.5 oz.)	.4	45
Russian:			
(USDA)	1 T. (.5 oz.)	.2	74
(Kraft)	1 T. (.5 oz.)	<.1	55
(Kraft) creamy	1 T. (.5 oz.)	<.1	68
(Marzetti) creamy	1 T. (.5 oz.)	<.1	75
(Wish-Bone)	1 T. (.5 oz.)	.1	54
(Saffola)	1 T. (.5 oz.)	.2	52
Salad Bowl (Kraft)	1 T. (.5 oz.)	<.1	53
Salad 'n Sandwich (Kraft)	1 T. (.5 oz.)	<.1	53
Salad Secret (Kraft)	1 T. (.5 oz.)	4.7	56
Sherry (Lawry's)	1 T. (.5 oz.)	.2	55
Slaw (Marzetti) regular or refrigerated	1 T. (.5 oz.)	.2	73
Spin Blend (Best Foods)	1 T. (.5 oz.)	<.1	56
Spin Blend Hellmann's)	1 T. (.5 oz.)	<.1	56
Supreme (McCormick)	.5 oz.	2.5	40
Sweet & Saucy (Marzetti)	1 T. (.5 oz.)	.1	70
Sweet 'n' Sour (Dutch Pantry)	1 T. (.6 oz.)	Tr.	76
Sweet & Sour (Kraft)	1 T. (.5 oz.)	<.1	28
Tahitian Isle (Wish-Bone)	1 T. (.5 oz.)	.1	54
Tang (Nalley's)	.5 oz.	.1	50
Tart & Creamy (Bama)	1 T. (.5 oz.)	.1	61
Thousand Island:			
(USDA)	1 T. (.6 oz.)	.1	80
(Bernstein's)	1 T. (.5 oz.)	.2	48
(Best Foods)	1 T. (.5 oz.)	.1	60
(Kraft)	1 T. (.5 oz.)	.2	71
(Kraft) Imperial	1 T. (.5 oz.)	.1	76
(Kraft) pourable	1 T. (.5 oz.)	.2	56
(Kraft) refrigerated	1 T. (.5 oz.)	.2	73
(Lawry's)	1 T. (.5 oz.)	.3	69
(Marzetti)	1 T. (.5 oz.)	.1	70
(Marzetti) refrigerated	1 T. (.5 oz.)	.1	71
(Wish-Bone)	1 T. (.5 oz.)	.2	71

(USDA): United States Department of Agriculture

Food and Description	Measure or Quantity	Protein (grams)	Calories
Tomato 'n' Spice (Dutch Pantry)	1 T. (.6 oz.)	Tr.	66
Vinaigrette (Bernstein's)	1 T. (.5 oz.)	.1	41

SALAD DRESSING, DIETETIC or LOW CALORIE:

Bleu or blue:

Low fat, 6% fat (USDA)	1 T. (.6 oz.)	.5	12
Low fat, 1% fat (USDA)	1 T. (.5 oz.)	.2	3
(Frenchette) chunky	1 T. (.5 oz.)	.6	20
(Kraft)	1 T. (.5 oz.)	.7	13
(Marzetti)	1 T. (.5 oz.)	.6	23
(Slim-ette)	1 T. (.5 oz.)	.6	12
(Tillie Lewis)	1 T. (.5 oz.)	.3	13
Caesar (Frenchette)	1 T. (.5 oz.)	.6	32
Catalina (Kraft)	1 T. (.5 oz.)	<.1	15
Cheese (Tillie Lewis)	1 T. (.5 oz.)	.3	12
Chef Style (Kraft)	1 T. (.5 oz.)	<.1	16
Chef's (Slim-ette)	1 T. (.5 oz.)	.1	14
Chef's (Tillie Lewis)	1 T. (.5 oz.)	Tr.	2
Coleslaw (Kraft)	1 T. (.5 oz.)	.2	28
Diet Mayo 7 (Bennett's)	1 T. (.5 oz.)	.3	23

French:

Low fat, 6% fat (USDA)	1 T. (.6 oz.)	<.1	15
Low fat, 1% fat, with artificial sweetener (USDA)	1 T. (.5 oz.)	<.1	2
Medium fat, with artificial sweetener (USDA)	1 T. (.5 oz.)	.1	23
(Bennett's)	1 T. (.5 oz.)	<.1	21
(Frenchette)	1 T. (.5 oz.)	Tr.	9
(Kraft)	1 T. (.5 oz.)	<.1	21
(Marzetti)	1 T. (.5 oz.)	<.1	11
(Tillie Lewis)	1 T. (.5 oz.)	.1	6
(Wish-Bone)	1 T. (.6 oz.)	<.1	23
Gourmet (Frenchette)	1 T. (.5 oz.)	.1	21
Green Goddess (Frenchette)	1 T. (.5 oz.)	.4	21
Green Goddess (Slim-ette)	1 T. (.5 oz.)	.1	12

Italian:

(USDA)	1 T. (.5 oz.)	<.1	8
(Bennett's)	1 T. (.5 oz.)	Tr.	7
(Bernstein's)	1 T. (.4 oz.)	<.1	4
(Bernstein's) with cheese	1 T. (.4 oz.)	.2	5
Italianette (Frenchette)	1 T. (.5 oz.)	Tr.	7

(USDA): United States Department of Agriculture

Food and Description	Measure or Quantity	Protein (grams)	Calories
(Kraft)	1 T. (.5 oz.)	<.1	10
(Marzetti)	1 T. (.5 oz.)	<.1	8
(Slim-ette)	1 T. (.5 oz.)	<.1	6
(Tillie Lewis)	1 T. (.5 oz.)	Tr.	<1
Mayonnaise, imitation:			
(USDA)	1 T. (.6 oz.)	.2	22
May-lo-naise (Tillie Lewis)	1 T. (.5 oz.)	1.0	16
Mayonette Gold (Frenchette)	1 T. (.5 oz.)	.1	32
Remoulade (Tillie Lewis)	1 T. (.5 oz.)	.2	10
Russian (Wish-Bone)	1 T. (.6 oz.)	.1	24
Slaw (Frenchette)	1 T. (.5 oz.)	Tr.	28
Slaw (Marzetti)	1 T. (.5 oz.)	Tr.	28
Thousand Island:			
(USDA)	1 T. (.5 oz.)	.1	27
(Frenchette)	1 T. (.5 oz.)	.2	22
(Kraft)	1 T. (.5 oz.)	.1	28
(Marzetti)	1 T. (.5 oz.)	.2	21
(Wish-Bone)	1 T. (.5 oz.)	.1	25
Vinaigrette (Bernstein's)	1 T. (.4 oz.)	Tr.	1
Whipped (Tillie Lewis)	1 T. (.5 oz.)	1.3	16

SALAD DRESSING MIX,

regular & low calorie:			
Bacon (Lawry's)	1 pkg. (.8 oz.)	3.3	69
Bleu or blue cheese:			
*(Good Seasons)	1 T. (.5 oz.)	.1	89
*(Good Seasons) thick, creamy	1 T. (.5 oz.)	.3	96
(Lawry's)	1 pkg. (.7 oz.)	3.9	79
Caesar garlic cheese (Lawry's)	1 pkg. (.8 oz.)	3.6	71
*Cheese garlic (Good Seasons)	1 T. (.5 oz.)	.1	84
*French (Good Seasons) old fashion	1 T. (.5 oz.)	.1	83
*French (Good Seasons) thick, creamy	1 T. (.5 oz.)	.2	97
French (Lawry's) old fashion	1 pkg. (.8 oz.)	1.1	72
*French, Riviera (Good Seasons)	1 T. (.6 oz.)	.1	90
*Garlic (Good Seasons)	1 T. (.5 oz.)	.1	84
Green Goddess (Lawry's)	1 pkg. (.8 oz.)	2.5	69
Italian:			
*(Good Seasons)	1 T. (.5 oz.)	.1	84
*(Good Seasons) cheese	1 T. (.5 oz.)	.1	89

(USDA): United States Department of Agriculture
*Prepared as Package Directs

Food and Description	Measure or Quantity	Protein (grams)	Calories
*(Good Seasons) mild	1 T. (.5 oz.)	.1	89
*(Good Seasons) thick, creamy	1 T. (.5 oz.)	.2	94
(Lawry's)	1 pkg. (.8 oz.)	1.1	44
(Lawry's) cheese	1 pkg. (.8 oz.)	2.6	69
*Low calorie (Good Seasons)	1 tsp.	.1	3
*Onion (Good Seasons)	1 T. (.5 oz.)	.1	84
*Thousand Island (Good Seasons) thick, creamy	1 T. (.5 oz.)	.2	80
SALAD SEASONING:			
(Durkee)	1 tsp. (4 grams)	.2	4
Salad Lift (French's)	1 tsp. (4 grams)	.2	6
Salad Mate (Durkee)	1 tsp. (4 grams)	.5	7
With cheese (Durkee)	1 tsp. (3 grams)	.5	10
SALAMI:			
Dry (USDA)	1 oz.	6.7	128
Cooked (USDA)	1 oz.	5.0	88
Cotto, all meat (Oscar Mayer)	.8-oz. slice (10 per ½ lb.)	3.4	55
Cotto, pure beef (Oscar Mayer)	.8-oz. slice	3.4	55
Dilusso Genoa (Hormel)	1 oz.	6.0	120
For beer (Oscar Mayer)	.8-oz. slice	3.4	52
Hard (Hormel) Dairy	1 oz.	5.7	120
Hard, all meat (Oscar Mayer)	1 slice (.4 oz.)	2.1	41
Machiaeh Brand, pure beef cooked (Oscar Mayer)	.8-oz. slice	3.2	60
SALISBURY STEAK:			
Canned (Morton House) with mushroom gravy	⅓ of 12-oz. can	8.7	160
Frozen:			
(Banquet) buffet	2 lb. pkg.	102.5	1524
(Banquet) cooking bag	5-oz. bag	16.0	239
(Swanson) *Hungry Man*	17-oz. dinner	45.4	943
(Swanson) with potato	6-oz. pkg.	16.0	360
Dinner (Banquet)	11-oz. dinner	20.8	335
Dinner (Morton)	11-oz. dinner	15.6	343
Dinner (Morton) 3-course	11-oz. dinner	22.0	642
Dinner (Swanson) 3-course	16-oz. dinner	27.1	517

(USDA): United States Department of Agriculture
*Prepared as Package Directs

Food and Description	Measure or Quantity	Protein (grams)	Calories
SALMON:			
Atlantic (USDA):			
Raw, whole	1 lb. (weighed whole)	66.3	640
Raw, meat only	4 oz.	25.5	246
Canned, solids & liq., including bones	4 oz.	24.6	230
Chinook or King (USDA):			
Raw, steak	1 lb. (weighed with bones)	76.2	886
Raw, meat only	4 oz.	21.7	252
Canned, solids & liq., including bones	4 oz.	22.2	238
Chum, canned, solids & liq., including bones (USDA)	4 oz.	24.4	158
Coho, canned, solids & liq. (USDA)	4 oz.	23.6	174
Pink or Humpback:			
Raw, steak (USDA)	1 lb. (weighed with bones)	79.8	475
Raw, meat only (USDA)	4 oz.	22.7	135
Canned, solids & liq.:			
(USDA)	4 oz.	23.2	160
(Del Monte)	7¾-oz. can	43.3	268
(Del Monte)	1 cup (8 oz.)	44.7	277
Sockeye or Red or Blueback:			
Canned, solids & liq.:			
Including bones (USDA)	4 oz.	23.0	194
(Bumble Bee)	1 cup (6 oz.)	33.6	286
(Del Monte)	7¾-oz. can	35.9	304
(Del Monte)	1 cup (8 oz.)	37.0	313
Unspecified kind of salmon, baked or broiled with vegetable shortening:			
(USDA)	4″ × 3″ × ½″ (4.2 oz.)	32.4	218
(USDA)	6¾″ × 2½″ × 1″ (5.1 oz.)	39.2	264
SALMON RICE LOAF, home recipe (USDA)	4 oz.	13.6	138

(USDA): United States Department of Agriculture

Food and Description	Measure or Quantity	Protein (grams)	Calories
SALMON, SMOKED:			
(USDA)	4 oz.	24.5	200
Lox, drained (Vita)	4-oz. jar	18.7	136
Nova, drained (Vita)	4-oz. jar	21.3	221
SALSIFY (USDA):			
Raw, without tops	1 lb. (weighed untrimmed)	11.4	51
Boiled, drained	4 oz.	2.9	14
SALT:			
Butter flavored, imitation (Durkee)	1 tsp. (4 grams)	Tr.	3
Garlic (French's)	1 tsp. (6 grams)	.1	4
Garlic, parslied (French's)	1 tsp. (4 grams)	.1	6
Garlic (Lawry's)	2.9-oz. pkg.	5.5	116
Garlic (Lawry's)	1 tsp. (4 grams)	.2	5
Hickory smoke (French's)	1 tsp. (4 grams)	.1	1
Lite Salt (Morton's)	1 tsp. (6 grams)	0.	0
Onion (French's)	1 tsp. (5 grams)	.1	5
Onion (Lawry's)	3-oz. pkg.	2.2	106
Onion (Lawry's)	1 tsp. (3 grams)	< .1	4
Seasoned (Lawry's)	3-oz. pkg.	2.0	21
Seasoned (Lawry's)	1 tsp. (5 grams)	.1	1
Seasoning (French's)	1 tsp. (4 grams)	Tr.	3
Substitute (Adolph's)	1 tsp. (6 grams)	Tr.	Tr.
Substitute, seasoned (Adolph's)	1 tsp. (5 grams)	Tr.	5
Substitute (Morton)	1 tsp. (6 grams)	0.	Tr.
Substitute, seasoned (Morton)	1 tsp. (6 grams)	< .1	3
Table (USDA)	1 tsp. (6 grams)	0.	0.
Table (Morton)	1 tsp. (7 grams)	0.	0.
SALT PORK, raw (USDA):			
With skin	1 lb. (weighed with skin)	17.0	3410
Without skin	1 oz.	1.1	222
SALT STICK (see **BREAD STICK**)			
SAND DAB, raw (USDA):			
Whole	1 lb. (weighed whole)	25.0	118

(USDA): United States Department of Agriculture

Food and Description	Measure or Quantity	Protein (grams)	Calories
Meat only	4 oz.	18.9	90
SANDWICH SPREAD:			
(USDA)	1 cup (8.7 oz.)	1.7	932
(USDA)	1 T. (.5 oz.)	.1	57
(USDA) low calorie	1 T. (.5 oz.)	.2	17
(Bama)	1 T. (.5 oz.)	.1	51
(Bennett's)	1 T. (.5 oz.)	< .1	44
(Best Foods)	1 T. (.5 oz.)	.2	62
(Hellmann's)	1 T. (.5 oz.)	.2	62
(Kraft)	1 oz.	.2	105
(Kraft) Salad Bowl	1 oz.	.2	101
(Nalley's)	1 oz.	.1	100
(Oscar Mayer)	1 oz.	2.6	60
Chicken salad (Carnation)	⅕ can (1.5 oz.)	4.8	99
Corned beef (Carnation)	⅕ can (1.5 oz.)	5.6	96
Ham & cheese (Carnation)	⅕ can (1.5 oz.)	5.4	80
Ham salad (Carnation)	⅕ can (1.5 oz.)	4.0	90
Pimento (Kraft)	1 oz.	3.4	117
Tuna salad (Carnation)	⅕ can (1.5 oz.)	5.1	84
Turkey salad (Carnation)	⅕ can (1.5 oz.)	5.5	92
SAPODILLA, fresh (USDA):			
Whole	1 lb. (weighed with skin & seeds)	1.8	323
Flesh only	4 oz.	.6	101
SAPOTES or MARMALADE PLUM, fresh (USDA):			
Whole	1 lb. (weighed with skin & seeds)	6.2	431
Flesh only	4 oz.	2.0	142
SARDINE:			
Atlantic, canned in oil (USDA);			
Solids & liq.	3¾-oz. can	21.8	330
Drained solids	3¾-oz. can	22.0	187
Atlantic, canned in tomato sauce, solids & liq. (Del Monte)	1½ large sardines (3.5 oz.)	17.8	138

(USDA): United States Department of Agriculture

Food and Description	Measure or Quantity	Protein (grams)	Calories
Norwegian, canned:			
(Snow)	1 oz.	2.3	66
In mustard sauce			
(Underwood)	3¾-oz. can	15.8	196
In oil, drained (Underwood)	3¾-oz. can	22.0	232
In tomato sauce (Underwood)	3¾-oz. can	15.8	169
Pacific (USDA):			
Raw	4 oz.	21.8	181
Canned, in brine or mustard, solids & liq.	4 oz.	21.3	222
Canned, in tomato sauce, solids & liq.	4 oz.	21.2	223
SAUCE, regular & dietetic:			
A1	1 T. (.6 oz.)	.2	12
Barbecue:			
(USDA)	½ cup (4.4 oz.)	1.9	114
(USDA)	1 T. (.6 oz.)	.2	15
(Contadina) oven	1 fl. oz. (1.2 oz.)	.4	29
(French's)	1 T.	.2	9
(French's) mild	1 T.	.2	17
(French's) smoky	1 T.	.2	15
(General Foods) hickory smoke, *Open Pit*	1 T. (.6 oz.)	.2	27
(General Foods) hot 'n spicy, *Open Pit*	1 T. (.6 oz.)	.3	27
(General Foods) original flavor, *Open Pit*	1 T. (.6 oz.)	.2	26
(General Foods) original flavor with onions, *Open Pit*	1 T. (.6 oz.)	.2	27
(Heinz) with onions, hickory smoke	1 T.	.2	19
(Heinz) with onions, regular	1 T.	.2	18
(Kraft)	1 oz.	.4	34
(Kraft) garlic	1 oz.	.4	32
(Kraft) hickory smoke	1 oz.	.4	34
(Kraft) hot	1 oz.	.4	31
(Kraft) mustard flavored	1 oz.	.7	28
(Kraft) onion	1 oz.	.4	41
Cheese (Kraft) *Deluxe Dinner*	1 oz.	4.2	77
Chili (see **CHILI SAUCE**)			
Creole (Contadina)	1 fl. oz. (1.1 oz.)	.5	21

(USDA): United States Department of Agriculture

Food and Description	Measure or Quantity	Protein (grams)	Calories
Escoffier Sauce Robert	1 T. (.6 oz.)	.3	20
Famous (Durkee) 6½-oz. bottle	1 T. (.6 oz.)	.6	72
Famous (Durkee) 10-oz. bottle	1 T. (.6 oz.)	.6	69
57 (Heinz)	1 T.	.4	14
Horseradish (Marzetti)	1 T. (.5 oz.)	<.1	58
H.P. Steak Sauce (Lea & Perrins)	1 T. (1 oz.)	Tr.	20
Marinara (Buitoni)	4 oz.	3.3	67
Marinara (Chef Boy-Ar-Dee)	¼ of 15-oz. can	1.6	68
Meat loaf (Contadina)	1 fl. oz. (1.1 oz.)	.6	16
Mushroom (Contadina)	1 fl. oz. (1.1 oz.)	.2	22
Newburg, canned (Snow)	1 oz.	.8	32
Savory (Heinz)	1 T.	.2	20
Seafood (Bernstein's)	1 T. (.5 oz.)	Tr.	19
Seafood cocktail (Del Monte)	1 T. (.6 oz.)	.3	20
Sloppy Joe (Contadina)	1 fl. oz. (1.1 oz.)	.4	18
Sloppy Joe, chili (Contadina)	1 fl. oz. (1.1 oz.)	.2	11
Sloppy Joe, pizza (Contadina)	1 fl. oz. (1.1 oz.)	.6	16
Soy (USDA)	1 oz.	1.6	19
Spaghetti (see **SPAGHETTI SAUCE**)			
Steak Supreme (Heublein)	1 T. (.6 oz.)	.2	20
Stroganoff (Contadina)	1 fl. oz. (1.1 oz.)	.4	17
Sweet 'n sour (Contadina)	1 fl. oz. (1.2 oz.)	.1	37
Sweet & sour (Kraft)	1 oz.	<.1	55
Swiss steak (Contadina)	1 fl. oz. (1.1 oz.)	.3	11
Tartar:			
(USDA) regular	1 T. (.5 oz.)	.2	74
(USDA) low calorie	1 T. (.5 oz.)	<.1	31
(Bama)	1 T. (.5 oz.)	.1	63
(Bennett's)	1 T. (.5 oz.)	.2	80
(Best Foods)	1 T. (.5 oz.)	.1	71
(Hellmann's)	1 T. (.5 oz.)	.1	71
(Kraft)	1 oz.	.3	145
(Marzetti)	1 T. (.5 oz.)	<.1	69
(Mrs. Paul's)	4.2-oz. pkg.	1.4	668
Tomato (see **TOMATO SAUCE**)			
White, home recipe (USDA):			
Thin	1 cup (8.8 oz.)	9.8	302
Medium	1 cup (9 oz.)	9.9	413
Thick	1 cup (8.7 oz.)	9.9	489

(USDA): United States Department of Agriculture

Food and Description	Measure or Quantity	Protein (grams)	Calories
Worcestershire:			
(French's)	1 T.	.1	6
(Heinz)	1 T.	.2	11
(Lea & Perrins)	1 T. (.6 oz.)	Tr.	12
SAUCE MIX:			
A la King (Durkee)	1.1-oz. pkg.	2.1	132
*A la King (Durkee)	¼ cup (2 oz.)	.6	34
*Barbecue (Kraft)	1 oz.	.3	32
*Bordelaise (Betty Crocker)	¼ cup	.3	33
Cheese:			
*(Betty Crocker)	¼ cup	4.4	87
(Durkee)	1.1-oz. pkg.	3.3	44
*(Durkee)	¼ cup	5.4	84
(French's)	1¼-oz. pkg.	8.8	165
*(French's)	¼ cup	4.3	81
(McCormick)	1¼-oz. pkg.	9.5	170
*(McCormick)	2-oz. serving	4.5	80
*Cheddar (Kraft)	1 oz.	2.2	52
Hollandaise:			
*(Betty Crocker)	¼ cup	1.2	84
(Durkee)	1-oz. pkg.	8.9	139
*(Durkee)	⅓ cup (3 oz.)	5.1	78
(French's)	1⅛-oz. pkg.	4.5	192
*(French's)	1 T.	.4	16
*(Kraft)	1 oz.	1.1	54
*(McCormick)	2-oz. serving	1.0	73
Miracle (Mrs. Paul's)	½-oz. pkg.	.1	10
*Mushroom (Betty Crocker)	¼ cup	.4	36
*Mushroom (Golden Grain) *Noodle-Roni*	½ cup	2.5	140
*Newburg (Betty Crocker)	¼ cup	2.6	68
Sloppy Joe (see **SLOPPY JOE MIX**)			
Sour cream:			
(Durkee)	1-oz. pkg.	3.9	109
*(Durkee)	⅓ cup (2.5 oz.)	4.5	108
(French's)	1¼-oz. pkg.	2.1	192
*(French's)	1 T.	.6	25
*(Kraft)	1 öz.	1.9	61
*(McCormick)	2-oz. serving	1.0	40

*Prepared as Package Directs

Food and Description	Measure or Quantity	Protein (grams)	Calories
Spaghetti (see **SPAGHETTI SAUCE MIX**)			
Stroganoff (French's)	1¾-oz. pkg.	6.3	192
*Stroganoff (French's)	⅓ cup	4.4	104
*Sweet-sour (Durkee)	1 cup (2-oz. pkg.)	1.1	230
Tartar (Lawry's)	.6-oz. pkg.	2.9	64
*Teri-yaki (Durkee)	⅔ cup (1¼-oz. pkg.)	.6	66
White:			
(Durkee) medium	1-oz. pkg.	1.9	155
*(Durkee) medium	¼ cup	2.6	79
*(Kraft)	1 oz.	1.2	44
*Supreme (McCormick)	2-oz. serving	.4	22
SAUERKRAUT, canned:			
Solids & liq. (USDA)	1 cup (8.3 oz.)	2.4	42
Drained solids (USDA)	1 cup (5 oz.)	2.0	31
Solids & liq. (Del Monte)	1 cup (8 oz.)	1.8	36
Solids & liq. (Steinfeld's Western Acres)	8-oz. can	2.2	50
Solids & liq. (Stokely-Van Camp)	1 cup (7.8 oz.)	2.2	40
SAUERKRAUT JUICE, canned (USDA)	½ cup (4.3 oz.)	.8	12
SAUGER, raw (USDA):			
Whole	1 lb. (weighed whole)	28.4	133
Meat only	4 oz.	20.3	95
SAUSAGE (see also individual kinds):			
Breakfast (Hormel)	8-oz. can	27.5	838
Breakfast, smoked, all meat (Oscar Mayer)	1 link (7 to 5-oz. pkg.)	2.8	67
Brown & serve:			
Before browning (USDA)	1 oz.	3.8	111
After browning (USDA)	1 oz.	4.7	120
After browning (Swift)	1 link (10 to 8-oz. pkg.)	3.3	88
Brown 'n Serve (Hormel)	1 piece (.8 oz.)	3.2	78
New England Brand, all meat (Oscar Mayer)	.8-oz. slice	3.9	36

(USDA): United States Department of Agriculture
*Prepared as Package Directs

Food and Description	Measure or Quantity	Protein (grams)	Calories
In sauce, canned (Prince)	1 oz.	10.5	187
SCALLION (see **ONION, GREEN**)			
SCALLOP:			
Raw, muscle only (USDA)	4 oz.	17.4	92
Steamed (USDA)	4 oz.	26.3	127
Frozen:			
Breaded, fried, reheated (USDA)	4 oz.	20.4	220
Breaded, fried (Mrs. Paul's)	7-oz. pkg.	25.9	411
Crisps (Gorton)	½ of 7-oz. pkg.	15.0	155
SCHAV SOUP (Manischewitz)	8 oz. (by wt.)	.5	11
***SCOTCH BROTH,** canned (Campbell)	1 cup	5.2	84
SCRAPPLE:			
(USDA)	4 oz.	10.0	244
(Oscar Mayer)	4 oz.	10.2	202
SCUP (see **PORGY**)			
SEABASS, WHITE, raw, meat only (USDA)	4 oz.	24.3	109
SEAFOOD CHOWDER, New England (Snow)	8 oz.	10.8	144
SEAFOOD PLATTER, breaded, fried, with potato puffs, frozen (Mrs. Paul's)	9-oz. pkg.	23.5	518
SEAFOOD SEASONING (French's)	1 tsp. (5 grams)	.1	2
SESAME SEED:			
Dry, whole (USDA)	1 oz.	5.3	160
Dry, hulled (USDA)	1 oz.	5.2	165
Liquid, Tahini (A. Sahadi)	1 T. (8 grams)	2.0	57

(USDA): United States Department of Agriculture
*Prepared as Package Directs

Food and Description	Measure or Quantity	Protein (grams)	Calories
SHAD (USDA):			
Raw, whole	1 lb. (weighed whole)	40.5	370
Raw, meat only	4 oz.	21.1	193
Cooked, home recipe:			
Baked with butter or margarine & bacon slices	4 oz.	26.3	278
Creole, made with tomatoes, onion, green pepper, butter & flour	4 oz.	17.0	172
Canned, solids & liq.	4 oz.	19.2	172
SHAD, GIZZARD raw (USDA):			
Whole	1 lb. (weighed whole)	25.7	299
Meat only	4 oz.	19.5	227
SHAKE'N BAKE, seasoned mixes:			
Chicken-coating	2⅜-oz. pkg.	7.2	276
Fish-coating	2-oz. pkg.	6.5	224
Hamburger-coating	2-oz. pkg.	7.2	158
Pork-coating	2⅜-oz. pkg.	7.2	260
SHALLOT, raw (USDA):			
With skin	1 oz.	.6	18
With skin removed	1 oz.	.7	20
SHEEFISH (see **INCONNU**)			
SHEEPSHEAD, Atlantic, raw (USDA):			
Whole	1 lb. (weighed whole)	29.0	159
Meat only	4 oz.	23.4	128
SHERBET (see also individual brands):			
Orange (USDA)	1 cup (6.8 oz.)	1.7	259
Orange (USDA)	¼ pint (3.4 oz.)	.9	130
Any flavor (Borden)	¼ pint (3 oz.)	1.0	120
Any flavor (Dean) 1.7% fat	¼ pint (3.3 oz.)	.7	137
Orange (Sealtest)	¼ pint (3.1 oz.)	1.3	120

(USDA): United States Department of Agriculture

Food and Description	Measure or Quantity	Protein (grams)	Calories
SHORTENING (see **FATS**)			
SHREDDED OATS, cereal, includes protein & other added nutrients (USDA)	1 oz.	5.3	107
SHREDDED WHEAT, cereal:			
(USDA) plain	1 cup (1.2 oz.)	3.5	124
(USDA) with malt & sugar	1 cup (2.1 oz.)	5.5	220
(Kellogg's) cinnamon or sugar, frosted, *Mini-Wheats*	4 biscuits (1 oz.)	2.5	107
(Nabisco)	1 biscuit (.9 oz.)	2.4	86
(Nabisco) *Spoon Size*	⅔ cup (1 oz.)	3.0	107
(Nabisco) *Spoon Size*	1 piece (1 gram)	.1	4
(Quaker)	2 biscuits (1⅓ oz.)	3.6	135
SHRIMP:			
Raw (USDA):			
Whole	1 lb. (weighed in shell)	56.7	285
Meat only	4 oz.	20.5	103
Canned:			
Dry pack or drained (USDA)	1 cup (22 large or 76 small, 4.5 oz.)	30.1	148
Wet pack, solids & liq. (USDA)	4 oz.	18.4	91
Solids & liq. (Bumble Bee)	4½-oz. can	18.3	90
Cooked, french-fried, dipped in egg, breadcrumbs & flour or in batter (USDA)	4 oz.	23.0	255
Frozen:			
Raw:			
Breaded, not more than 50% breading (USDA)	4 oz.	13.9	158
Breaded (Gorton)	¼ of 1-lb. pkg.	14.0	158
Recipe Shrimp (Henderson's)	¼ of 20-oz. pkg.	15.6	71
Cooked (Sau-Sea)	4 oz.	17.4	73
Cooked (Weight Watchers)	¼ of 1-lb. pkg.	13.2	67
Fried (Mrs. Paul's)	6-oz. pkg.	20.2	316
Scampi (Gorton)	½ of 7½-oz. pkg.	13.0	285

(USDA): United States Department of Agriculture

[293]

Food and Description	Measure or Quantity	Protein (grams)	Calories
SHRIMP CAKE:			
Frozen, fried, breaded (Mrs. Paul's)	1 cake (3 oz.)	2.8	157
Frozen, thins (Mrs. Paul's)	1 cake (2½ oz.)	7.5	158
SHRIMP COCKTAIL:			
(Sau-Sea)	4-oz. jar	6.0	107
(Sea Snack)	4-oz. jar	8.2	110
SHRIMP DINNER, frozen:			
(Morton)	7¾-oz. dinner	15.4	374
(Swanson)	8-oz. dinner	18.4	358
SHRIMP PASTE, canned (USDA)	1 oz.	5.9	51
SHRIMP PUFF, frozen (Durkee)	1 piece (.5 oz.)	2.0	44
SHRIMP SOUP, Cream of:			
*Canned (Campbell)	1 cup (8 oz.)	6.8	171
Frozen:			
Condensed (USDA)	8 oz. (by wt.)	9.1	302
*Prepared with equal volume water (USDA)	1 cup (8.5 oz.)	4.8	158
*Prepared with equal volume milk (USDA)	1 cup (8.6 oz.)	9.3	243
SKATE, raw, meat only (USDA)	4 oz.	24.4	111
SLENDER (Carnation):			
Dry, all flavors	1 pkg. (1 oz.)	9.5	104
Liquid, all flavors	1 pkg. (1 oz.)	16.2	225
SLIM JIM:			
Polish sausage, all beef	1 piece (1¾ oz.)	5.6	108
Sausage	1 piece (¼ oz.)	3.3	83
SLOPPY JOE:			
Canned, with beef (Morton House)	15-oz. can	33.2	706
Frozen (Banquet) cooking bag	5-oz. bag	14.8	251
Mix, including seasoning mix:			
(Durkee)	1½-oz. pkg.	1.0	97

(USDA): United States Department of Agriculture
*Prepared as Package Directs

Food and Description	Measure or Quantity	Protein (grams)	Calories
*With tomato paste & meat			
(Durkee)	3 cups (1½-oz. pkg.)	87.0	1453
(French's)	1½-oz. pkg.	2.6	117
*(Kraft)	1 oz.	3.6	47
(Lawry's)	1½-oz. pkg.	3.5	139
*(Wyler's)	6 fl. oz.	.9	25
Pizza flavor (Durkee)	1-oz. pkg.	.8	97
*Pizza flavor (Durkee)	3 cups	89.7	1490
SMELT, Atlantic, jack & bay (USDA):			
Raw, whole	1 lb. (weighed whole)	46.4	244
Raw, meat only	4 oz.	21.1	111
Canned, solids & liq.	4 oz.	20.9	227
SMOKIE SAUSAGE:			
(Eckrich):			
Smoked, skinless	2 oz.	3.5	94
Smokee, from 12-oz. pkg.	1.2-oz. link	4.0	110
Smokee, from 1-lb. pkg.	1.6-oz. link	6.0	152
Smokette	1 piece	3.0	79
Smok-Y-Links, with maple flavor or skinless	1 piece	3.0	79
(Hormel)	1 piece (.8 oz.)	3.2	75
(Oscar Mayer):			
8 links per ¾ lb.	1 link (1.5 oz.)	5.6	131
7 links per 5 oz.	1 link (.7 oz.)	2.6	61
Cheese	1.5-oz. link	5.6	131
Little Smokies, all meat	1 link (16 per 5 oz.)	1.2	29
Smoky Snax	1 link (4 oz.)	14.7	350
(Wilson)	1 oz.	4.0	84
SMOKY SNAX SPREAD (Oscar Mayer)	1 oz.	3.4	101
SNACK (see CRACKER, POPCORN, POTATO CHIP, etc.)			
SNAIL, raw:			
(USDA)	4 oz.	18.3	102
Giant African (USDA)	4 oz.	11.2	83

(USDA): United States Department of Agriculture
*Prepared as Package Directs

Food and Description	Measure or Quantity	Protein (grams)	Calories
SNAPPER (see **RED SNAPPER**)			
SNO BALL (Hostess)	1 cake (1.5 oz.)	1.4	162
SOFT SWIRL (Jell-O):			
*All flavors except chocolate	½ cup (3.8 oz.)	3.1	168
*Chocolate	½ cup (4 oz.)	4.1	190
SOLE:			
Raw, whole (USDA)	1 lb. (weighed whole)	25.0	118
Raw, meat only (USDA)	4 oz.	18.9	90
Frozen:			
(Gorton)	⅓ of 1-lb. pkg.	25.0	120
Dinner (Weight Watchers)	18-oz. dinner	43.9	279
& cauliflower (Weight Watchers)	9½-oz. luncheon	23.1	190
In lemon butter (Gorton)	⅓ of 9-oz. pkg.	11.0	147
SORGHUM, grain (USDA)	1 oz.	3.1	94
SORREL (see **DOCK**)			
SOUP (see individual listing by kind)			
SOUP BASE, beef or chicken (Wyler's)	1 tsp. (6 grams)	.6	29
SOURSOP, raw, (USDA):			
Whole	1 lb. (weighed with skin & seeds)	3.1	200
Flesh only	4 oz.	1.1	74
SOUSE (USDA)	1 oz.	3.7	51
SOYBEAN (USDA):			
Young seeds:			
Raw	1 lb. (weighed in pods)	26.2	322
Boiled, drained	4 oz.	11.1	134
Canned, solids & liq.	4 oz.	7.4	85
Canned, drained solids	4 oz.	10.2	117

(USDA): United States Department of Agriculture
*Prepared as Package Directs

Food and Description	Measure or Quantity	Protein (grams)	Calories
Mature seeds, dry:			
Raw	1 lb.	154.7	1828
Raw	1 cup (7.4 oz.)	71.6	846
Cooked	4 oz.	12.5	147
Roasted, *Soy Town*	1 oz.	10.8	145
SOYBEAN CURD or TOFU (USDA)	4.2-oz. cake (2¾" × 2½" × 1")	9.4	86
SOYBEAN FLOUR (see **FLOUR**)			
SOYBEAN GRITS, high fat (USDA)	1 cup (4.9 oz.)	56.9	524
SOYBEAN MILK (USDA):			
Fluid	4 oz.	3.9	37
Powder	1 oz.	11.9	122
Sweetened:			
Dry powder	1 oz.	5.8	128
Liquid concentrate	4 oz. (by wt.)	5.4	143
SOYBEAN PROTEIN (USDA)	1 oz.	21.2	91
SOYBEAN PROTEINATE (USDA)	1 oz.	22.9	88
SOYBEAN SPROUT (see **BEAN SPROUT**)			
SOY SAUCE (see **SAUCE**)			
SPAGHETTI. Plain spaghetti products are essentially the same in caloric value and protein content on the same weight basis. The longer the cooking, the more water is absorbed and this affects the nutritive value (USDA):			
Dry	1 oz.	3.5	105
Dry, broken	1 cup (2.5 oz.)	8.9	262

(USDA): United States Department of Agriculture

Food and Description	Measure or Quantity	Protein (grams)	Calories
Cooked:			
8-10 minutes, "al dente"	1 cup (5.1 oz.)	7.3	216
8-10 minutes, "al dente"	4 oz.	5.7	168
14-20 minutes, tender	1 cup (4.9 oz.)	4.8	155
14-20 minutes, tender	4 oz.	3.9	126
SPAGHETTI DINNER:			
*With meat balls (Chef Boy-Ar-Dee)	8¾-oz. pkg.	15.9	362
*With meat sauce (Chef Boy-Ar-Dee)	7-oz. pkg.	10.7	285
*With meat sauce (Kraft) *Deluxe*	4 oz.	6.4	151
*With mushroom sauce (Chef Boy-Ar-Dee)	7-oz. pkg.	9.5	273
Frozen, with meat balls:			
(Banquet)	11½-oz. dinner	14.7	450
(Morton)	11-oz. dinner	12.5	390
(Swanson)	12-oz. dinner	14.3	323
SPAGHETTI & FRANKFURTERS in TOMATO SAUCE, canned: *SpaghettiO's*			
(Franco-American)	1 cup	8.9	253
(Heinz)	8½-oz. can	10.8	308
SPAGHETTI & GROUND BEEF in TOMATO SAUCE, canned:			
(Buitoni)	8 oz.	9.3	266
(Chef Boy-Ar-Dee)	½ of 15-oz. can	7.9	192
(Franco-American)	1 cup	11.1	272
(Nalley's)	8 oz.	9.2	220
SPAGHETTI & MEATBALLS in TOMATO SAUCE:			
Home recipe (USDA)	1 cup (8.7 oz.)	18.6	332
Canned:			
(USDA)	1 cup (8.8 oz.)	12.2	258
(Buitoni)	8 oz.	13.9	258
(Chef Boy-Ar-Dee)	⅕ of 40-oz. can	7.9	213
(Franco-American)	1 cup	11.1	264

(USDA): United States Department of Agriculture
*Prepared as Package Directs

Food and Description	Measure or Quantity	Protein (grams)	Calories
SpaghettiO's			
(Franco-American)	1 cup	10.7	215
(Hormel)	15-oz. can	17.8	348
(Van Camp)	1 cup (7.8 oz.)	10.6	226
Frozen (Buitoni)	8 oz.	13.3	262
Frozen (Morton)	8-oz. casserole	13.6	288
Frozen (Morton)	20-oz. casserole	34.0	720
SPAGHETTI with MEAT SAUCE:			
Canned (Heinz)	8½-oz. can	10.2	207
Frozen:			
(Banquet) cooking bag	8-oz. bag	13.8	311
(Banquet) buffet	2 lbs.	61.8	1324
(Banquet) entree	8 oz.	13.8	310
(Kraft)	12½-oz. pkg.	12.4	343
(Morton)	8-oz. casserole	13.9	289
(Morton)	20-oz. casserole	34.8	722
SPAGHETTI MIX (Kraft):			
*American style	4 oz.	3.7	121
*Italian style	4 oz.	4.5	119
SPAGHETTI SAUCE:			
Clam, red (Buitoni)	4 oz.	8.8	106
Clam, white (Buitoni)	4 oz.	11.2	140
Italian, frozen (Celeste)	½ cup (4 oz.)	2.3	68
Italian, canned (Contadina)	½ cup	2.0	76
Meat:			
(Buitoni)	4 oz.	4.0	111
(Chef Boy-Ar-Dee)	¼ of 15-oz. can	3.5	93
(Heinz)	½ cup	3.7	110
Meatball (Chef Boy-Ar-Dee)	⅓ of 15-oz. can	8.5	202
With ground meat (Chef Boy-Ar-Dee)	⅐ of 29-oz. jar	6.0	136
Meatless or plain:			
(Buitoni)	4 oz.	3.5	76
(Chef Boy-Ar-Dee)	¼ of 16-oz. jar	1.5	73
(Heinz)	½ cup	2.7	98
Mushroom:			
(Buitoni)	4 oz.	3.4	69
(Chef Boy-Ar-Dee)	¼ of 15-oz. can	1.4	68
(Heinz)	½ cup	2.6	97

*Prepared as Package Directs

Food and Description	Measure or Quantity	Protein (grams)	Calories
With meat (Heinz)	½ cup	3.7	106
SPAGHETTI SAUCE MIX:			
(Durkee)	1½-oz. pkg.	1.4	85
*(Durkee)	½ cup	1.4	45
*(Kraft)	4 oz.	2.0	60
*(McCormick)	4-oz. serving	2.0	80
Italian (French's)	1½-oz. pkg.	3.1	108
*Italian (French's)	½ cup	1.8	90
*Prepared with oil (Spatini)	½ cup (4.2 oz.)	3.0	76
Meatball (Lawry's)	3½-oz. pkg.	11.9	330
Mushroom (Durkee)	1.2-oz. pkg.	1.6	69
*Mushroom (Durkee)	⅔ cups (1.2-oz. pkg.)	7.4	208
Mushroom (French's)	1⅜-oz. pkg.	4.3	122
*Mushroom (French's)	½ cup	2.6	73
Mushroom (Lawry's)	1½-oz. pkg.	6.5	147
*(Wyler's)	½ cup	.5	21
SPAGHETTI WITH TOMATO SAUCE:			
Twists (Buitoni)	8 oz.	7.5	142
(Van Camp)	1 cup (7.8 oz.)	4.8	168
With cheese:			
Home recipe (USDA)	1 cup (8.8 oz.)	8.8	260
Canned:			
(USDA)	1 cup (8.8 oz.)	5.5	190
(Chef Boy-Ar-Dee)	⅕ of 40-oz. can	4.5	156
(Franco-American)	1 cup	7.3	185
SpaghettiO's (Franco-American)	1 cup	6.1	183
Italian Style (Franco-American)	1 cup	8.0	174
(Heinz)	8½-oz. can	4.8	164
SPAM (Hormel), canned:			
Regular	3 oz.	11.9	260
& cheese	3 oz.	12.9	260
Spread	1 oz.	4.3	80

(USDA): United States Department of Agriculture
*Prepared as Package Directs

Food and Description	Measure or Quantity	Protein (grams)	Calories
SPANISH MACKEREL, raw (USDA)			
Whole	1 lb. (weighed whole)	54.0	490
Meat only	4 oz.	22.1	201
SPANISH-STYLE VEGETABLES, frozen (Birds Eye)	⅓ of 10-oz. pkg.	1.1	85
SPECIAL K, cereal (Kellogg's)	1¼ cups (1 oz.)	6.1	107
SPICE CAKE MIX:			
*(Betty Crocker)	1/12 of cake	3.0	202
*(Duncan Hines)	1/12 of cake (2.7 oz.)	3.0	199
*(Pillsbury) streusel	1/12 of cake	4.0	350
*Apple with raisins (Betty Crocker)	1/12 of cake	3.1	206
Honey:			
(USDA)	1 oz.	1.2	126
*Prepared with eggs, water, caramel icing (USDA)	2 oz.	2.3	200
SPINACH:			
Raw (USDA):			
Untrimmed	1 lb. (weighed with large stems & roots)	10.5	85
Trimmed or packaged	1 lb.	14.5	118
Trimmed, whole leaves	1 cup (1.2 oz.)	1.1	9
Trimmed, chopped	1 cup (1.8 oz.)	1.7	14
Boiled, whole leaves, drained (USDA)	1 cup (5.5 oz.)	4.7	36
Canned, regular pack:			
Solids & liq. (USDA)	½ cup (4.1 oz.)	2.3	22
Drained solids (USDA)	½ cup (4 oz.)	3.0	27
Drained liq. (USDA)	4 oz.	.6	7
Drained solids (Del Monte)	½ cup (4 oz.)	3.4	24
Solids & liq. (Stokely-Van Camp)	½ cup (3.9 oz.)	2.2	21
Canned, dietetic pack, low-sodium:			
Solids & liq. (USDA)	4 oz.	2.8	24

(USDA): United States Department of Agriculture
*Prepared as Package Directs

[301]

Food and Description	Measure or Quantity	Protein (grams)	Calories
Drained solids (USDA)	4 oz.	3.6	29
Drained liq. (USDA)	4 oz.	.6	9
Solids & liq. (Blue Boy)	4 oz.	2.1	22
Frozen:			
Chopped:			
Not thawed (USDA)	4 oz.	3.5	27
Boiled, drained (USDA)	4 oz.	3.4	26
(Birds Eye)	⅓ of 10-oz. pkg.	2.8	23
Deviled, with cheddar cheese, casserole (Green Giant)	⅓ of 10-oz. pkg.	4.3	70
In cream sauce (Green Giant)	⅓ of 10-oz. pkg.	2.4	57
Leaf:			
Not thawed (USDA)	4 oz.	3.4	28
Boiled, drained (USDA)	4 oz.	3.3	27
(Birds Eye)	⅓ of 10-oz. pkg.	2.8	23
Creamed (Birds Eye)	⅓ of 9-oz. pkg.	2.3	61
In butter sauce (Green Giant)	⅓ of 10-oz. pkg.	2.4	48

SPINACH, NEW ZEALAND (see **NEW ZEALAND SPINACH**)

SPINACH SOUFFLE, frozen (Stouffer's)	12-oz. pkg.	20.0	484

SPINY LOBSTER (see **CRAYFISH**)

SPLEEN, raw (USDA):			
Beef & calf	4 oz.	20.5	118
Hog	4 oz.	19.4	121
Lamb	4 oz.	21.3	130

SPONGE CAKE, home recipe (USDA)	1/12 of 10″ cake (2.3 oz.)	5.0	196

SPOT, fillets (USDA):			
Raw	1 lb.	79.8	993
Baked	4 oz.	25.9	335

(USDA): United States Department of Agriculture

Food and Description	Measure or Quantity	Protein (grams)	Calories
SQUAB, pigeon, raw (USDA):			
Dressed	1 lb. (weighed with feet, inedible viscera & bones)	38.0	569
Meat & skin	4 oz.	21.0	333
Meat only	4 oz.	19.8	161
Light meat only, without skin	4 oz.	23.1	142
Giblets	1 oz.	5.6	44
SQUASH SEEDS, dry (USDA):			
In hull	4 oz.	24.4	464
Hulled	1 oz.	8.2	157
SQUASH, SUMMER:			
Fresh (USDA):			
Crookneck & Straightneck, yellow:			
Whole	1 lb. (weighed untrimmed)	5.3	89
Boiled, drained, diced	½ cup (3.6 oz.)	1.0	15
Boiled, drained, slices	½ cup (3.1 oz.)	.9	13
Scallop, white & pale green:			
Whole	1 lb. (weighed untrimmed)	4.0	93
Boiled, drained, mashed	½ cup (4.2 oz.)	.8	19
Zucchini & Cocozelle, green:			
Whole	1 lb. (weighed untrimmed)	5.2	73
Boiled, drained slices	½ cup (2.7 oz.)	.8	9
Canned, zucchini in tomato sauce (Del Monte)	½ cup (4.1 oz.)	1.3	25
Frozen:			
Not thawed (USDA)	4 oz.	1.6	24
Boiled, drained (USDA)	4 oz.	1.6	24
Fried, breaded, zucchini (Mrs. Paul's)	9-oz. pkg.	11.3	567
Parmesan, zucchini (Mrs. Paul's)	12-oz. pkg.	18.4	259
Summer squash, slices (Birds Eye)	½ cup (3.3 oz.)	1.4	20
Zucchini (Birds Eye)	⅓ of 10-oz. pkg.	.9	20

(USDA): United States Department of Agriculture

Food and Description	Measure or Quantity	Protein (grams)	Calories
SQUASH, WINTER:			
Fresh (USDA):			
Acorn:			
Whole	1 lb. (weighed with skin & seeds)	5.2	152
Baked, flesh only, mashed	½ cup (3.6 oz.)	1.9	56
Boiled, mashed	½ cup (4.1 oz.)	1.4	39
Butternut:			
Whole	1 lb. (weighed with skin & seeds)	4.4	171
Baked, flesh only	4 oz.	2.0	77
Boiled, flesh only	4 oz.	1.2	46
Hubbard:			
Whole	1 lb. (weighed with skin & seeds)	4.2	117
Baked, flesh only	4 oz.	2.0	57
Baked, flesh only, mashed	½ cup (3.6 oz.)	1.8	51
Boiled, flesh only, diced	½ cup (4.2 oz.)	1.3	35
Frozen:			
Not thawed (USDA)	4 oz.	1.4	43
Heated (USDA)	½ cup (4.2 oz.)	1.4	46
(Birds Eye)	⅓ of 12-oz. pkg.	1.4	43
SQUID, raw, meat only (USDA)	4 oz.	18.6	
STARCH (see **CORNSTARCH**)			
START, instant breakfast drink	½ cup (4.7 oz.)	Tr.	60
STOCKPOT SOUP, canned (Campbell)	1 cup	5.1	90
STOMACH, PORK, scalded (USDA)	4 oz.	18.7	172
STRAINED FOOD (see **BABY FOOD**)			
STRAWBERRY:			
Fresh, whole (USDA)	1 lb. (weighed with caps & stems)	3.0	161
Fresh, whole, capped (USDA)	1 cup (5.1 oz.)	1.0	53

(USDA): United States Department of Agriculture
*Prepared as Package Directs

Food and Description	Measure or Quantity	Protein (grams)	Calories
Canned, unsweetened, or low calorie:			
Water pack, solids & liq.			
(USDA)	4 oz.	.5	25
Solids & liq. (Blue Boy)	4 oz.	1.0	16
Low calorie, solids & liq.			
(S and W) *Nutradiet*	4 oz.	.6	23
Frozen:			
Sweetened, whole, not thawed:			
(USDA)	16-oz. can	1.8	418
(USDA)	½ cup (4.5 oz.)	.5	116
Sweetened, sliced, not thawed (USDA)	10-oz. pkg.	1.4	310
Sweetened, sliced, not thawed (USDA)	½ cup (4.5 oz.)	.6	138
Whole (Birds Eye)	¼ of 1-lb. pkg.	.6	101
Halves (Birds Eye)	½ cup (5.3 oz.)	.9	162
Quick-thaw (Birds Eye)	½ cup (5 oz.)	.7	122
STRAWBERRY CAKE:			
*Mix (Duncan Hines)	1/12 of cake (2.7 oz.)	3.4	207
*Mix (Pillsbury)	1/12 of cake	3.0	190
Frozen, Shortcake (Mrs. Smith's)	1/6 of 9" cake	1.8	379
STRAWBERRY DRINK (Hi-C)			
Cooler	6 fl. oz. (6.6 oz.)	.7	98
STRAWBERRY ICE CREAM,			
(Sealtest)	¼ pt. (2.3 oz.)	1.9	133
STRAWBERRY PIE:			
Home recipe (USDA)	1/6 of 9" pie (5.6 oz.)	3.0	313
Creme (Tastykake)	4-oz. pie	3.8	356
Frozen:			
Cream (Banquet)	2½-oz. serving	1.8	187
Cream (Morton)	1/6 of 16-oz. pie	1.7	183
Cream (Mrs. Smith's)	1/6 of 8" pie (4.2 oz.)	1.2	221
STRAWBERRY PIE FILLING:			
(Comstock)	½ cup (5.4 oz.)	.4	159
(Lucky Leaf)	8 oz.	.8	248

(USDA): United States Department of Agriculture
*Prepared as Package Directs

Food and Description	Measure or Quantity	Protein (grams)	Calories
(Wilderness)	21-oz. can	2.7	738
STRAWBERRY PRESERVE or JAM:			
Sweetened:			
(Bama)	1 T. (.7 oz.)	.1	54
(Smucker's)	1 T. (.7 oz.)	.1	51
Dietetic or low calorie:			
(Diet Delight)	1 T. (.6 oz.)	.1	23
(Kraft)	1 oz.	.1	36
(S and W) *Nutradiet*	1 T. (.5 oz.)	<.1	12
(Slenderella)	1 T. (.7 oz.)	Tr.	25
(Smucker's)	1 T.(.5 oz.)	<.1	2
(Tillie Lewis)	1 T. (.5 oz.)	.1	12
STRAWBERRY RENNET MIX:			
Powder:			
Dry (Junket)	1 oz.	<.1	115
*(Junket)	4 oz.	3.7	108
Tablet:			
Dry (Junket)	1 tablet (<1 gram)	Tr.	1
& sugar (Junket)	4 oz.	3.7	101
STRAWBERRY-RHUBARB PIE:			
(Tastykake)	4-oz. pie	3.5	399
Frozen:			
(Mrs. Smith's)	⅙ of 8″ pie (4.2 oz.)	2.0	319
(Mrs. Smith's)	⅛ of 10″ pie (5.6 oz.)	2.6	419
(Mrs. Smith's) natural juice	⅙ of 8″ pie (4.2 oz.)	2.5	316
STRAWBERRY-RHUBARB PIE FILLING, canned (Lucky Leaf)	8 oz.	.6	258
STRAWBERRY SOFT DRINK (Yoo-Hoo) High Protein	6 fl. oz. (6.4 oz.)	6.0	100
STRAWBERRY TURNOVER, frozen (Pepperidge Farm)	1 turnover (3.3 oz.)	2.9	326
STRUDEL, frozen (Pepperidge Farm):			
Apple	⅙ of strudel (2.5 oz.)	2.1	202
Blueberry	⅙ of strudel (2.5 oz.)	2.1	240

*Prepared as Package Directs

Food and Description	Measure or Quantity	Protein (grams)	Calories
Cherry	⅙ of strudel (2.5 oz.)	2.3	204
Pineapple-cheese	⅙ of strudel (2.3 oz.)	4.4	209
STURGEON (USDA):			
Raw, section	1 lb. (weighed with skin & bones)	69.8	362
Raw, meat only	4 oz.	20.5	107
Smoked	4 oz.	35.4	169
Steamed	4 oz.	28.8	181
SUCCOTASH, frozen:			
Not thawed (USDA)	4 oz.	4.9	110
Boiled, drained (USDA)	½ cup (3.4 oz.)	4.0	89
(Birds Eye)	½ cup (3.3 oz.)	4.1	87
SUCKER, CARP, raw (USDA):			
Whole	1 lb. (weighed whole)	34.0	196
Meat only	4 oz.	21.8	126
SUCKER, including **WHITE AND MULLET,** raw (USDA):			
Whole	1 lb. (weighed whole)	40.2	203
Meat only	4 oz.	23.4	118
SUET, raw (USDA)	1 oz.	.4	242
SUGAR	1 T. (.4 oz.)	0.	46
SUGAR APPLE, raw (USDA):			
Whole	1 lb. (weighed with skin & seeds)	3.7	192
Flesh only	4 oz.	2.0	107
SUGAR CHEX, cereal (Ralston Purina)	⅞ cup (1 oz.)	1.5	119
SUGAR FROSTED FLAKES, cereal:			
(Kellogg's)	¾ cup (1 oz.)	1.4	108
(Ralston Purina)	¾ cup (1 oz.)	1.3	113

(USDA): United States Department of Agriculture

[307]

Food and Description	Measure or Quantity	Protein (grams)	Calories
SUGAR JETS, cereal (General Mills)	1 cup (1 oz.)	2.1	111
SUGAR POPS, cereal (Kellogg's)	1 cup (1 oz.)	1.4	109
SUGAR SMACKS, cereal (Kellogg's)	1 cup (1 oz.)	1.8	111
SUGAR SPARKLED TWINKLES, cereal (General Mills)	1 cup (1 oz.)	1.9	112
SUGAR SUBSTITUTE (Adolph's)	1 tsp. (4 grams)	0.	14
SUKI-YAKI MIX: (Durkee)	1¾-oz. pkg.	.8	82
*With meat & vegetables (Durkee)	1 cup	14.9	293
SUNFLOWER SEED: In hulls (USDA)	4 oz. (weighed in hull)	14.7	343
Hulled (USDA)	1 oz.	6.8	159
Hulled (Planters)	1 oz.	6.5	164
Dry roasted (Flavor House)	1 oz.	7.7	178
Dry roasted (Planters)	1 oz.	6.5	159
Toasted, salted (Planters)	1 oz.	5.8	170
SUNFLOWER SEED FLOUR (see **FLOUR**)			
SUPER ORANGE CRISP WHEAT PUFFS, cereal (Post)	1 cup (1 oz.)	1.6	109
SUPER SUGAR CRISP WHEAT PUFFS, cereal (Post)	⅞ cup (1 oz.)	1.7	107
SURINAM CHERRY (see **PITANGA**)			
SUZY Q (Hostess)	1 cake (2¼ oz.)	2.2	268

(USDA): United States Department of Agriculture
*Prepared as Package Directs

Food and Description	Measure or Quantity	Protein (grams)	Calories
SWAMP CABBAGE (USDA):			
Raw, whole	1 lb. (weighed untrimmed)	11.0	107
Boiled, trimmed, drained	4 oz.	2.5	24
SWEETBREADS (USDA):			
Beef, raw	1 lb.	66.2	939
Beef, braised	4 oz.	29.4	363
Calf, raw	1 lb.	80.7	426
Calf, braised	4 oz.	37.0	191
Hog (see **PANCREAS**)			
Lamb, raw	1 lb.	64.0	426
Lamb, braised	4 oz.	31.9	198
SWEET POTATO:			
Raw (USDA):			
All kinds, unpared	1 lb. (weighed whole)	6.2	419
All kinds, pared	4 oz.	1.9	129
Firm-fleshed, Jersey types, pared	4 oz.	2.0	116
Soft-fleshed, Puerto Rico variety, pared	4 oz.	1.9	133
Baked, peeled after baking (USDA)	3.9-oz. sweet potato (5″ × 2″)	2.3	155
Boiled, peeled after boiling (USDA)	5-oz. sweet potato (5″ × 2″)	2.5	168
Candied, home recipe (USDA)	6.2-oz. sweet potato (3½″ × 2¼″)	2.3	294
Canned, regular pack:			
In syrup, solids & liq. (USDA)	4 oz.	1.1	129
Vacuum or solid pack (USDA)	½ cup (3.8 oz.)	2.2	118
Heavy syrup, solids & liq. (Del Monte)	½ cup (4.2 oz.)	.8	138
Vacuum pack (Taylor's)	½ cup	1.9	135
Canned, dietetic pack (USDA)	4 oz.	.8	52
Dehydrated flakes, dry (USDA)	½ cup (2 oz.)	2.4	220
*Dehydrated flakes, prepared with water (USDA)	½ cup (4.4 oz.)	1.3	120
Frozen:			
Candied (Mrs. Paul's)	⅓ of 12-oz. pkg.	1.0	186

(USDA): United States Department of Agriculture
*Prepared as Package Directs

Food and Description	Measure or Quantity	Protein (grams)	Calories
Sweets & Apples, candied			
(Mrs. Paul's)	⅓ of 12-oz. pkg.	1.2	145
Candied yams (Birds Eye)	⅓ of 12-oz. pkg.	1.4	215
With brown sugar, pineapple			
glaze (Birds Eye)	½ cup (3.3 oz.)	1.4	145
SWEET POTATO PIE:			
Home recipe (USDA)	⅙ of 9″ pie (5.4 oz.)	6.8	324
(Tastykake)	4-oz. pie	5.9	359
SWEETSOP (see **SUGAR APPLE**)			
*SWISS BURGER, mix, dinner			
(Jeno's) Add 'n Heat	⅙ of 30-oz. pkg.	18.4	250
SWISS STEAK, frozen:			
(Stouffer's)	10-oz. pkg.	54.0	569
Dinner (Swanson)	10-oz. dinner	20.2	361
SWORDFISH (USDA):			
Raw, meat only	1 lb.	87.1	535
Broiled with butter or			
margarine	3″ × 3″ × ½″		
	steak (4.4 oz.)	35.0	218
Canned, solids & liq.	4 oz.	19.8	116

T

TABASCO (McIlhenny)	¼ tsp. (1 gram)	<.1	<1
TACO SEASONING MIX:			
(Durkee)	1⅛-oz. pkg.	1.8	67
*(Durkee)	½ cup.	20.8	321
(French's)	1¾-oz. pkg.	4.4	123
(Lawry's)	1¼-oz. pkg.	3.6	120
TAMALE:			
Canned:			
(Armour Star)	15½-oz. can	20.2	620
(Hormel) beef	1 tamale (2.1 oz.)	1.9	80
(Wilson)	15½-oz. can	19.8	601

(USDA): United States Department of Agriculture
*Prepared as Package Directs

Food and Description	Measure or Quantity	Protein (grams)	Calories
Frozen:			
(Banquet) buffet	2-lb. pkg.	45.2	1488
(Banquet) cooking bag	2 tamales with sauce (3 oz. each)	8.5	279
TAMARIND, fresh (USDA):			
Whole	1 lb. (weighed with pods & seeds)	6.1	520
Flesh only	4 oz.	3.2	271
TANDY TAKE (Tastykake):			
Chocolate	.7-oz. cake	1.9	147
Choc-o-mint or Dandy Kake	.6-oz. cake	.8	102
Orange	.6-oz. cake	1.4	103
*TANG, instant breakfast drink	½ cup (4.7 oz.)	Tr.	61
TANGELO, fresh (USDA):			
Juice from whole fruit	1 lb. (weighed with peel, membrane & seeds)	1.3	104
Juice	½ cup (4.4 oz.)	.6	51
TANGERINE or MANDARIN ORANGE, fresh:			
Whole (USDA)	1 lb. (weighed with peel, membrane & seeds)	2.7	154
Whole (USDA)	4.1-oz. tangerine (2⅜″ dia.)	1.7	39
Peeled (Sunkist)	1 large tangerine (4.1 oz.)	1.0	39
Sections, without membranes (USDA)	1 cup (6.8 oz.)	1.5	89
TANGERINE JUICE:			
Fresh (USDA)	½ cup (4.4 oz.)	.6	53
Canned, unsweetened (USDA)	½ cup (4.4 oz.)	.6	53
Canned, sweetened (USDA)	½ cup (4.4 oz.)	.6	62
Frozen concentrate, unsweetened:			
Undiluted (USDA)	6-oz. can	3.6	340

(USDA): United States Department of Agriculture
*Prepared as Package Directs

[311]

Food and Description	Measure or Quantity	Protein (grams)	Calories
*Prepared with 3 parts water by volume (USDA)	½ cup (4.4 oz.)	.6	57
Frozen concentrate, sweetened:			
*(Minute Maid)	½ cup (4.2 oz.)	.6	57
*(Snow Crop)	½ cup (4.2 oz.)	.6	57
TAPIOCA, dry, quick cooking, granulated:			
(USDA)	1 cup (5.4 oz.)	.9	535
(USDA)	1 T. (10 grams)	<.1	35
(Minute Tapioca)	1 T.	Tr.	40
TAPIOCA PUDDING:			
Apple, home recipe (USDA)	½ cup (4.4 oz.)	.2	146
Cream, home recipe (USDA)	½ cup (2.9 oz.)	4.1	110
Chilled (Sealtest)	4 oz.	3.0	130
Canned (Hunt's)	5-oz. can	3.6	166
Mix:			
*All flavors (Jell-O)	½ cup (5.1 oz.)	4.6	166
*Chocolate (Royal)	½ cup (5.1 oz.)	4.9	186
*Fluffy recipe (Minute Tapioca)	½ cup (4.4 oz.)	5.9	150
*Vanilla (Royal)	½ cup (5.1 oz.)	4.3	169
TARO, raw (USDA):			
Tubers, whole	1 lb. (weighed with skin)	7.2	373
Tubers, skin removed	4 oz.	2.2	111
Leaves & stems	1 lb.	13.6	181
TASTE AMERICA (Green Giant):			
Florida style	⅓ of 10-oz. pkg.	.9	48
New Orleans style	⅓ of 10-oz. pkg.	2.8	76
Northwest style	⅓ of 10-oz. pkg.	2.8	83
Pennsylvania Dutch style	⅓ of 10-oz. pkg.	1.9	90
San Francisco style	⅓ of 10-oz. pkg.	.9	31
TAUTOG or BLACKFISH, raw:			
Whole (USDA)	1 lb. (weighed whole)	31.2	149
Meat only (USDA)	4 oz.	21.1	101

(USDA): United States Department of Agriculture
*Prepared as Package Directs

Food and Description	Measure or Quantity	Protein (grams)	Calories
TEAM, cereal	1⅓ cups (1 oz.)	1.7	107
TEA MIX, iced:			
*All flavors (Salada)	1 cup (.5 oz. dry)	.2	57
*Low calorie, lemon (Tender Leaf)	1 cup	Tr.	12
TEMPTYS (Tastykake):			
Butter creme	⅔-oz. cake	.7	94
Chocolate	⅔-oz. cake	1.0	95
Lemon	⅔-oz. cake	.7	95
TENDERGREEN (see MUSTARD SPINACH)			
TERRAPIN, DIAMOND BACK, raw (USDA):			
In shell	1 lb. (weighed in shell)	17.7	106
Meat only	4 oz.	21.1	126
TEXTURED VEGETABLE PROTEIN:			
Breakfast links, *Morningstar Farms*	8-oz. link	3.2	54
Breakfast patties, *Morningstar Farms*	1.3-oz. pattie	6.0	98
Breakfast slices, *Morningstar Farms*	1-oz. slice	4.5	50
Burger Builder (Betty Crocker)	¼ cup	7.0	60
Chili seasoning (Williams)	4-oz. pkg.	45.5	428
Hamburger seasoning (Williams)	4-oz. pkg.	40.7	385
*Hamburger, seasoning (Williams)	4 oz.	18.4	216
Meatloaf seasoning (Williams)	4-oz. pkg.	40.7	393
*Meatloaf seasoning (Williams)	4 oz.	18.0	212
Pathmark Plus:			
Dry	⅓ oz.	5.0	28
*Prepared	4 oz.	27.6	178
Sloppy Joe seasoning (Williams)	4-oz. pkg.	35.7	382
*Sloppy Joe Seasoning (Williams)	4 oz.	14.2	186

(USDA): United States Department of Agriculture
*Prepared as Package Directs

[313]

Food and Description	Measure or Quantity	Protein (grams)	Calories
Spaghetti sauce (Williams)	4-oz. pkg.	25.6	366
*Spaghetti sauce (Williams)	4 oz.	6.2	118
Taco seasoning (Williams)	4-oz. pkg.	39.7	378
*Taco seasoning (Williams)	4 oz.	15.4	194
THICK & FROSTY (General Foods)	1 cup (8.3 oz.)	5.5	314
THURINGER, sausage:			
(USDA)	1 oz.	5.3	87
(Hormel) Old Smokehouse	1 oz.	4.6	100
Summer sausage, meat (Oscar Mayer)	.8-oz. slice	3.7	74
Summer sausage, beef (Oscar Mayer)	.8-oz. slice	3.7	68
TILEFISH (USDA):			
Raw, whole	1 lb. (weighed whole)	40.5	183
Baked, meat only	4 oz.	27.8	156
TOASTER CAKE:			
Corn Treats (Arnold)	1.1-oz. piece	1.9	111
Toastee (Howard Johnson's):			
Blueberry	1 piece (1¼ oz.)	2.1	121
Cinnamon raisin	1 piece (1.1 oz.)	2.2	114
Corn	1 piece (1¼ oz.)	2.0	112
Orange or pound	1 piece (1 oz.)	1.9	113
Toastette (Nabisco):			
Apple	1 piece (1⅔ oz.)	1.8	184
Blueberry	1 piece (1⅔ oz.)	1.8	184
Brown sugar, cinnamon	1 piece (1⅔ oz.)	2.0	189
Cherry	1 piece (1⅔ oz.)	1.8	182
Orange marmalade	1 piece (1⅔ oz.)	2.1	181
Peach	1 piece (1⅔ oz.)	2.0	183
Strawberry	1 piece (1⅔ oz.)	1.8	184
Toast-r-Cake (Thomas):			
Bran	1 piece (1.2 oz.)	2.4	116
Corn or orange	1 piece (1.2 oz.)	2.7	118
TOASTERINO, frozen (Buitoni):			
Cheese, grilled	4 oz.	12.8	286
Pizzaburger	4 oz.	16.9	298

(USDA): United States Department of Agriculture
*Prepared as Package Directs

Food and Description	Measure or Quantity	Protein (grams)	Calories
Sloppy Joe	4 oz.	13.4	295
TOASTER SANDWICH:			
Cheese zesty (Borden)	1 sandwich (1.8 oz.)	5.8	166
Grilled cheese (Borden)	1 sandwich (1.8 oz.)	5.8	166
Pizza (Borden)	1 sandwich (1.8 oz.)	5.0	157
TODDLER FOOD (see **BABY FOOD**)			
TOFFEE KRUNCH BAR (Sealtest)	3 fl. oz. (1.7 oz.)	1.9	149
TOFU (see **SOYBEAN CURD**)			
TOMATO:			
Fresh, green, whole, untrimmed (USDA)	1 lb. (weighed with core & stem ends)	5.0	99
Fresh, green, trimmed, unpeeled (USDA)	4 oz.	1.4	27
Fresh, ripe (USDA):			
Whole, eaten with skin	1 lb.	5.0	100
Whole, peeled	1 lb. (weighed with skin, stem ends & hard core)	4.4	88
Whole, peeled	1 med. (2″ × 2½″, 5.3 oz.)	1.6	33
Whole, peeled	1 small (1¾″ × 2¼″, 3.9 oz.)	1.2	24
Sliced, peeled	½ cup (3.2 oz.)	1.0	20
Fresh, ripe, cherry (HEW/FAO)	4 pieces (2.3 oz.)	.6	14
Boiled (USDA)	½ cup (4.3 oz.)	1.6	31
Canned, regular pack:			
Whole, solids & liq. (USDA)	½ cup (4.2 oz.)	1.2	25
(Contadina)	½ cup (4 oz.)	1.2	28
Solids & liq. (Del Monte)	½ cup (4.2 oz.)	1.0	30
Diced, in puree (Contadina)	½ cup (4 oz.)	1.2	44
Sliced (Contadina)	½ cup (4 oz.)	.8	36
Stewed (Contadina)	½ cup (4 oz.)	1.2	36
Stewed (Del Monte)	½ cup (4.2 oz.)	1.0	30

(HEW/FAO): Health, Education and Welfare/
Food and Agricultural Organization
(USDA): United States Department of Agriculture   [315]

Food and Description	Measure or Quantity	Protein (grams)	Calories
Wedges, solids & liq. (Del Monte)	½ cup (4.1 oz.)	1.0	28
Whole, peeled (Hunt's)	½ cup (4.2 oz.)	1.2	24
Whole, solids & liq. (Stokely-Van Camp)	½ cup (4.1 oz.)	1.1	23
Canned, dietetic pack, low sodium:			
Solids & liq. (USDA)	4 oz.	1.1	23
Solids & liq. (Blue Boy)	4 oz.	1.2	25
Solids & liq. (Tillie Lewis)	½ cup (4.3 oz.)	1.2	25
Stewed, solids & liq. (Tillie Lewis)	½ of 8¼-oz. can	1.4	29
Whole, peeled (Diet Delight)	½ cup (4.3 oz.)	1.1	39
Whole, unseasoned (S and W) *Nutradiet*	4 oz.	1.1	24
TOMATO JUICE:			
Canned, regular pack:			
(USDA)	6 fl. oz. (6.4 oz.)	1.6	35
(USDA)	½ cup (4.3 oz.)	1.1	23
(Campbell)	6 fl. oz.	1.2	33
(Del Monte)	½ cup (4.3 oz.)	1.0	22
(Heinz)	5½-fl.-oz. can	1.3	34
(Hunt's)	5½-fl.-oz. can	1.4	30
(Sacramento)	5½-fl.-oz. can	1.3	33
Peppy (Sacramento)	5½-fl.-oz. can	1.4	39
(Stokely-Van Camp)	½ cup (4.3 oz.)	1.0	28
Canned, dietetic pack, low sodium:			
(USDA)	4 oz. (by wt.)	.9	22
(Blue Boy)	4 oz. (by wt.)	1.1	24
(Diet Delight)	½ cup (3.9 oz.)	.9	24
(S and W) *Nutradiet*, unseasoned	4 oz. (by wt.)	1.0	25
Concentrate, canned (USDA)	4 oz. (by wt.)	3.9	86
*Concentrate, canned, diluted with 3 parts water by volume (USDA)	4 oz. (by wt.)	1.0	23
Dehydrated (USDA)	1 oz.	3.3	86
*Dehydrated (USDA)	½ cup (4.3 oz.)	1.0	24
TOMATO JUICE COCKTAIL:			
(USDA)	4 oz. (by wt.)	.8	24

(USDA): United States Department of Agriculture
*Prepared as Package Directs

Food and Description	Measure or Quantity	Protein (grams)	Calories
Snap-E-Tom	6 fl. oz. (6.5 oz.)	1.3	36
Tomato Plus (Sacramento)	5½-fl.-oz. can	1.5	49
TOMATO PASTE, canned:			
(USDA)	6-oz. can	5.8	139
(USDA)	½ cup (4.6 oz.)	4.4	106
(USDA)	1 T. (.6 oz.)	.5	13
(Contadina)	1 T. (.5 oz.)	.5	12
(Del Monte)	1 T. (.6 oz.)	.6	14
(Hunt's)	½ cup (4.6 oz.)	4.5	108
(Hunt's)	1 T. (.6 oz.)	.6	14
(Stokely-Van Camp)	½ cup (4.6 oz.)	4.4	106
TOMATO PRESERVE			
(Smucker's)	1 T. (.7 oz.)	.2	52
TOMATO PUREE:			
Canned, regular pack: (USDA)	1 cup (8.8 oz.)	4.2	98
(Contadina)	1 cup	4.0	96
(Hunt's)	1 cup (8.8 oz.)	4.0	92
Canned, dietetic pack (USDA)	8 oz.	3.9	88
TOMATO SALAD, jellied			
(Contadina)	½ cup	1.6	60
TOMATO SAUCE, canned:			
(Contadina)	1 cup	3.2	80
(Del Monte) plain	1 cup (8.8 oz.)	3.5	60
(Del Monte) with mushrooms	1 cup (8.8 oz.)	3.8	75
(Del Monte) with onions	1 cup (8.8 oz.)	4.0	90
(Del Monte) with tomato tidbits	1 cup (8.8 oz.)	2.8	90
(Hunt's) plain	1 cup (8.7 oz.)	3.4	78
(Hunt's) herb	1 cup (8.8 oz.)	4.2	186
(Hunt's) special	1 cup (8.7 oz.)	3.3	94
(Hunt's) with bits	1 cup (8.7 oz.)	3.5	82
(Hunt's) with cheese	1 cup (8.8 oz.)	5.0	107
(Hunt's) with mushrooms	1 cup (8.8 oz.)	3.5	83
(Hunt's) with onions	1 cup (8.8 oz.)	4.0	102
TOMATO SOUP:			
Canned, regular pack: Condensed (USDA)	8 oz. (by wt.)	3.6	163

(USDA): United States Department of Agriculture

[317]

Food and Description	Measure or Quantity	Protein (grams)	Calories
*Prepared with equal volume water (USDA)	1 cup (8.6 oz.)	2.0	88
*Prepared with equal volume milk (USDA)	1 cup (8.8 oz.)	6.5	172
*(Campbell)	1 cup	1.6	79
*(Heinz) California	1 cup (8.5 oz.)	1.4	81
(Heinz) Great American	1 cup (8¾ oz.)	3.0	168
*(Manischewitz)	1 cup	1.4	60
*Beef (Campbell) Noodle-O's	1 cup	4.3	109
*Bisque (Campbell)	1 cup	2.5	115
*Rice, old fashioned (Campbell)	1 cup	1.8	99
*Rice (Manischewitz)	1 cup	1.5	78
With vegetables (Heinz) Great American	1 cup (8¾ oz.)	2.4	126
Canned, dietetic pack:			
Low sodium (Campbell)	7½-oz. can	2.2	97
*With rice (Claybourne)	8 oz.	1.6	73
*With rice (Slim-ette)	8 oz. (by wt.)	.9	34
(Tillie Lewis)	1 cup (8 oz.)	2.5	70
TOMATO SOUP MIX:			
(Lipton) Cup-a-Soup	1 pkg. (.8 oz.)	1.4	79
Cream of (Lipton) Cup-a-Soup	1 pkg. (.9 oz.)	1.8	98
Cream of (Wyler's)	1 pkg. (1 oz.)	2.9	96
With noodles (USDA)	1 oz.	2.5	98
*With noodles (USDA)	1 cup (8 oz.)	1.4	62
*With noodles (Lipton)	1 cup (8 oz.)	1.8	68
TOMCOD, ATLANTIC, raw (USDA):			
Whole	1 lb. (weighed whole)	30.4	136
Meat only	4 oz.	19.5	87
TONGUE (USDA):			
Beef, medium fat, raw, untrimmed	1 lb.	56.5	714
Beef, medium, fat braised	4 oz.	24.4	277
Calf, raw, untrimmed	1 lb.	64.6	454
Calf, braised	4 oz.	27.1	181
Hog, raw, untrimmed	1 lb.	57.9	741
Hog, braised	4 oz.	24.9	287

(USDA): United States Department of Agriculture
*Prepared as Package Directs

Food and Description	Measure or Quantity	Protein (grams)	Calories
Lamb, raw, untrimmed	1 lb.	46.0	659
Lamb, braised	4 oz.	23.2	288
Sheep, raw, untrimmed	1 lb.	45.4	877
Sheep, braised	4 oz.	22.5	366
TONGUE, CANNED:			
Pickled (USDA)	1 oz.	5.5	76
Potted or deviled (USDA)	1 oz.	5.3	82
(Hormel)	1 oz. (12-oz. can)	5.5	67
TOPPING (see also **CHOCOLATE SYRUP**):			
Sweetened:			
Butterscotch (Kraft)	1 oz.	.9	84
Butterscotch (Smucker's)	1 T. (.7 oz.)	.3	66
Caramel:			
(Smucker's)	1 T. (.7 oz.)	.3	65
Chocolate (Kraft)	1 oz.	1.5	84
Peanut butter (Smucker's)	1 T. (.6 oz.)	1.4	61
Vanilla (Kraft)	1 oz.	1.9	84
Cherry (Smucker's)	1 T. (.6 oz.)	<.1	53
Chocolate or chocolate flavored:			
(Kraft)	1 oz.	1.1	73
Fudge (Hershey's)	1 oz.	.6	54
Fudge (Kraft)	1 oz.	1.4	74
Fudge (Smucker's)	1 T. (.7 oz.)	.5	54
Fudge, mint (Smucker's)	1 T. (.7 oz.)	.5	55
Milk chocolate (Smucker's)	1 T. (.7 oz.)	1.7	68
Marshmallow creme (Kraft)	1 oz.	.3	90
Pecan (Kraft)	1 oz.	1.0	122
Pecan in syrup (Smucker's)	1 T. (.7 oz.)	.8	61
Pineapple (Kraft)	1 oz.	.1	80
Pineapple (Smucker's)	1 T. (.6 oz.)	<.1	53
Spoonmallow (Kraft)	1 oz.	.3	83
Strawberry (Kraft)	1 oz.	.1	80
Strawberry (Smucker's)	1 T. (.6 oz.)	<.1	49
Walnut (Kraft)	1 oz.	1.4	113
Walnut in syrup (Smucker's)	1 T. (.6 oz.)	1.0	61
Dietetic, chocolate:			
(Diet Delight)	1 T. (.6 oz.)	.3	20
(Tillie Lewis)	1 T. (.6 oz.)	.4	8

(USDA): United States Department of Agriculture

Food and Description	Measure or Quantity	Protein (grams)	Calories
TOPPING, WHIPPED:			
(USDA) pressurized	1 cup (2.5 oz.)	1.0	190
(USDA) pressurized	1 T. (4 grams)	Tr.	10
(Birds Eye) *Cool Whip*	1 T. (4 grams)	Tr.	16
(Kraft)	1 oz.	.1	79
(Lucky Whip)	1 T. (4 grams)	Tr.	12
(Sealtest) *Big Top*	1.5 fl. oz. (.5 oz.)	Tr.	15
(Sealtest) *Zip Whipt*, real cream	1.5 fl. oz. (.4 oz.)	.2	26
TOPPING, WHIPPED, MIX:			
*(D-Zerta)	1 T.	.1	7
*(Dream Whip)	1 T. (.2 oz.)	.2	14
(Lucky Whip)	1 oz.	1.2	159
*(Lucky Whip)	1 T. (4 grams)	.2	10
TORTILLA (USDA)	.7-oz. tortilla (5″)	1.2	42
TOTAL, cereal (General Mills)	1¼ cups (1 oz.)	2.5	100
TOWEL GOURD, raw (USDA):			
Unpared	1 lb. (weighed with skin)	3.1	69
Pared	4 oz.	.9	20
TREET (Armour)	1 oz.	4.0	84
TRIPE, beef (USDA):			
Commercial	4 oz.	21.7	113
Pickled	4 oz.	13.4	70
TRIX, cereal (General Mills)	1 cup (1 oz.)	1.5	110
TROUT:			
Brook, fresh, whole (USDA)	1 lb. (weighed whole)	42.7	224
Brook, fresh, meat only (USDA)	4 oz.	21.8	115
Lake (see **LAKE TROUT**)			
Rainbow (USDA):			
Fresh, meat with skin	4 oz.	24.4	221
Canned	4 oz.	23.4	237
Frozen (1000 Springs):			
Boned	5-oz. trout	20.4	135

(USDA): United States Department of Agriculture
*Prepared as Package Directs

Food and Description	Measure or Quantity	Protein (grams)	Calories
Dressed	5-oz. trout	24.8	164
TUNA:			
Raw, bluefin, meat only (USDA)	4 oz.	28.6	164
Raw, yellowfin, meat only (USDA)	4 oz.	28.0	151
Canned in oil:			
Solids & liq:			
(USDA)	6½-oz. can	44.5	530
(Breast O'Chicken)	6½-oz. can	44.2	540
Chunk, light (Chicken of the Sea)	6½-oz. can	43.8	405
Chunk, light (Del Monte)	6½-oz. can	43.1	478
Chunk, light (Del Monte)	1 cup (4.7 oz.)	31.1	346
White albacore (Del Monte)	6½-oz. can	47.5	543
White albacore (Del Monte)	1 cup (4.7 oz.)	34.3	392
Drained solids:			
(USDA)	6½-oz. can	45.2	309
(Bumble Bee)	1 cup (6 oz.)	49.0	334
Albacore (Del Monte)	1 cup (5.6 oz.)	43.8	306
Chunk, light (Chicken of the Sea)	6½-oz. can	45.0	294
Chunk, light (Del Monte)	1 cup (5.6 oz.)	38.9	374
Canned in water:			
Solids & liq., (USDA)	6½-oz. can	51.5	234
Solids & liq. (Breast O' Chicken)	6½-oz. can	51.5	230
Solids & liq., white, solid (Chicken of the Sea)	7-oz. can	50.0	240
Drained, solid, light (Chicken of the Sea)	6½-oz. can	48.4	224
Drained, solid, white (Chicken of the Sea)	6½-oz. can	50.2	216
Canned, dietetic, drained, chunk, white (Chicken of the Sea)	6½-oz. can	44.1	200
TUNA CAKE, frozen, thins (Mrs. Paul's)	10-oz. pkg.	28.0	689

(USDA): United States Department of Agriculture

Food and Description	Measure or Quantity	Protein (grams)	Calories
TUNA PIE, frozen:			
(Banquet)	8-oz. pie	14.1	479
(Morton)	8-oz. pie	15.9	384
TUNA SALAD, home recipe, made with tuna, celery, mayonnaise, pickle, onion & egg (USDA)	4 oz.	16.6	193
TURBOT, GREENLAND, raw (USDA):			
Whole	1 lb. (weighed whole)	38.7	344
Meat only	4 oz.	18.6	166
Frozen (Weight Watchers)	18-oz. dinner	32.5	426
Frozen, with apple (Weight Watchers)	9½-oz. luncheon	21.3	277
TURKEY:			
Raw, ready-to-cook (USDA)	1 lb. (weighed with bones)	66.6	722
Raw, light meat (USDA)	4 oz.	27.9	132
Raw, dark meat (USDA)	4 oz.	23.7	145
Raw, skin only (USDA)	4 oz.	13.7	459
Roasted (USDA):			
Flesh, skin & giblets	From 13½-lb. raw ready-to-cook turkey	993.6	9678
Flesh & skin	From 13½-lb. raw ready-to-cook turkey	1126.1	7872
Flesh & skin	4 oz.	36.2	253
Meat only:			
Chopped	1 cup (5 oz.)	44.4	268
Diced	1 cup (4.8 oz.)	42.5	256
Light	4 oz.	37.3	200
Light	1 slice (4″ × 2″ × ¼″, 3 oz.)	14.0	75
Dark	4 oz.	34.0	230
Dark	1 slice (2½″ × 1⅝″ × ¼″, .7 oz.)	6.4	43
Skin only	1 oz.	4.8	128
Giblets, simmered (USDA)	2 oz.	11.7	132

(USDA): United States Department of Agriculture

Food and Description	Measure or Quantity	Protein (grams)	Calories
Smoked, cooked, pressed (Oscar Mayer)	.8-oz. slice	4.9	23
Canned, boned:			
(USDA)	4 oz.	23.7	229
Solids & liq. (Lynden Farms)	5-oz. jar	36.9	239
(Swanson) with broth	5-oz. can	34.0	217
Canned, roast (Wilson) *Tender Made*	4 oz.	23.9	117

TURKEY DINNER:

Canned, noodle (Lynden Farms)	15-oz. can	21.2	463
Frozen:			
Sliced turkey, mashed potato, peas (USDA)	12 oz.	28.6	381
(Banquet)	11-oz. dinner	23.4	293
(Morton) sliced	11-oz. dinner	25.0	346
(Morton) 3-course	16¾-oz. dinner	33.2	613
(Swanson)	11½-oz. dinner	28.5	401
(Swanson) 3-course	16-oz. dinner	32.5	501
With gravy, dressing & potato (Swanson)	8¾-oz. pkg.	21.9	292
(Weight Watchers)	18-oz. dinner	40.4	302

TURKEY FRICASSEE, canned

(Lynden Farms)	14.5-oz. can	32.9	366

TURKEY GIZZARD (USDA):

Raw	4 oz.	23.0	178
Simmered	4 oz.	30.4	222

TURKEY PIE:

Home recipe, baked (USDA)	⅓ of 9″ pie (8.2 oz.)	24.1	550
Frozen:			
Commercial, unheated (USDA)	8 oz.	13.2	447
(Banquet)	8-oz. pie	14.5	415
(Banquet)	2-lb. 4-oz. pie	74.0	1327
(Morton)	8-oz. pie	18.2	411
(Swanson)	8-oz. pie	16.4	422
(Swanson) deep-dish	1-lb. pie	30.7	746

TURKEY, POTTED (USDA)

	1 oz.	5.0	70

(USDA): United States Department of Agriculture

Food and Description	Measure or Quantity	Protein (grams)	Calories
TURKEY SOUP, canned:			
(Campbell) *Chunky*	1 cup	8.1	112
Broth (Lynden Farms)	1 cup (8 oz.)	2.3	14
Noodle:			
Condensed (USDA)	8 oz. (by wt.)	8.2	147
*Prepared with equal volume			
water (USDA)	1 cup (8.8 oz.)	4.5	82
*(Campbell)	1 cup	3.8	72
*(Heinz)	1 cup (8½ oz.)	3.5	83
(Heinz) *Great American*	1 cup (8¾ oz.)	5.4	97
Low sodium (Campbell)	7½-oz. can	3.2	78
Rice, with mushrooms (Heinz)			
Great American	1 cup (8½ oz.)	5.4	101
*Vegetable (Campbell)	1 cup	3.1	73
Vegetable (Heinz) *Great*			
American	1 cup (8½ oz.)	5.3	98
*TURKEY SOUP MIX, noodle			
(Lipton)	1 cup (8 oz.)	3.1	61
TURKEY TETRAZZINI, frozen:			
(Morton)	11-oz. dinner	15.6	396
(Stouffer's)	12-oz. pkg.	32.0	694
TURNIP (USDA):			
Fresh, without tops	1 lb. (weighed		
	with skins)	3.9	117
Fresh, pared, diced	½ cup (2.4 oz.)	.7	20
Fresh, pared, slices	½ cup (2.3 oz.)	.6	19
Boiled, drained, diced	½ cup (2.8 oz.)	.6	18
Boiled, drained, mashed	½ cup (4 oz.)	.9	26
TURNIP GREENS leaves &			
stems:			
Fresh (USDA)	1 lb. (weighed		
	untrimmed)	11.4	107
Boiled, in small amount water,			
short time, drained (USDA)	½ cup (2.5 oz.)	1.6	14
Boiled, in large amount water,			
long time, drained (USDA)	½ cup (2.5 oz.)	1.6	14
Canned, solids & liq.:			
(USDA)	½ cup (4.1 oz.)	1.7	21
(Stokely-Van Camp)	½ cup (3.9 oz.)	1.6	20

(USDA): United States Department of Agriculture
*Prepared as Package Directs

Food and Description	Measure or Quantity	Protein (grams)	Calories
Frozen:			
Not thawed (USDA)	4 oz.	2.9	26
Boiled, drained (USDA)	½ cup (2.9 oz.)	2.0	19
Chopped (Birds Eye)	½ cup (3.3 oz.)	2.4	22
TURNOVER (see individual kinds)			
TURTLE, GREEN (USDA):			
Raw, in shell	1 lb. (weighed in shell)	21.6	97
Raw, meat only	4 oz.	22.5	101
Canned	4 oz.	26.5	120
TV DINNER (see individual listing such as BEEF DINNER, CHICKEN DINNER, CHINESE DINNER, ENCHILADA DINNER, etc.)			
TWINKIE (Hostess):			
Devil's food	1 cake (1½ oz.)	1.7	170
Gold sponge	1 cake (1½ oz.)	1.7	162

V

Food and Description	Measure or Quantity	Protein (grams)	Calories
VANILLA CAKE, frozen (Pepperidge Farm)	⅛ of cake (3.1 oz.)	2.5	332
VANILLA ICE CREAM (see also individual brand names):			
(Borden) 10.5% fat	¼ pt. (2.3 oz.)	2.1	132
Lady Borden 14% fat	¼ pt. (2.5 oz.)	1.8	162
(Sealtest) Party Slice	¼ pt. (2.3 oz.)	2.3	133
(Sealtest) 10.2% fat	¼ pt. (2.3 oz.)	2.3	133
(Sealtest) 12.1% fat	¼ pt. (2.3 oz.)	2.4	144
French (Prestige)	¼ pt. (2.6 oz.)	3.0	183
Fudge royale (Sealtest)	¼ pt. (2.3 oz.)	2.3	132
VANILLA ICE MILK (Borden)			
Lite-line	¼ pt.	3.4	108

(USDA): United States Department of Agriculture

Food and Description	Measure or Quantity	Protein (grams)	Calories
VANILLA PIE FILLING MIX (see **VANILLA PUDDING MIX**)			
VANILLA PUDDING:			
Blancmange, home recipe, with starch base (USDA)	½ cup (4.5 oz.)	4.5	142
Canned (Betty Crocker)	½ cup	2.7	170
Canned (Del Monte)	5-oz. can	3.1	190
Canned (Hunt's)	5-oz. can	2.4	238
Canned (Royal) *Creamerina*	5-oz. can	3.4	217
Canned (Thank You)	½ cup (4.5 oz.)	2.5	169
Chilled (Breakstone)	5-oz. container	2.0	252
Chilled (Sanna)	4¼ oz. container	3.1	172
Chilled (Sealtest)	4 oz.	3.0	125
VANILLA PUDDING or PIE FILLING MIX: Sweetened;			
*Regular, plain or French (Jell-O)	½ cup (5.2 oz.)	4.3	173
*Instant, plain or French (Jell-O)	½ cup (5.3 oz.)	4.3	178
*Regular (Royal)	½ cup (5.1 oz.)	4.3	163
*Instant (Royal)	½ cup (5.1 oz.)	4.3	176
*Low-calorie (D-Zerta)	½ cup (4.6 oz.)	4.8	107
VANILLA RENNET MIX: Powder:			
Dry (Junket)	1 oz.	<.1	116
*(Junket)	4 oz.	3.7	108
Tablet:			
Dry (Junket)	1 tablet (<1 gram)	Tr.	1
*& sugar (Junket)	4 oz.	3.7	101
VANILLA SOFT DRINK (Yoo-Hoo) High Protein	6 fl. oz. (6.4 oz.)	6.0	100
VEAL, medium fat (USDA):			
Chuck, raw	1 lb. (weighed with bone)	70.4	628
Chuck, braised, lean & fat	4 oz.	31.6	266

[326]

(USDA): United States Department of Agriculture
*Prepared as Package Directs

Food and Description	Measure or Quantity	Protein (grams)	Calories
Flank, raw	1 lb. (weighed with bone)	74.1	1410
Flank, stewed, lean & fat	4 oz.	26.3	442
Foreshank, raw	1 lb. (weighed with bone)	46.5	368
Foreshank, stewed, lean & fat	4 oz.	32.5	245
Loin, raw	1 lb. (weighed with bone)	72.3	681
Loin, broiled, medium done, chop, lean & fat	4 oz.	29.9	265
Plate, raw	1 lb. (weighed with bone)	65.6	828
Plate, stewed, lean & fat	4 oz.	29.6	344
Rib, raw, lean & fat	1 lb. (weighed with bone)	65.7	723
Rib, roasted, medium done, lean & fat	4 oz.	30.8	305
Round & rump, raw	1 lb. (weighed with bone)	68.1	573
Round & rump, broiled, steak or cutlet, lean & fat	4 oz. (weighed without bone)	30.7	245
VEAL DINNER, frozen: Breaded veal with spaghetti in tomato sauce (Swanson)	8¼-oz. pkg.	17.3	272
Parmigian (Banquet)	11-oz. dinner	20.6	421
Parmigiana (Kraft)	11-oz. dinner	26.5	474
Parmigiana (Swanson)	12¼-oz. dinner	23.6	492
Parmigiana (Weight Watchers)	9½-oz. luncheon	21.5	260
VEGETABLE BOUILLON CUBE: (Herb-Ox)	1 cube (4 grams)	.6	6
(Steero)	1 cube (4 grams)	.6	4
(Wyler's)	1 cube (4 grams)	.5	7
VEGETABLE FAT (see **FAT**)			
VEGETABLE JUICE COCKTAIL, canned: (USDA)	4 oz. (by wt.)	1.0	19

(USDA): United States Department of Agriculture

[327]

Food and Description	Measure or Quantity	Protein (grams)	Calories
Unseasoned (S and W)			
Nutradiet	4 oz. (by wt.)	.9	24
V-8 (Campbell)	¾ cup	1.2	31
VEGETABLES, MIXED:			
Canned regular pack:			
(Veg-All)	½ cup (4 oz.)	1.5	39
Drained liq. (Del Monte)	4 oz.	1.1	22
Drained solids (Del Monte)	½ cup (2.8 oz.)	2.6	41
Solids & liq. (Del Monte)	½ cup (4 oz.)	2.7	46
Solids & liq. (Stokely-Van Camp)	½ cup (3.8 oz.)	3.6	71
Canned, Chop Suey (Hung's)	4 oz.	1.7	20
Frozen:			
Not thawed (USDA)	4 oz.	3.7	74
Boiled, drained (USDA)	½ cup (3.3 oz.)	2.9	58
(Birds Eye)	½ cup (3.3 oz.)	3.0	50
Chinese (Birds Eye)	⅓ of 10-oz. pkg.	2.5	65
In butter sauce (Green Giant)	⅓ of 10-oz. pkg.	2.2	67
Jubilee (Birds Eye)	⅓ of 10-oz. pkg.	3.0	138
With onion sauce (Birds Eye)	½ cup (2.7 oz.)	2.9	117
VEGETABLE OYSTER (see SALSIFY)			
VEGETABLE SOUP:			
Canned, regular pack:			
*(Campbell)	1 cup	3.1	77
(Campbell) *Chunky*	1 cup	2.9	105
*(Campbell) old fashioned	1 cup	3.1	70
Beef, condensed (USDA)	9 oz. (by wt.)	9.5	147
*Beef, prepared with equal volume water (USDA)	1 cup (8.6 oz.)	5.1	78
*Beef (Campbell)	1 cup	5.7	75
*Beef (Heinz)	1 cup (8½ oz.)	3.7	66
Beef (Heinz) *Great American*	1 cup (8¾ oz.)	11.1	126
With beef broth, condensed (USDA)	8 oz. (by wt.)	5.0	145
*With beef broth, prepared with equal volume water (USDA)	1 cup (8.8 oz.)	2.8	80
With beef broth (Heinz) *Great American*	1 cup (8¾ oz.)	8.3	144

(USDA): United States Department of Agriculture
*Prepared as Package Directs

Food and Description	Measure or Quantity	Protein (grams)	Calories
With beef stock (Heinz)	1 cup (8½ oz.)	2.9	83
With ground beef (Heinz)			
Great American	1 cup (8¾ oz.)	10.6	139
ᘐ Noodle-O's (Campbell)	1 cup	2.3	71
Vegetarian:			
Condensed (USDA)	8 oz. (by wt.)	4.1	145
*Prepared with equal			
volume water (USDA)	1 cup (8.6 oz.)	2.2	78
*(Campbell)	1 cup	2.0	71
*(Heinz)	1 cup (8¾ oz.)	1.9	83
(Heinz) *Great American*	1 cup (8½ oz.)	4.8	125
*(Manischewitz)	1 cup	1.7	63
Canned, dietetic pack:			
Low sodium (Campbell)	7½-oz. can	2.3	83
Beef, low sodium (Campbell)	7½-oz. can	4.5	78
*(Claybourne)	8 oz.	3.6	104
*(Slim-ette)	8 oz. (by wt.)	1.0	46
(Tillie Lewis)	1 cup (8 oz.)	1.8	62
Frozen:			
With beef, condensed			
(USDA)	8 oz. (by wt.)	12.2	159
*With beef, prepared with			
equal volume water			
(USDA)	8 oz. (by wt.)	6.1	79
VEGETABLE SOUP MIX:			
*(Wyler's)	1 cup	2.0	56
*Beef (Lipton)	1 cup (8 oz.)	3.4	58
*Chicken (Wyler's)	1 cup	1.7	37
Italian-style (Lipton)	1 pkg. (2.1 oz.)	7.2	210
*& noodle (Lipton) Country	1 cup	2.5	75
Spring (Lipton) *Cup-a-Soup*	1 pkg. (½ oz.)	1.4	42
VEGETABLE STEW, canned			
(Hormel) *Dinty Moore*	8 oz.	5.0	170
"VEGETARIAN FOODS":			
Canned or dry:			
Beans, rich brown (Loma			
Linda)	½ cup (3.7 oz.)	6.0	122
Bean, soy:			
Boston style (Loma Linda)	½ cup (3.7 oz.)	14.6	141

(USDA): United States Department of Agriculture
*Prepared as Package Directs

Food and Description	Measure or Quantity	Protein (grams)	Calories
Green, drained (Loma Linda)	½ cup (3.7 oz.)	10.5	109
Tomato sauce (Loma Linda)	½ cup (3.7 oz.)	7.8	118
Big franks, drained (Loma Linda)	1 frank (1.6 oz.)	8.0	86
Breading meal (Loma Linda)	1 cup (3.7 oz.)	21.9	364
Breading meal (Worthington)	¼ cup (1.1 oz.)	7.0	116
Burger Aid (Worthington)	1 oz.	16.0	102
Cheze-O-Soy (Worthington)	½" slice (2.5 oz.)	8.9	125
Chili (Worthington)	¼ can (5 oz.)	14.0	166
Chili with beans (Loma Linda)	½ cup (4.7 oz.)	10.0	158
Choplet (Worthington)	1 choplet (2.2 oz.)	12.0	72
Choplet burger (Worthington)	⅓ cup (3.2 oz.)	13.0	118
Cutlet (Worthington)	1 cutlet (2.2 oz.)	12.0	71
Dinner bits, drained (Loma Linda)	1 bit (.5 oz.)	2.3	23
Dinner cuts, drained (Loma Linda)	1 cut (1.6 oz.)	7.4	45
Dinner cuts, no salt added, drained (Loma Linda)	1 cup (1.6 oz.)	7.4	46
Fry stick (Worthington)	1 piece (2.3 oz.)	12.9	103
Garbanzo (Loma Linda)	½ cup (4.1 oz.)	8.9	145
Granburger (Worthington)	1 oz.	17.0	96
Granola (Loma Linda)	½ cup (2 oz.)	8.1	246
Gravy Quik, brown (Loma Linda)	⅙ pkg. (6 grams)	.6	15
J-7901 or J-7901A (Worthington)	1 oz.	17.9	95
Jell Quik, dry (Loma Linda)	1 oz.	<.1	106
Kaffir Tea (Worthington)	1 bag (1 gram)	.1	3
Lentils, drained (Loma Linda)	¾ cup (3.5 oz.)	6.7	91
Linketts, drained (Loma Linda)	1 link (1.3 oz.)	7.4	76
Little links, drained (Loma Linda)	1 link (.8 oz.)	3.8	45
Madison burger (Worthington)	⅓ cup (2 oz.)	14.0	86
Meat-like loaf (Loma Linda): Beef, drained	¼" slice (2 oz.)	10.9	120

Food and Description	Measure or Quantity	Protein (grams)	Calories
Chicken	¼" slice (2 oz.)	10.9	114
Luncheon	¼" slice (2 oz.)	11.1	122
Turkey	¼" slice (2 oz.)	11.3	125
Meat loaf mix (Worthington)	2 oz.	19.0	236
Multigen powder (Loma Linda)	½ cup (2.1 oz.)	14.1	227
Non-meatballs (Worthington)	1 piece (.6 oz.)	3.3	35
Not meat with tomato (Worthington)	2⅓ oz.	9.2	214
Numete (Worthington)	½" slice (2.3 oz.)	9.0	163
Nuteena (Loma Linda)	½" slice (2.5 oz.)	8.8	152
Oven-cooked wheat (Loma Linda)	½ cup (2.6 oz.)	11.0	263
Peanuts & soya (USDA)	4 oz.	13.3	269
Prime vegetable burger (Worthington)	½" slice (2.1 oz.)	15.1	74
Proteena (Loma Linda)	½" slice (2.6 oz.)	15.8	141
Protose, canned (Worthington)	½" slice (2.7 oz.)	18.8	175
Rediburger (Loma Linda)	½" slice (2.5 oz.)	14.2	153
Redi-loaf mix, beef (Loma Linda)	⅓ cup (1.2 oz.)	13.7	160
Redi-loaf mix, chicken (Loma Linda)	⅓ cup (1.2 oz.)	14.6	159
Ruskets, biscuits (Loma Linda)	1 biscuit (.6 oz.)	2.2	65
Ruskets, flakes (Loma Linda)	1 cup (1 oz.)	3.5	101
Sandwich spread (Loma Linda)	1 T. (.6 oz.)	1.2	23
Sandwich spread (Worthington)	3 T. (1.2 oz.)	4.8	79
Saucette (Worthington)	1 link (.6 oz.)	3.3	39
Savita (Worthington)	1 tsp. (.4 oz.)	4.1	20
Savorex (Loma Linda)	1 oz.	11.6	52
Seasoning, chicken (Loma Linda)	1 T. (3 grams)	.5	8
Soyagen:			
A.P. & Malt powder (Loma Linda)	¼ cup (1.4 oz.)	9.0	184
Carob, powder (Loma Linda)	¼ cup (1.4 oz.)	9.2	184
Liquid (Loma Linda)	1 cup (8.6 oz.)	7.3	146

(USDA): United States Department of Agriculture

Food and Description	Measure or Quantity	Protein (grams)	Calories
Soyalac, concentrate, liquid (Loma Linda)	1 fl. oz. (1.1 oz.)	1.3	41
Soyalac, infant powder (Loma Linda)	1 oz.	6.2	136
Soyalac, ready-to-use (Loma Linda)	1 oz.	.6	19
Soyamel, any kind (Worthington)	1 oz.	6.7	129
Soyameat (Worthington:			
Diced beef	1 oz.	2.8	30
Sliced beef	1 slice (1 oz.)	6.5	69
Diced chicken	1 oz.	3.1	30
Fried chicken	1 piece (1.2 oz.)	4.5	64
Sliced chicken	1 slice (1.1 oz.)	3.5	34
Salisbury steak	1 slice (2.5 oz.)	12.0	157
Soy flour (Loma Linda)	¼ cup (.9 oz.)	10.7	115
Stew pac, drained (Loma Linda)	½ cup (3 oz.)	13.5	138
Stripple Zips (Worthington)	1 oz.	14.4	131
Tamales (Worthington)	1 oz.	1.6	23
Tastee cuts (Loma Linda)	1 cup (1.5 oz.)	7.5	42
Tenderbit (Loma Linda)	1 piece (.9 oz.)	2.8	28
Vegeburger (Loma Linda)	½ cup (3.9 oz.)	21.1	126
Vegeburger, no salt added (Loma Linda)	½ cup (3.9 oz.)	21.1	132
Vegechee (Loma Linda)	½" slice (2.5 oz.)	8.3	106
Vegelona (Loma Linda)	½" slice (3.3 oz.)	19.7	154
Vegetable skallop (Worthington)	1 piece (.7 oz.)	5.0	26
Vegetable steaks (Worthington)	1 piece (.7 oz.)	3.0	21
Vegetarian burger (Worthington)	⅓ cup (2.5 oz.)	13.0	104
Veja-links (Worthington)	1 link (1.2 oz.)	4.0	72
Wham, sliced or loaf (Worthington)	1 slice (1 oz.)	4.5	46
Wheat germ, natural or toasted (Loma Linda)	1 T. (.4 oz.)	3.2	40
Wheat protein (USDA)	4 oz.	18.5	124
Wheat protein, nuts or peanuts (USDA)	4 oz.	23.0	240
Wheat protein, vegetable oil (USDA)	4 oz.	21.7	214

(USDA): United States Department of Agriculture

Food and Description	Measure or Quantity	Protein (grams)	Calories
Wheat & soy protein (USDA)	4 oz.	18.3	118
Wheat & soy protein, soy or other vegetable oil (USDA)	4 oz.	18.3	170
Worthington 209	1 slice (.5 oz.)	1.2	18
Yum (Worthington)	1 serving (1.5 oz.)	10.0	89
Frozen (Worthington):			
Beef pie	1 pie (7.9 oz.)	11.2	506
Beef style, loaf or sliced	1 slice (1 oz.)	5.6	52
Chicken pie	1 pie (7.9 oz.)	10.8	450
Chicken style, diced, roll or sliced	1 oz. or 1 slice	6.0	72
Chic-Ketts	1 oz.	7.1	55
Corned beef, loaf or sliced	1 slice (.5 oz.)	2.5	36
Croquettes	1 croquette (1 oz.)	5.0	69
FriPats	1 pat (2.6 oz.)	13.9	198
Holiday roast	1 slice (2 oz.)	11.4	142
Non-meatballs	1 piece (.6 oz.)	4.7	55
Prosage	⅜″ slice (1.2 oz.)	6.0	81
Salisbury steak	1 slice (2 oz.)	9.4	111
Smoked beef, roll or sliced	1 slice (7 grams)	1.6	15
Stripples	1 slice (7 grams)	1.6	17
Turkey, smoked, loaf or sliced	1 slice (.7 oz.)	3.6	46
Wham, diced, sliced or loaf	1 slice (1 oz.)	6.0	53
VENISON, raw, lean meat only (USDA)	4 oz.	23.8	143
VIENNA SAUSAGE, canned:			
(USDA)	1 oz.	4.0	68
(USDA)	1 sausage (from 5-oz. can, .6 oz.)	2.2	38
(Armour Star)	5-oz. can	17.0	393
(Armour Star)	1 sausage (.6 oz.)	2.0	45
(Hormel)	1 sausage (.6 oz.)	1.3	42
(Van Camp)	1 oz.	4.0	68
(Wilson)	1 oz.	3.4	85
VINEGAR (USDA)	½ cup (4.2 oz.)	Tr.	17

(USDA): United States Department of Agriculture

Food and Description	Measure or Quantity	Protein (grams)	Calories
VINESPINACH or BASELLA, raw (USDA)	4 oz.	2.0	22

W

WAFER (see **COOKIE** or **CRACKER**)

WAFFLE:			
Home recipe (USDA)	2.6-oz. waffle (7″ dia.)	7.0	209
Frozen:			
(USDA)	1.6-oz. waffle (8 in 13-oz. pkg.)	3.3	116
(USDA)	.8-oz. waffle (6 in 5-oz. pkg.)	1.7	61
Buttermilk (Aunt Jemima)	1 section (¾ oz.)	1.2	57
Original (Aunt Jemima)	1 section (¾ oz.)	1.2	57
WAFFLE MIX (USDA) (see also **PANCAKE & WAFFLE MIX**):			
Dry, complete mix	1 oz.	1.8	130
*Prepared with water	2.6-oz. waffle (½″ × 4½″ × 5½″, 7″ dia.)	3.6	229
Dry, incomplete mix	1 oz.	2.4	101
*Prepared with egg & milk	2.6-oz. waffle (7″ dia.)	6.6	206
*Prepared with egg & milk	7.1-oz. waffle (9″ × 9″ × ⅝″, 1⅛ cups batter)	17.6	550
WALNUT:			
Black:			
In shell, whole (USDA)	1 lb. (weighed in shell)	20.5	627
Shelled, whole (USDA)	4 oz.	23.2	712
Chopped (USDA)	½ cup (2.1 oz.)	12.3	377
Kernels (Hammons)	4 oz.	31.3	746
English or Persian:			
In shell, whole (USDA)	1 lb. (weighed in shell)	30.2	1329
Shelled, whole (USDA)	4 oz.	16.8	738

(USDA): United States Department of Agriculture
*Prepared as Package Directs

Food and Description	Measure or Quantity	Protein (grams)	Calories
Chopped (USDA)	½ cup (2.1 oz.)	8.9	391
Chopped (USDA)	1 T. (8 grams)	1.1	49
Halves (USDA)	½ cup (1.8 oz.)	7.4	326
(Diamond)	3-oz. bag (3.4 cups)	12.8	564
(Diamond)	15 halves (.5 oz.)	2.2	99
*WALNUT CAKE MIX, Black, (Betty Crocker)	¹⁄₁₂ of cake	2.7	202
WATER CHESTNUT, CHINESE, raw (USDA):			
Whole	1 lb. (weighed unpeeled)	4.9	272
Peeled	4 oz.	1.6	90
WATERCRESS, raw (USDA):			
Untrimmed	½ lb. (weighed untrimmed)	4.6	40
Trimmed	½ cup (.6 oz.)	.4	3
WATERMELON, fresh (USDA):			
Whole	1 lb. (weighed with rind)	1.0	54
Wedge	2-lb. wedge (4″ × 8″, measured with rind)	2.1	111
Slice	½ slice (12.2 oz., ¾″ × 10″)	.8	41
Diced	1 cup (5.6 oz.)	.8	42
WAX GOURD, raw (USDA):			
Whole	1 lb. (weighed with skin & cavity contents)	1.3	41
Flesh only	4 oz.	.5	15
WEAKFISH (USDA):			
Raw, whole	1 lb. (weighed whole)	35.9	263
Broiled, meat only	4 oz.	27.9	236
WELSH RAREBIT:			
Home recipe (USDA)	1 cup (8.3 oz.)	18.8	415

(USDA): United States Department of Agriculture
*Prepared as Package Directs

Food and Description	Measure or Quantity	Protein (grams)	Calories
Canned (Snow)	4 oz.	9.0	171
WEST INDIAN CHERRY (see **ACEROLA**)			
WHALE MEAT, raw (USDA)	4 oz.	23.4	177
WHEAT CHEX, cereal (Ralston Purina)	⅔ cup (1 oz.)	2.6	115
* **WHEATENA,** cereal	½ cup (.9 oz. dry)	2.8	88
WHEAT FLAKES, cereal, crushed (USDA)	1 cup (2.5 oz.)	7.1	248
WHEAT GERM, cereal:			
(USDA)	¼ cup (1 oz.)	8.5	110
Regular (Kretschmer)	¼ cup (1 oz.)	8.9	106
Caramel apple or cinnamon raisin (Kretschmer)	¼ cup (1 oz.)	6.0	110
Sugar & honey (Kretschmer)	¼ cup (1 oz.)	7.0	114
Toasted (Pillsbury)	¼ cup	9.0	120
WHEAT GERM, crude, commercial, milled (USDA)	1 oz.	7.5	103
WHEATIES, cereal (General Mills)	1¼ cups (1 oz.)	2.5	101
WHEAT OATA, cereal, dry	¼ cup (1 oz.)	4.5	109
WHEAT, ROLLED (USDA):			
Uncooked	1 cup (3.1 oz.)	8.6	296
Cooked	1 cup (7.7 oz.)	4.8	163
WHEAT, SHREDDED, cereal (see **SHREDDED WHEAT**)			
WHEY (USDA):			
Dried	1 oz.	3.7	99
Fluid	1 cup (8.6 oz.)	2.2	63
* **WHIP 'N CHILL** (Jell-O):			
All flavors except chocolate	½ cup (3 oz.)	2.8	135

(USDA): United States Department of Agriculture
*Prepared as Package Directs

Food and Description	Measure or Quantity	Protein (grams)	Calories
Chocolate	½ cup (3 oz.)	3.2	144
WHITEFISH, LAKE (USDA):			
Raw, whole	1 lb. (weighed whole)	40.3	330
Raw, meat only	4 oz.	21.4	176
Baked, stuffed, made with bacon, butter, onion, celery & breadcrumbs, home recipe	4 oz.	17.2	244
Smoked	4 oz.	23.7	176
WHITEFISH & PIKE (see **GEFILTE FISH**)			
WIENER (see **FRANKFURTER**)			
WILD BERRY, fruit drink (Hi-C)	6 fl. oz. (6.3 oz.)	.6	90
WILD RICE, raw:			
(USDA)	½ cup (2.9 oz.)	11.6	289
(Gourmet House)	⅛ cup (¾ oz.)	3.0	75
WORCESTERSHIRE SAUCE, (see **SAUCE**, Worcestershire)			
WRECKFISH, raw, meat only (USDA)	4 oz.	20.9	129

Y

YAM (USDA):			
Raw, whole	1 lb. (weighed with skin)	8.2	394
Raw, flesh only	4 oz.	2.4	115
Canned & frozen (see **SWEET POTATO**)			
YAM BEAN, raw (USDA):			
Unpared tuber	1 lb. (weighed unpared)	5.7	225
Pared tuber	4 oz.	1.6	62

(USDA): United States Department of Agriculture

Food and Description	Measure or Quantity	Protein (grams)	Calories
YEAST:			
Baker's:			
Compressed (USDA)	1 oz.	3.4	24
Compressed (Fleischmann's)	⅗-oz. cake	2.7	19
Dry (USDA)	1 oz.	10.5	80
Dry (USDA)	1 pkg. (7 grams)	2.6	20
Dry (Fleischmann's)	¼ oz. (pkg. or jar)	2.8	24
Brewer's dry, debittered (USDA)	1 oz.	11.0	80
Brewer's dry, debittered (USDA)	1 T. (8 grams)	3.1	23
YELLOWTAIL, raw, meat only (USDA)	4 oz.	23.8	156
YODEL (Drake's) chocolate	1 roll (.9 oz.)	1.3	116
YOGURT:			
Made from whole milk (USDA)	½ cup (4.3 oz.)	3.7	76
Made from partially skimmed milk, plain or vanilla:			
(USDA)	½ cup (4.3 oz.)	4.1	61
(USDA)	8-oz. container	7.7	113
Plain:			
(Borden) Swiss style	5-oz. container	7.0	82
(Borden) Swiss style	8-oz. container	11.1	131
(Breakstone)	8-oz. container	11.6	141
(Breakstone)	1 T. (.5 oz.)	.7	9
(Dannon)	8-oz. container	10.9	136
(Dean)	8-oz. container	11.7	143
Apple, Dutch (Axelrod)	8-oz. container	10.9	228
Apple, Dutch (Dannon)	8-oz. container	7.8	258
Apricot:			
(Breakstone)	8-oz. container	9.3	220
(Breakstone) *Swiss Parfait*	8-oz. container	9.8	249
(Dannon)	8-oz. container	7.8	258
Black cherry (Breakstone) *Swiss Parfait*	8-oz. container	9.8	256
Blueberry:			
(Axelrod)	8-oz. container	11.0	226
(Breakstone)	8-oz. container	9.3	252

(USDA): United States Department of Agriculture

Food and Description	Measure or Quantity	Protein (grams)	Calories
(Breakstone) *Swiss Parfait*	8-oz. container	9.5	286
(Dannon)	8-oz. container	7.8	258
(Dean)	8-oz. container	9.9	259
(Sealtest) *Light n' Lively*	8-oz. container	9.3	257
(SugarLo)	8-oz. container	9.5	117
Boysenberry (Dannon)	8-oz. container	7.8	258
Cherry:			
(Dannon)	8-oz. container	7.8	258
(Dean)	8-oz. container	10.0	245
Black (SugarLo)	8-oz. container	9.5	117
Cinnamon apple (Breakstone)	8-oz. container	9.1	229
Coffee (Dannon)	8-oz. container	9.1	198
Danny (Dannon):			
Cuplet, any flavor	4-oz. container	3.9	129
Frozen pop	2½-oz. pop	2.5	127
Honey (Breakstone) *Swiss Parfait*	8-oz. container	9.5	277
Lemon:			
(Breakstone) *Swiss Parfait*	8-oz. container	9.5	254
(Dannon)	8-oz. container	9.1	198
(Sealtest) *Light n' Lively*	8-oz. container	9.8	229
Lime (Breakstone) *Swiss Parfait*	8-oz. container	10.0	243
Mandarin orange:			
(Broden) Swiss Style	5-oz. container	5.3	142
(Borden) Swiss Style	8-oz. container	8.4	227
(Breakstone) *Swiss Parfait*	8-oz. container	9.5	263
Orange (Dean)	8-oz. container	10.2	311
Peach:			
(Axelrod)	8-oz. container	11.1	220
(Borden) Swiss Style	5-oz. container	5.2	138
(Borden) Swiss Style	8-oz. container	8.4	221
(Breakstone) *Swiss Parfait*	8-oz. container	9.1	254
(Dean)	8-oz. container	11.9	259
Melba (Breakstone) *Swiss Parfait*	8-oz. container	9.1	268
(Sealtest) *Light n' Lively*	8-oz. container	9.5	252
(SugarLo)	8-oz. container	9.5	118
Pineapple:			
(Breakstone)	8-oz. container	9.1	220
(Dean)	8-oz. container	13.3	265
(Sealtest) *Light n' Lively*	8-oz. container	9.3	241
(SugarLo)	8-oz. container	9.5	118

[339]

Food and Description	Measure or Quantity	Protein (grams)	Calories
Pineapple-cherry (Axelrod)	8-oz. container	10.6	229
Pineapple-orange (Dannon)	8-oz. container	7.8	258
Prune whip (Breakstone)	8-oz. container	9.3	231
Prune whip (Dannon)	8-oz. container	7.8	258
Raspberry:			
(Axelrod)	8-oz. container	11.2	227
(Borden) Swiss Style	5-oz. container	5.5	147
(Borden) Swiss Style	8-oz. container	8.9	236
(Breakstone)	8-oz. container	9.3	249
(Dannon)	8-oz. container	7.8	258
Red (Breakstone) *Swiss Parfait*	8-oz. container	10.2	263
Red (Dean)	8-oz. container	10.6	272
Red (Sealtest) *Light n' Lively*	8-oz. container	10.2	225
(SugarLo)	8-oz. container	9.5	118
Strawberry:			
(Axelrod)	8-oz. container	10.9	224
(Borden) Swiss Style	5-oz. container	5.5	142
(Borden) Swiss Style	8-oz. container	8.9	227
(Breakstone)	8-oz. container	9.1	225
(Breakstone) *Swiss Parfait*	8-oz. container	9.5	259
(Dannon)	8-oz. container	7.8	258
(Dean)	8-oz. container	12.2	256
(Sealtest) *Light n' Lively*	8-oz. container	10.0	234
(SugarLo)	8-oz. container	9.5	111
Vanilla:			
(Borden) Swiss Style	5-oz. container	5.9	148
(Borden) Swiss Style	8-oz. container	9.5	235
(Breakstone)	8-oz. container	10.2	195
(Dannon)	8-oz. container	9.1	198

YOUNGBERRY, fresh (see **BLACKBERRY,** fresh)

Z

ZING, cereal beverage, 0.4% alcohol	12 fl. oz. (12 oz.)	.6	62
ZITI, baked, with sauce, frozen (Buitoni)	4 oz.	7.0	128

Food and Description	Measure or Quantity	Protein (grams)	Calories
ZUCCHINI (see **SQUASH, SUMMER**)			
ZWIEBACK:			
(USDA)	1 oz.	3.0	120
(Nabisco)	1 piece (7 grams)	.9	31

BIBLIOGRAPHY

Dawson, Elsie H., Gilpin, Gladys L., and Fulton, Lois H. *Average weight of a measured cup of various foods.* U.S.D.A. ARS 61–6, February 1969. 19 pp.

Food and Agriculture Organization of the United Nations. *Energy and protein requirements.* Report of a Joint FAO/WHO Ad Hoc Expert Committee. 1973. 118 pp.

Irwin, M. Isabel and Hegsted, D. Mark. *A conspectus of research on protein requirements of man.* Jour. of Nutrition 101 (3):385–430. 1971.

Kofrányi, Ernst, Jekat, F., and Muller-Wecker, H. *The minimum protein requirement of humans, tested with mixtures of whole egg plus potato and maize plus beans.* Hoppe-Seyler's Z. Physiol. Chem. Bd. 351, s. 1485–1493, December 1970.

Leung, W. T. W., Busson, F. and Jardin, C. *Food Composition Table for Use in Africa.* U.S. Department of Health, Education and Welfare and Food and Agriculture Organization of the United Nations. 1968. 306 pp.

Leung, W. T. W., Butrum, R. V. and Chang, F. H. *Food Composition Table for Use in East Asia.* U. S. Department of Health, Education and Welfare and Food and Agriculture Organization of the United Nations. December 1972. 334 pp.

Merrill, A. L. and Watt, B. K. *Energy value of foods—basis and derivation.* U.S.D.A. Handb. 74. 1955. 105 pp.

Orr, M. L. and Watt, B. K. *Amino acid content of foods.* U.S.D.A. Home Economics Research Rept. No. 4. 1957 and 1968. 82 pp.

Pecot, Rebecca K., Jaeger, Carol M., and Watt, B. K. *Proximate composition of beef from carcass to cooked meat: Method of derivation and tables of values.* U.S.D.A. Home Economics Research Report 31, 1965. 32 pp.

Pecot, Rebecca K. and Watt, Bernice K. *Food yields: Summarized by different stages of preparation.* U.S.D.A. Handb. 102. 1956. 93 pp.

[343]

BIBLIOGRAPHY

    U.S.D.A. Nutritive value of foods. Home and Garden Bul. 72, 1964, 36 pp.; and revised edition, 1970. 41 pp.

    U.S.D.A. Unpubl. Data, 1969 and 1974.

    Watt, Bernice K., Merrill, Annabel L., et al. *Composition of foods: Raw, processed, prepared.* U.S.D.A. Handb. 8. 1963. 190 pp.

WITHDRAWAL